W0080554

Instrumental Methods of Analysis

D Muralidhara Rao
Assistant Professor, Department of Biotechnology
Sri Krishnadevaraya University
Anantapur 515 055, Andhra Pradesh, India

AVN Swamy
Professor, Department of Chemical Engineering
JNT University, Anantapur
Anantapur 515 002, Andhra Pradesh, India

D Dharaneeswara Reddy
Assistant Professor, Department of Biotechnology
Sree Vidyanikethan Engineering College (Autonomous)
A Rangampet, Tirupati 517 102, Andhra Pradesh, India

CBSPD

CBS Publishers & Distributors Pvt Ltd

New Delhi • Bengaluru • Chennai • Kochi • Kolkata • Lucknow • Mumbai
Hyderabad • Jharkhand • Nagpur • Patna • Pune • Uttarakhand

Instrumental Methods
of Analysis

ISBN: 978-81-239-2327-7

Copyright © Publisher

First Edition: 2013
Reprint: 2019, 2024, **2025**

All rights reserved. No part of this book may be reproduced or transmitted in any form or by any means, electronic or mechanical, including photocopying, recording, or any information storage and retrieval system without permission, in writing, from the author and the publishers.

Published by **Satish Kumar Jain** and produced by **Varun Jain** for

CBS Publishers & Distributors Pvt Ltd

4819/XI Prahlad Street, 24 Ansari Road, Daryaganj, New Delhi 110 002, India.
Ph: 011-23289259, 23266838   Website: www.cbspd.com
                                                          e-mail: delhi@cbspd.com

Corporate Office: 204 FIE, Industrial Area, Patparganj, Delhi 110 092
Ph: 011-4934 4934   Fax: 011-4934 4935
                                         e-mail: publishing@cbspd.com; publicity@cbspd.com

Branches

- **Bengaluru:** Seema House 2975, 17th Cross, KR Road, Banasankari 2nd Stage, Bengaluru 560 070, Karnataka, India
  Ph: +91-80-26771678/79   Fax: +91-80-26771680   e-mail: bangalore@cbspd.com
- **Chennai:** 7, Subbaraya Street, Shenoy Nagar, Chennai 600 030, Tamil Nadu, India
  Ph: +91-44-26680620, 26681266   Fax: +91-44-42032115   e-mail: chennai@cbspd.com
- **Kochi:** 42/1325, 1326, Power House Road, Opp KSEB, Power House, Ernakulum Kochi 682 018, Kerala, India
  Ph: +91-484-4059061-65,67   Fax: +91-484-4059065   e-mail: kochi@cbspd.com
- **Kolkata:** 147, Hind Ceramics Compound, 1st Floor, Nilgunj Road, Belghoria, Kolkata-700056, West Bengal, India
  Ph: +033-25633055, 033-25633056
- **Lucknow:** Basement, Khushnuma Complex, 7 Meerabai Marg (Behind Jawahar Bhawan), Lucknow-226001, UP, India
  Ph: +0522-4000032   e-mail: tiwari.lucknow@cbspd.com
- **Mumbai:** PWD Shed, Gala no 25/26, Ramchandra Bhatt Marg, Next to JJ Hospital Gate no. 2, Opp. Union Bank of India, Noorbaug, Mumbai-400009, Maharashtra, India
  Ph: 022-66661880/89   e-mail: mumbai@cbspd.com

Representatives

• Hyderabad	0-9885175004	• Jharkhand	0-9811541605	• Nagpur	0-8692091830
• Patna	0-9334159340	• Pune	0-9664372571	• Uttarakhand	0-9716462459

Printed at Rashtriya Printers, Dilshad Garden, Delhi, India

डा. वै. वेंकटरामि रेड्डी
सदस्य
DR. Y. VENKATARAMI REDDY
MEMBER

Telefax (O) : 011-23070398
संघ लोक सेवा आयोग
नई दिल्ली - 110069
UNION PUBLIC SERVICE COMMISSION
Dholpur House, Shahjahan Road,
New Delhi-110069

Dated 18.02.2013

FOREWORD

Instrumental Methods of Analysis is a very important subject which is useful to all Processes of Engineering, Pharmaceutical sciences, Biotechnology, Food Technology, Drugs and Intermediates and Petrochemicals. This subject is taught as one of the core subjects of the above listed branches of Science and - Technology. The Control and Analysis of raw materials and finished products in process and pharmaceutical industries is carried out with the knowledge of instrumental methods of analysis. The various instruments, their working principle, and their application in analyzing various parameters with proper illustration, is dealt in this book of Instrumental Methods of Analysis.

I am glad that Prof. A.V.N. Swamy with his wide experience in process industries and teaching at JNTUA Anantapur, Dr. D. Muralidhar Rao Faculty of Bio-Technology at S.K. University Anantapur, with his experience ,in teaching biotechnology and Mr. D Dharaneeswara Reddy of Sree Vidyaniketan College, Tirupathi, have authored this book.

I am extremely happy to note that this publication would cater to the needs of Under Graduates and Post Graduates of Chemical Engineering, Pharmaceutical Sciences and Biotechnology. I also hope that this book would be useful to the working professionals of Pharmaceutical Industry, Chemical Industry and Biotechnology industry.

I wish all the three authors a grand success.

(Dr. Venkatarami Reddy Y)

PREFACE

While writing this book, the authors had primary objective to provide a new text-book of 'Instrumental Methods of Analysis' to meet the requirements of Technological universities, students of Biotechnology and Pharmacy grades. The scope of this book describes the concepts, principles and applications of biophysical and chemical techniques used in biotechnology. The book is covered in 8 chapters. The contents of this book are selected according to the different universities UG and PG level syllabi including syllabus of Jawaharlal Nehru Technological University and affiliated colleges in particular.

'Chapter-1' Presents introduction, history and role of analytical chemistry, types of analytical methods and their basic principles. It deals with the principles and applications of gravimetric, volumetric analysis, precipitation methods and modern analytical techniques and their data analysis.

'Chapter-2' This chapter provides, principles, instrumentation and applications of generalized microscopy and various types of microscopy techniques viz., bright field microscopy, dark field microscopy, fluorescence microscopy, phase contrast microscopy, confocal microscopy, electron microscopy and their types. It also highlights flow Cytometry.

'Chapter-3' Looks in to the theoretical, practical aspects, applications and types of centrifugation.

'Chapter-4' Introduces the fundamentals of radiation, energy, electromagnetic radiation, types of spectra and spectroscopy. This chapter is extended with more explanation on UV Visible spectrophotometer, Spectrofluorimeter, Atomic absorption and emission spectroscopy and Circular dichroism.

'Chapter-5' Further extension of chapter-4, provides an overview of infrared radiation, and includes Infrared spectroscopy, Mass spectroscopy and Electron spin resonance spectroscopy fundamentals, their designs, operation processes and their applications in various fields.

New analytical tools are able to provide fast, reliable and sensitive measurements with lower cost, many of them aimed at on-site analysis. These devices allow current field screening and monitoring methods. For control and monitoring, the development of various online monitoring and control devices are mentioned.

'Chapter-6' Deals with principles, working operations and applications of pH and oxygen electrode, sensors and X-ray crystallography techniques.

'Chapter-7' Introduces the principles, instrumentation and applications of high performance liquid chromatography, gas chromatography, ion exchange chromatography, affinity chromatography and membrane separation techniques bordering on ultra filtration and reverse osmosis.

'Chapter-8' Introduces the nuclear spin and magnetic moment of nuclei for detection of various chemical substances which provides an important nuclear magnetic resonance. Basic fundamentals, principle, instrumentation and applications of nuclear magnetic resonance spectroscopy are discussed in this chapter.

The book is generously illustrated with simple line drawings readily reproducible by the students. Most of the chapters are appended with suitable examples in order to convey various biophysical and chemical techniques in biotechnology. This book covers, present developments in biotechnology combined with the analytical techniques which are being used for the detection of various chemical and biological agents. The contents of this book will be useful for all graduate and postgraduate students of biotechnology, physical and chemical sciences, and pharmacy. This book should be useful for teachers and students. Any suggestions for improving the book will be gratefully received by the authors.

First and foremost, we would like to thank Dr. Y. Venkatarami Reddy, Member, Union Public Service Commission for writing the valuable foreword. In the preparation of this book, Dr. D. Muralidhara Rao is grateful to CSIR, New Delhi and Prof. K. Ramakrishna Reddy, the Hon. Vice-Chancellor of Sri Krishnadevaraya University, for his encouragement. Prof. A.V.N Swamy sincerely acknowledges Prof. K. Lal Kishore, the Hon. Vice Chancellor of the J.N.T.University, Anantapur, for his support. Mr. D. Dharaneeswara Reddy acknowledges deep sense of gratitude to Padmasri Dr. Mohan Babu, Chairman, Sree Vidyanikethan Educational Institutions and Dr. Gopal Rao, Special Officer, Sree Vidyanikethan Educational Institutions, Tirupati for providing an inspiring environment for teachers to undertake creative works of this kind. The authors are also grateful to M/S IKON Books for their authentic concern of both teachers and students, their wish to produce an educationally beneficial book. Finally, we would like to express our deep appreciation to our families, who patiently endured many evenings and weekends that were devoted to writing and who gave encouragement when it was most needed.

AUTHORS

Dr. D Muralidhara Rao

Dr. A V N Swamy

Mr. D Dharaneeswara Reddy

CONTENTS

Chapter 1
ANALYTICAL CHEMISTRY

1.0. INTRODUCTION

Analytical chemistry is a sub discipline of chemistry that has the broad mission of understanding the chemical composition of all matter. This is differs from other sub disciplines of chemistry, it is not intended to understand the physical basis as with physical chemistry and it is not necessarily intended to provide engineering tactics as are often used in materials science. Analytical chemistry has significant overlap with other branches of chemistry, especially those that are focused on a certain broad class of chemicals, such as organic chemistry, inorganic chemistry or biochemistry, as opposed to a particular way of understanding chemistry, such as theoretical chemistry. *Ex:* The field of bio-analytical chemistry is a growing area of analytical chemistry that addresses all analytical questions in biochemistry i.e. the chemistry of life. Analytical chemistry and experimental physical chemistry, however, have a unique relationship in that they are much unrelated in their mission but often share the most in common in the tools used in experiments.

Analytical chemistry is particularly concerned with the questions of what chemicals are present, what are their characteristics and in what quantities are they present? Analytical chemistry is a measurement science consisting of a set of powerful ideas and methods that are useful in all fields of science and medicine. Most of the major developments in analytical chemistry take place after 1900. During these periods instrumental analysis becomes progressively dominant in the field. In particular many of the spectrometric techniques were discovered in the early 20th century. Modern analytical chemistry is dominated by instrumental analysis. Many analytical chemists focus on a single type of instrument. Academics tend to either focus on new applications/ discoveries or on new methods of analysis. The late 20th century also saw an expansion of the application of analytical chemistry from somewhat academic chemical questions to forensic, environmental, industrial and medical questions.

The discovery of a chemical present in blood that increases the risk of cancer would be a discovery that an analytical chemist might be involved in. An effort to develop a new method might involve the use of a tunable laser to increase the specificity and sensitivity of a spectrometric method. Analytical chemistry is particularly true in industrial quality assurance (QA), forensic, and environmental applications and plays an increasingly important role in the pharmaceutical industry. It is used in discovery of new drug candidates and in clinical applications where understanding the interactions between the drug and the patient are critical.

1.0.1. Methods for detecting analytes

There are a limited number of ways to detect an analyte. The following general categories are used in analytical techniques.

1. Physical means: Mass, Color, Refractive index and Thermal conductivity

2. With electromagnetic radiation (Spectroscopy): Absorption, Emission and Scattering

3. By an electric charge: Electrochemistry and Mass spectrometry

1.0.2. Analytical targets

Analytical targets for analytical techniques are Bioanalytical chemistry, Material analysis, Chemical analysis, Environmental analysis and Forensics.

1.1. ANALYTICAL TYPES

Analytical chemistry has been split into two main types (1) Qualitative analysis and (2) Quantitative analysis. Both qualitative and quantitative information are required in an analysis.

1.1.1. Qualitative analysis

Qualitative Analysis is that branch of analytical chemistry which treats of the determination of the elements present in a chemical but does not deal with the quantities. The methods employed are usually the application of a number of tests to a solution of the substance to be examined, the tests being applied in a regular order, so that some of the constituents of the compound are obtained in an insoluble form. For example if hydrochloric acid is added to a solution of silver, lead and mercurous salts and chlorides of these metals are precipitated. The precipitate is boiled with water, when the lead chloride is dissolved, and filtration leaves only the silver and mercurous salts. Adding ammonia affects the solution of the silver, and leaves the mercurous chloride as a blackened insoluble mass. By filtering again, the complete separation of the three salts is effected. The filtrate obtained from the hydrochloric is again treated in an analogous manner until all the metals present are determined. In the case of organic chemistry the qualitative analysis consists of a number of separate tests to determine the presence of suspected elements.

    i. Qualitative inorganic analysis seeks to establish the presence of a given element or inorganic compound in a sample.

    ii. Qualitative organic analysis seeks to establish the presence of a given functional group or organic compound in a sample.

    iii. Qualitative analysis establishes the chemical identity of the species in the sample.

    iv. It is often an integral part of the separation step.

1.1.2. Quantitative analysis

    i. Quantitative analysis seeks to establish the amount of a given element or compound in a sample.

    ii. Quantitative analysis determines the relative amount of species or analytes in numerical terms.

    iii. Determining the identity of the analytes is an essential accessory to quantitative analysis.

The results are compute of a typical quantitative analysis from two measurements (1) The mass or the volume of sample being analyzed (2) The measurement of some quantity that is proportional to the amount of analyte in the sample such as mass, volume, intensity of light or electrical charge and this second measurement usually completes the analysis. The analytical methods are classified according to the nature of the final measurement.

    1. Gravimetric methods determine the mass of the analyte or some compound chemically related to it.

    2. In volumetric methods, the volume of a solution containing sufficient reagent to react completely with the analyte is measured.

    3. Electro-analytical methods involve the measurement of electrical

properties such as potential, current, resistance and quantity of electrical charge.

4. Spectroscopic methods are based on measurement of the interaction between electromagnetic radiation and analyte atoms or molecules or production of radiation by analyte.

1.1.2.1. Quantitative analysis steps

A typical quantitative analysis involves the following sequence of steps (fig.1.1).

1.1.2.1.1 Selection of method

The essential first step in any quantitative analysis is the selection of a method. One of the first question is to be considered in the selection process is the level of accuracy required. The selected method usually represents a compromise between the required accuracy and time for the analysis.

1.1.2.1.2 Acquire sample

The second step in quantitative analysis is to acquire the sample. To produce meaningful information, an analysis must be performed on sample that has the same composition as the bulk of material from which it was taken. When the bulk is large and heterogeneous, great effort is required to get a representative sample. Sampling is the process of collecting a small mass of a material whose composition accurately represents the bulk of the material being sampled.

The collection of specimens from biological sources represents a second type of sampling problem. Sampling of human blood for the determination of blood gases illustrates the difficulty of acquiring a representative sample from a complex biological system. The concentration of oxygen and carbon dioxide in blood depends on a variety of physiological and environmental variables. The procedures ensure that the sample is representative of the patient at the collected time and its integrity is preserved until the sample can be analyzed.

The sampling is simple or complex, the analyst must be sure that the laboratory sample is representative of the whole before proceeding with an analysis. Sampling is the most difficult step and the reliability of the final results of analysis will never be any greater than the reliability of the sampling step.

1.1.2.1.3. Process the sample

The third step in an analysis is to process the sample. Under certain circumstances, no sample processing is required prior to the measurements step. *Eg:* Water sample is withdrawn from a stream, a lake or an ocean, its pH can be measured directly. Under most circumstances, process the sample in any of a variety of the different following ways.

1.1.2.1.3.1. Preparing laboratory samples

A solid laboratory sample is griend to decrease the particle size, mixed to ensure homogeneity and stored for various lengths of time before analysis begins. To dry samples for loss or gain of water changes the chemical composition of solids just before starting an analysis. Liquid samples present a slightly different but related set of problems during preparation step. If samples are allowed to stand in open containers, the solvent may evaporate and change the concentration of the analyte. If the analyte is a gas dissolved in a liquid, the sample container must be kept inside a second sealed container, during the entire analytical procedure to prevent contamination by atmospheric gases.

1.1.2.1.3.2. Replicate the samples

Most chemical analyses are performed on replicate samples whose masses or volumes have been determined by careful measurements with an analytical balance or with precise volumetric device. Replication improves the quality of the results and provides a measure of reliability.

1.1.2.1.3.3. Sample solutions

Most analyses are performed on solutions of the sample made with a suitable solvent. The solvent should dissolve the entire sample including the analyte, rapidly and completely. The conditions of dissolution should be sufficiently mild so that loss of the analyte cannot occur. Many materials must be analyzed are insoluble in common solvent. *Eg:* Silicate minerals, High molecular weight polymers and Specimens of animal tissue.

1.1.2.1.3.4. Eliminate interferences

The sample in solution, have converted the analyte to an appropriate form for measurement, the next step is to eliminate substances from the sample that may interfere with measurement. Species other than the analyte that affect the final measurement are called interferences or interferents. To isolate the analytes from interferences before the final measurement is made.

1.1.2.1.4 Measure property 'X'

All analytical results depend on a final measurement 'X' of a physical or

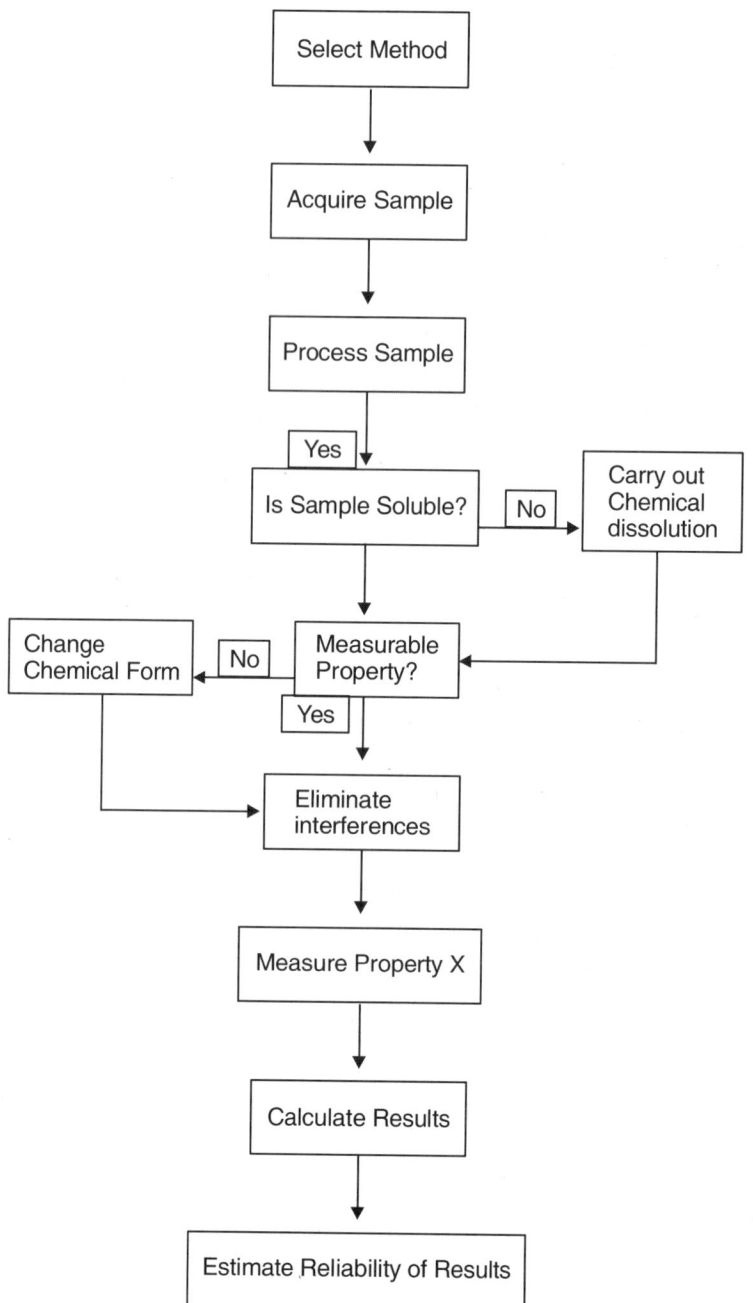

Fig. 1.1: The Steps in Quantitative Analysis

chemical property of the analyte. This property must vary in a known and reproducible way with concentration cA of the analyte. The measurement of the property is directly proportional to the concentration.

$$cA = kX \qquad \qquad \textit{... eq. 1.1}$$

Where k is proportionality constant, the process of determining k is an important step in most analyses this step is called a calibration.

1.1.2.1.5. Calculate results

Computing analyte concentrations from experimental data is usually relatively easy, particularly with modern calculators or computers. These computations are based on the raw experimental data collected in the measurement step, characteristics of the measurement instruments and stoichiometry of the analytical reaction.

1.1.2.2.6. Estimate reliability of results

Analytical results are incomplete without an estimate of their reliability. The experimenter must provide some measure of the uncertainties associated with computed results if the data are to have any value.

1.2. VOLUMETRIC ANALYSIS OR TITRATION

Volumetric analysis or Titration involves the addition of a reactant to a solution being analyzed until some equivalence point is reached. Often the amount of material in the solution being analyzed may be determined. Most familiar to those who have taken college chemistry is the acid-base titration involving a color changing indicator. There are many other types of titrations and may use different types of indicators to reach some equivalence point. *Ex:* Potentiometric titrations.

Volumetric or titrimetric analyses are quantitative analytical techniques which employ a titration in comparing an unknown with a standard. In a titration, a volume of a standardized solution containing a known concentration of reactant A is added to a sample containing an unknown concentration of reactant B. The titration proceeds until reactant B is just consumed (stoichiometric completion); this is known as an equivalence point. At this point the number of equivalents of A added to the unknown equals the number of equivalents of B originally present in the unknown. Volumetric methods have the potential for a precision of up to 0.1%. In almost all cases, a burette is used to meter out the titrant. When a titrant reacts directly with an analyte or with a reaction the product of the analyte and some intermediate compound, the procedure is termed a direct titration.

The alternative technique is called a back titration, here an intermediate reactant is added in excess of that required to exhaust the analyte, then the exact degree of excess is determined by subsequent titration of the unreacted intermediate with the titrant. Regardless of the type of titration, an indicator is always used to detect the equivalence point. Most common are the internal indicators, compounds added to the reacting solutions that undergo an abrupt change in a physical property usually absorbance or color at or near the equivalence point. Sometimes the analyte or titrant will serve auto indicating. External indicators, electrochemical devices such as pH meters, may also be used. Ideally, titrations should be stopped precisely at the equivalence point. However, the ever presence of random and systematic error, often results in a titration end point, the point at which a titration is stopped, that is not quite the same as the equivalence point. Fortunately, the systematic error or bias may be estimated by conducting a blank titration. In many cases the titrant is not available in a stable form of well defined composition. If this is true, the titrant must be standardized usually by volumetric analysis against a compound that is available in a stable, highly pure form i.e. a primary standard.

The following are the basic requirements or components of a volumetric method:

1. A standard solution (titrant) of known concentration which reacts with the analyte with a known and repeatable stoichiometry (acid/base, precipitation, redox, complexation).

2. A device to measure the mass or volume of sample *Ex:* pipet, graduated cylinder, volumetric flask and analytical balance.

3. A device to measure the volume of the titrant added (buret).

4. If the titrant-analyte reaction is not sufficiently specific, a pretreatment to remove interferon's.

5. A means by which the endpoint can be determined, this may be an internal indicator *Ex:* phenolphthalein or an external indicator *Ex:* pH meter.

Volumetric methods may be based on acid/base reactions, precipitation reactions, complexation reactions and redox reactions. The acid/base methods generally use a strong acid or base as a titrant with methyl orange/red (acid titration) or phenolphthalein (base titration) as the indicator. For all acidity and alkalinity determinations, the analyte must be separated from the major cations and anions prior to titration. Precipitate volumetric analysis relies on the formation of solid phase with a very low solubility product constant.

In environmental analysis, it may be used for chloride determination. Specific indicators are used to detect excess silver or mercury. Complex metric titrations often employ ethylenediaminetetraacetic acid (EDTA) [HOOCCH$_2$)$_2$NHCH$_2$CH$_2$NH (CH$_2$COOH)$_2$], this is a hexadentate ligand which binds very strongly to many metals. For calcium and total hardness determination, a couple of specific dyes are used to determine the presence of excess cation.

Many oxidation/reduction based volumetric methods employ the iodometric method. This involves the oxidation of iodide to iodine and subsequent titration with sodium thiosulfate using starch as an indicator. Many of these employ a series of redox reactions. The permanganate method for calcium is somewhat unique in that the calcium is precipitated as the oxalate and it is the solid phase oxalate group which participates in the redox reaction, not the calcium.

Alkalinity is a measure of water's ability to neutralize the strong acids. It reflects the water's buffer capacity or resistance to a drop in pH upon addition of acid. Conversely, acidity is a measure of water's ability to neutralize strong bases. Alkalinity is an important in assessing the need for additional buffering or pH control with pH sensitive operations. For example, the alkalinity of water must be known in order to calculate lime and soda ash doses for precipitative softening. Species responsible for either alkalinity or acidity can affect the rates of corrosion, metals and organic compounds, the rates of certain types of reactions and numerous biological processes. Alkalinity and acidity might also correlate with other properties of water such as hardness and TDS.

1.3. GRAVIMETRY

Gravimetric analysis involves determining the amount of material present by weighing the sample before and/or after some transformation. A common example is the determination of the amount of water in a hydrate by heating the sample to remove the water such that the difference in weight is due to the water lost. Gravimetry is used by geologists to map the subsurface features of the Earth's crust, such as underground masses of dense rock such as iron ore or light rock such as salt. Small variations in the gravitational field (gravimetric anomalies) can be caused by varying densities of rocks and structure beneath the surface. Such variations are measured by a device called a gravimeter or gravity meter, which consists of a weighted spring that is pulled further downwards where the gravity is stronger.

1.3.1. The Universal law of Gravitation

A popular story that Newton was sitting under an apple tree, an apple fell on his head and he suddenly thought of the Universal Law of Gravitation. As in all such legends, this is almost certainly not true in its details, but the story contains elements of what actually happened. Probably the more correct version of the story is that Newton, upon observing an apple fall from a tree, began to think along the following: The apple is accelerated, since its velocity changes from zero as it is hanging on the tree and moves toward the ground. Thus, by Newton's 2nd Law there must be a force that acts on the apple to cause this acceleration and this force is called as 'gravity', and the associated acceleration due to gravity. Then imagine the apple tree is twice as high. Again, we expect the apple to be accelerated toward the ground, so this suggests that this force that we call gravity reaches to the top of the tallest apple tree. If the force of gravity reaches to the top of the highest tree, might it not reach even further; in particular, might it not reach all the way to the orbit of the Moon. Then, the orbit of the Moon about the Earth could be a consequence of the gravitational force, because the acceleration due to gravity could change the velocity of the Moon in just such a way that it followed an orbit around the earth.

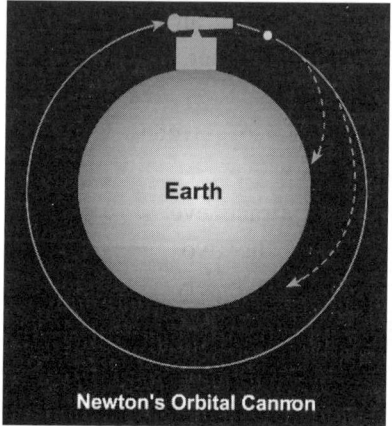

Fig. 1.2 : Newton's Orbital Cannon

Ex: Fire cannon horizontally from a high mountain; the projectile will eventually fall to earth, as indicated by the shortest trajectory (fig.-1.2) because of the gravitational force directed toward the center of the Earth and the associated acceleration, that acceleration is a change in velocity and that velocity is a vector, so it has both a magnitude and a direction.

Thus, acceleration occurs if either or both the magnitude and the direction of the velocity change. But as increase the muzzle velocity for imaginary cannon, the projectile will travel further and further before returning to earth. Finally, Newton reasoned that if the cannon projected the cannon ball with exactly the right velocity, the projectile would travel completely around the Earth, always falling in the gravitational field but never reaching the Earth, which is curving away at the same rate that the projectile falls. That is, *the cannon ball would have been put into orbit around the Earth.*

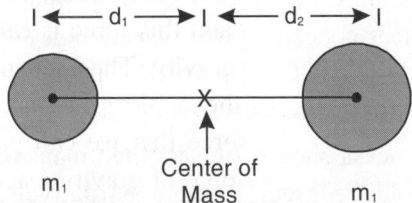

Fig. 1.3 : Universal attraction of objects directed along
the line of centers of two objects

Newton concluded that the orbit of the Moon was of exactly the same nature: the Moon continuously fell in its path around the Earth because of the acceleration due to gravity, thus producing its orbit. By such reasoning, Newton came to the conclusion that any two objects in the Universe exert gravitational attraction on each other, with the force having a universal form.

Law of Universal Gravitation

Every object in the Universe attracts every other object with a force directed along the line of centers for the two objects that is proportional to the product of their masses and inversely proportional to the square of the separation between the two objects (fig.-1.3).

$$Fg = G\ m_1m_2/r^2 \qquad\qquad ...\ eq.\ 1.2$$

Where Fg is the gravitational force, m_1 and m_2 are the masses of the two objects, r is the separation between the objects and G is the universal gravitational constant.

The constant of proportionality G is known as the *universal gravitational constant*. It is termed a universal constant because it is thought to be the same at all places and all times, and thus universally characterizes the intrinsic strength of the gravitational force.

1.4. PRECIPITATION

Precipitation is accomplished by combining a selected ion(s) in solution with a suitable counter ion in sufficient concentrations to exceed the solubility of the resulting compound and produce a supersaturated solution. Nucleation occurs and growth of the crystalline substance then proceeds in an orderly manner to produce the precipitate. The precipitate is collected from the solvent by a physical method, such as filtration or centrifugation. A cation (Sr^{+2}) will precipitate from an aqueous solution in the presence of a carbonate anion, forming the insoluble compound, strontium carbonate ($SrCO_3$), when sufficient concentrations of each ion are present in solution to exceed the solubility of $SrCO_3$. The method is used to isolate and collect strontium from water for radio analysis.

A precipitation process should satisfy three main requirements (1) The targeted species should be precipitated quantitatively. (2) The resulting precipitate should be in a form suitable for subsequent handling; it should be easily filterable and should not creep. (3) If it is used as part of a quantitative scheme, the precipitate should be pure or known purity at the time of weighing for gravimetric analysis. Precipitation processes are useful in several different kinds of laboratory operations, particularly gravimetric yield determinations as a separation technique and for preconcentration to eliminate interfering ions or for coprecipitation.

Ideally, separation of common ions from foreign ions in solution by precipitation will result in a pure solid that is easy to filter. This method should ensure the production of a precipitate to meet these criteria as closely as possible. The physical process of the formation of a precipitate is quite complex, and involves both nucleation and crystal growth. Nucleation is the formation within a supersaturated solution of the smallest particles of a precipitate (nuclei) capable of spontaneous growth.

The importance of nucleation is the nucleation processes govern the nature and purity of the resulting precipitates. If the precipitation is carried out in such a manner as to produce numerous nuclei, precipitation will be rapid, individual crystals will be small, filtration and washing difficult, and purity low. On the other hand, if precipitation is carried out so that only a few nuclei are formed, precipitation will be slower, crystals larger, filtration easier and purity higher. Hence, control of nucleation processes is of considerable significance in analytical chemistry. Once the crystal nuclei are formed, crystal growth proceeds through diffusion of the ions to the surface of the growing crystal and deposition of those ions on the surface. This crystal growth continues until super saturation of the precipitating material is eliminated and equilibrium solubility is attained. Thus, the goal is to produce fewer nuclei during precipitation so that the process will occur slowly, within reasonable limits and larger crystals will be formed.

Impurities result from three mechanisms: (1) inclusion, either by isomorphous replacement (isomorphic inclusion), replacement of a common ion in the crystal structure by foreign ions of similar size and charge to form a mixed crystal, or by solid solution formation (non-isomorphic inclusion), simultaneous crystallization of two or more solids mixed together (2) surface absorption of foreign ions. (3) Occlusion, the subsequent entrapment of adsorbed ions as the crystal grows. Slow growth gives the isomorphous ion time to be replaced by a common ion that fits the crystal structure perfectly, producing a more stable crystal. It also promotes establishment of equilibrium conditions for the formation of the crystal structure so that adsorbed impurities are more likely to desorbs and be replaced by a common ion rather than becoming entrapped. In addition, for a given weight of the solid that is forming, a small number of large crystals present an overall smaller surface area than a large number of small crystals. The large crystals provide less surface area for impurities to adsorb.

1.4.1. Conditions for precipitation

An analytical precipitate for gravimetric analysis should consist of perfect crystals large enough to be easily washed and filtered. The perfect crystal would be free from impurities and be large enough so that it presented a minimum surface area onto which foreign ions could be adsorbed. The precipitate should also be insoluble i.e. slight solubility that loses from dissolution would be minimal. During the precipitation the particle size of precipitates is inversely proportional to the relative super saturation of the solution.

Relative super saturation = (Q – S)/S   *... eq. 1.3*

Where Q is the molar concentration of the mixed reagents before any precipitation occurs and S is the molar solubility of the product (precipitate) when the system has reached equilibrium. For the best possible results, conditions need to be adjusted such that Q will be as low as possible and S will be relatively large. The following conditions are used for analytical precipitation.

1. Precipitation from dilute solution. Slow addition of precipitating reagent with effective stirring. This keeps Q low, stirring prevents local high concentrations of the precipitating agent.

2. Precipitation at a pH, near the acidic end of the pH range in which the precipitate is quantitative. Many precipitates are more soluble at the lower (more acidic) pH values and so the rate of precipitation is slower.

3. Precipitation from hot solution. The solubility S of precipitates increases with temperature and so an increase in S decreases the super saturation.

4. Digestion of the precipitate. The digestion period results in some improvement in the internal perfection of the crystal structure called ripening here some internal foreign atoms may be expelled.

1.4.2. Factors affecting the Precipitation

Several factors affect the nature and purity of the crystals formed during precipitation. These factors permits the selection and application of laboratory procedures that increase the effectiveness of precipitation as a technique for the separation and purification of ions and for the formation of precipitates that are easily isolated. These factors include the following:

1. Rate of precipitation: Formation of large, well shaped crystals is encouraged through slow precipitation because fewer nuclei form and they have time to grow into larger crystals to the detriment of smaller crystals present. Solubility of the larger crystals is less than that of smaller crystals because smaller crystals expose more surface area to the solution. Larger crystals also provide less surface area for the absorption of foreign ions. Slow precipitation can be accomplished by adding a very dilute solution of the precipitant gradually, with stirring to a medium in which the resulting precipitate initially has a moderate solubility.

2. Concentration of Ions and Solubility of Solids: The rate of precipitation depends on the concentration of ions in solution and the solubility of the solids formed during the equilibrium process. A solution containing a low concentration of ions but sufficient concentration to form a precipitate will slow the process resulting in larger crystal formation. At the same time increasing the solubility of the solid either by selecting the counter ion for precipitation or by altering the precipitating conditions for slow precipitation. Many radionuclide's form insoluble solids with a variety of ions and the choice of precipitating agent will affect the solubility of the precipitate.

3. Temperature: Precipitation at higher temperature slows the nucleation and crystal growth because of the increased thermal motion of the particles in solution. Therefore larger crystals form, reducing the amount of adsorption and occlusion. However, most solids are more soluble at elevated temperatures, effectively reducing precipitate yield and an optimum temperature balances these opposing factors.

4. Digestion: Extremely small particles with a radius on the order of one micron are more soluble than larger particles because of their larger surface area compared to their volume (weight). Therefore, when a precipitate is heated over time (digestion) the small crystals dissolve and larger crystals grow. Effectively the small crystals are recrystallized, allowing the escape of impurities (occluded ions) and growth of larger crystals. This process reduces the surface area for adsorption of foreign ions and at the same time replaces the impurities with common ions that properly fit the crystal lattice. Recrystallization perfects the crystal lattice producing a purer precipitate.

5. Degree of Super-saturation: A relatively high degree of super-saturation is required for spontaneous nucleation and degree of super-saturation is the main factor in determining the physical character of a precipitate. Generally, the higher the super-saturation required flocculated colloid will precipitate because more nuclei form under conditions of higher super-saturation and crystal growth is faster. In contrast, the lower super-saturation required, the more likely a crystalline precipitate will form because fewer nuclei form under these conditions and crystal growth is slower. Most perfect crystals are formed therefore from supersaturated solutions that require lower ion concentrations to reach the necessary degree of super saturation and as a result inhibit the rate of nucleation and crystal growth. Degree of super-saturation ultimately depends on physical properties of the solid that affect its formation. Choice of counter-ion will determine the type of solid formed from a radionuclide which in turn determines the degree of saturation required for precipitation. Many radionuclides form insoluble solids with a variety of ions and the choice of precipitating agent will affect the nature of the precipitate.

6. Solvent: The nature of the solvent affects the solubility of an ionic solid (precipitate) in the solvent. The polarity of water can be reduced by the addition of other miscible solvents such as alcohols

thereby reducing the solubility of precipitates. Strontium chromate ($SrCrO_4$) is soluble in water but it is insoluble in a methyl alcohol (CH_3OH) water mixture and can be effectively precipitated from the solution. In some procedures, precipitation is achieved by adding alcohol to an aqueous solution but the dilution effect might reduce the yield because it lowers the concentration of ions in solution.

7. Ion Concentration: The common ion effect causes precipitation to occur when the concentration of ions exceeds the solubility product constant. In some cases, however excess presence of common ions increases the solubility of the precipitate by decreasing the activity of the ions in solution as they become more concentrated in solution and deviate from ideal behavior. An increase in concentration of the ions is necessary to reach the activity of ions for precipitate formation.

8. Stirring: Stirring the solution during precipitation increases the motion of particles in solution and decreases the localized buildup of concentration of ions by keeping the solution thoroughly mixed. Both of these properties slow nucleation and crystal growth thus promoting larger and purer crystals. This approach also promotes re-crystallization because the smaller crystals with their net larger surface area are more soluble under these conditions. Virtually all radiochemical laboratories employ stirring with a magnetic stirrer during precipitation reactions.

9. pH: pH is made more alkaline by hydrolysis of urea (NH_2CONH_2) in boiling aqueous solution. The ammonia slowly liberated raises the pH of the solution homogeneously causing metal ions that form insoluble hydroxides or hydrous oxides to precipitate. In the precipitation of aluminium from homogenous solution urea is added to an acidic solution of aluminium containing sulphuric or succinic acid. No precipitation occurs until the solution has been boiled long enough for the ammonia to raise the pH to the necessary value. In this procedure aluminium precipitates as the basic sulphate or the basic succinate not as aluminium hydroxide. The precipitate obtained in the way is much denser from impurities than are aluminium precipitates formed by the conventional addition of ammonia to aluminium solutions. Another example is the precipitation from homogeneous solution of barium chromate. Chromate is added to barium in solution which is acidic enough to prevent precipitation.

Urea is added and the solution boiled. The ammonia released raises the solution pH and barium chromate slowly precipitates out.

1.5. TRADITIONAL ANALYTICAL TECHNIQUES

Modern analytical chemistry is dominated by sophisticated instrumentation, the roots of analytical chemistry and principles used in modern instruments are from traditional techniques many of which are still used today. These techniques also tend to form the backbone of most undergraduate analytical chemistry educational labs (Table 1.1 and 1.2). Eg: Instrumental analysis - *Spectroscopy- Mass Spectrometry, Spectrophotometry and Calorimetry; Chromatography and Electrophoresis, Crystallography, Microscopy and Electrochemistry.*

Table 1.1: Useful common signals for analytical signals.

S.No	Physical Property Measured	Analytical methods based on measurement of property
1.	Absorption of Radiation	UV Visible, Infra Red, Nuclear Magnetic Resonance and Electron Spin Resonance Spectroscopy techniques
2.	Emission of Radiation	Emission spectroscopy, Flame photometry and Fluorescence spectroscopy
3.	Diffraction of Radiation	X-ray electron diffraction methods
4.	Scattering of Radiation	Turbidimetry, Nephelometry and Raman Spectroscopy
5.	Mass	Gravimetric methods
6.	Volume	Volumetric methods

Table 1.2: Classification of Instrumental methods

S. No.	Method	Phenomenon underlying the method	Property/Quantity measured
Optical methods:			
These are based on the relation between optical properties of a system and its composition.			
1.	Emission Spectroscopy	Appearance of emission spectra because of high temperatures.	Position and intensity of spectral lines.
2.	X-ray Spectroscopy	Emission of X-ray spectrum by atoms under the influence of X-rays.	Position and intensity of spectral lines.

3.	Raman Spectroscopy	Absorption of monochromatic radiation by matter and subsequent emission of new radiation differing from that absorbed by the wavelength.	Position and intensity of spectral lines
4.	Atomic Absorption Spectroscopy	Absorption by atoms in the source of excitation of monochromatic radiation.	Intensity of absorption
5.	Absorption spectrophotometry	Absorption of poly and monochromatic radiant energy by molecules and ions present in solution.	Optical density of the solution.
6.	Turbidimetry	Absorption and scattering of a light beam by non-homogeneous or turbid media.	Optical density of the medium.
7.	Nephelometry	Scattering and reflection of a light beam by non-homogeneous media.	Optical density of the medium.

Electro-chemical methods

These methods are based on the interdependence of electrochemical properties and composition of the system.

8.	Potentiometry	Change in the electrode potentials of a system in the course of a chemical reaction.	Electrode Potential
9.	Electro-Gravimetry	Deposition of matter on an electrode under electrolysis conditions.	Electrode potential
10.	Polarography	Electrode polarisation	Current voltage
11.	Conductometry	Change in the electicconductivity of a solution in the course of a chemical reaction.	Electrical conductivity, Electric resistance.

Radiometric methods

Some methods are used for Quantitative analysis are isotopic dilution mass spectroscopy and NMR etc., these are called as Radiometric Methods.

12.	Nuclear Magnetic Resonance Spectroscopy	Nuclear Magnetism (Resonance absorption of electro magnetic irradiations by matter in a constant magnetic field).	Position and intensity of lines of the NMR spectrum.
13.	Mass spectroscopy	Isolation of atoms, molecules and ions under the combined action of electric and magnetic fields and the appearance of mass spectra.	Position and Intensity of signals in mass spectrum.

1.5.1. Spectroscopy

Spectroscopy measures the interaction of molecules with electromagnetic radiation. Spectroscopy consists of many different applications such as UV visible spectrophotometer, infra red spectroscopy, atomic absorption spectroscopy, atomic emission spectroscopy, X-ray fluorescence spectroscopy, Infrared spectroscopy, Raman spectroscopy, Nuclear magnetic resonance spectroscopy and Photoemission spectroscopy.

1.5.2. Mass spectroscopy

Mass spectrometry measures mass-to-charge ratio of molecules using electric and magnetic fields. There are several ionization methods: electron impact, chemical ionization, electro spray, fast atom bombardment, matrix assisted laser desorption ionization etc. Also, mass spectrometry is categorized by approaches of mass analyzers: magnetic-sector, quadrupole mass analyzer, quadrupole ion trap, Time-of-flight and Fourier transform ion cyclotron resonance.

1.5.3. Crystallography

Crystallography is a technique that characterizes the chemical structure of materials at the atomic level by analyzing the diffraction patterns of X-rays that have been deflected by atoms in the material. From the raw data the relative placement of atoms in space may be determined.

1.5.4. Electrochemical analysis

Electro-analytical methods measure the potential (volts) and/or current (amps) in an electrochemical cell containing the analyte. These methods can be categorized according to which aspects of the cell are controlled and which are measured. The three main categories are potentiometry - the difference in electrode potentials is measured, caulometry - the cell's current is measured over time, and voltammetry-the cell's current is measured while actively altering the cell's potential.

1.5.5. Thermal analysis

Calorimetric and thermo-gravimetric analysis measure the interaction of a material and heat.

1.5.6. Separation techniques

Separation processes are used to decrease the complexity of material mixtures. These are used for removal of unwanted materials from the sample and to purify the sample and used in various industrial processes for isolating metals organic compounds etc. Separation techniques are chromatography, ion-exchange, liquid extraction etc.

1.5.7. Hyphenated techniques

A hyphenated separation technique refers to a combination of two or more techniques to detect and separate chemicals from solutions. Hyphenated techniques are widely used in chemistry and biochemistry. A slash is sometimes used instead of hyphen, especially if the name of one of the methods contains a hyphen itself. Several examples are in popular use today and new hybrid techniques are under development. Ex: Gas chromatography-mass spectrometry, LC-MS, HPLC-MS, GC-IR, LC-NMR, CE-MS, CE-UV and GC-MS.

1.5.8. Microscopy

The visualization of single molecules, single cells, biological tissues and nano- and micro materials is very important and attractive approach in analytical science. Also, hybridization with other traditional analytical tools is revolutionizing analytical science. Microscopy can be categorized into three different fields: optical microscopy, electron microscopy, and scanning probe microscopy. Recently this field is rapidly progressing because of the rapid development of computer and camera industries.

1.6. ROLE OF ANALYTICAL CHEMISTRY

Analytical chemistry has played critical roles in the understanding of basic science to a variety of practical applications, such as biomedical applications, environmental monitoring, quality control of industrial manufacturing and forensic science. Many chemists', biochemists and medicinal chemists devote much time in the laboratory gathering quantitative information about systems that are important and interesting to them. Quantitative analytical measurements are also play a vital role in many research areas in chemistry, biochemistry, biology, geology, physics and other sciences. The central role of analytical chemistry in this enterprise and many other is illustrated in Fig. 1.4. Analytical chemistry has a similar function with respect to the many other scientific fields listed in the Fig. 1.4. The central location of analytical chemistry in Fig. 1.4 signifies its importance and the breadth of its interactions with many other disciplines. Chemistry is often called the central science; its top center position and the central position of analytical chemistry in the figure emphasize this importance. The interdisciplinary nature of chemical analysis makes it a vital tool in medical, industrial, government and academic laboratories throughout the world.

EXAMPLES

1. The concentrations of oxygen and carbon dioxide are determined in millions of blood samples every day and used to diagnose and treat illnesses.

2. Quantities of hydrocarbons, nitrogen oxides and carbon monoxide present in automobile exhaust gases are measured to assess the effectiveness of smog control devices.

3. Quantitative measurements of ionized calcium in blood serum help diagnose parathyroid disease in humans.

4. Quantitative determination of nitrogen in foods established their protein content and thus their nutritional value.

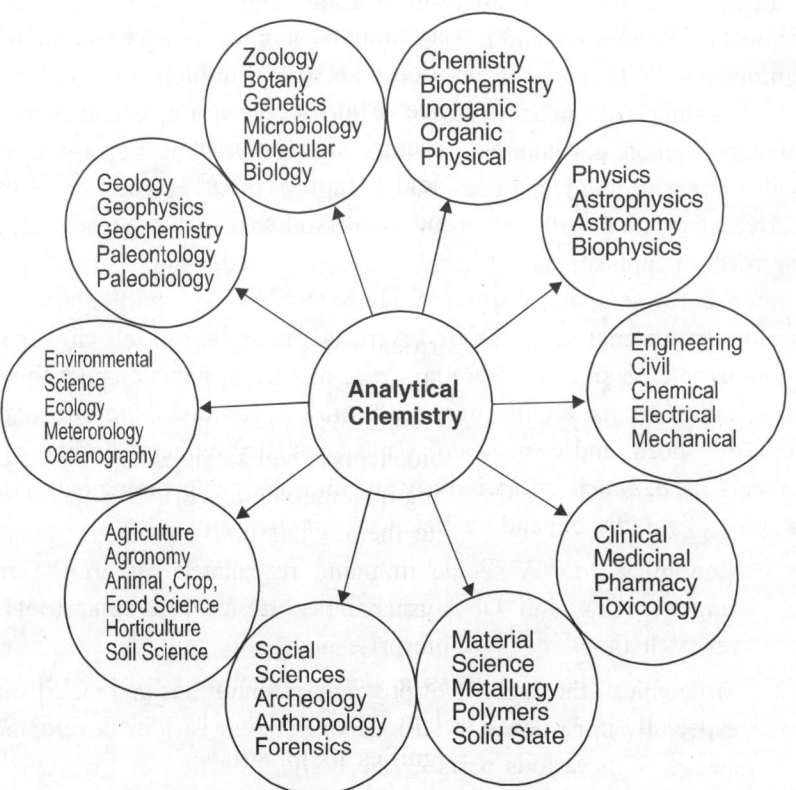

Fig. 1.4: The relationship between analytical chemistry, other branches of chemistry and other sciences

5. Farmers tailor fertilization and irrigation schedules to meet changing plant needs during the growing season, gauging these needs from quantitative analyses of the plant and the soil in which they grow.
6. Measurements of potassium, calcium and sodium ions in the body fluids of animals permit physiologists to study the role of these ions in nerve signal conduction as well as muscle contraction and relaxation.
7. Materials scientists rely heavily on quantitative analyses of crystalline germanium and silicon in their studies of semiconductor devices.

Analytical chemistry research is largely driven by the performance i.e. sensitivity, selectivity, robustness, linear range, accuracy, precision speed and cost i.e. purchase, operation, training, time and space. Among the main branches of contemporary analytical atomic spectrometry the most widespread and universal are optical and mass spectrometry.

In the direct elemental analysis of solid samples, the new leaders are laser induced breakdown and laser ablation mass spectrometry and the related techniques with transfer of the laser ablation products into inductively coupled plasma. Advances in design of diode lasers and optical parametric oscillators promote developments in fluorescence and ionization spectrometry and also in absorption techniques where uses of optical cavities for increased effective absorption path length are expected to expand. Steady progress and growth in applications of plasma- and laser-based methods are noticeable. An interest towards the absolute analysis has revived, particularly in the emission spectrometry. A lot of effort is put in shrinking the analysis techniques to chip size, there are numbers of such systems competitive with traditional analysis techniques, potential advantages include size or portability, speed, and cost. Micro scale chemistry reduces the amounts of chemicals used. Much effort is also put into analyzing biological systems. Examples of rapidly expanding fields in this area are:

1. Genomics - DNA sequencing and its related research. Genetic fingerprinting and DNA microarray are very popular tools and research fields.

2. Proteomics - the analysis of protein concentrations and modifications especially in response to various stressors at various developmental stages or in various parts of the body.

3. Metabolomics - similar to proteomics but dealing with metabolites.

4. Transcriptomics- mRNA and its associated field.

5. Lipidomics - lipids and its associated field.

6. Peptidomics - peptides and its associated field.

7. Metalomics - similar to proteomics and metabolomics but dealing with metal concentrations and especially with their binding to proteins and other molecules.

The recent developments of computer automation and information technologies have innervated analytical chemistry to initiate a number of new biological fields. For example, automated DNA sequencing machines were the basis to complete human genome projects leading to the birth of genomics. Protein identification and peptide sequencing by mass spectrometry opened a new field of proteomics. Also, analytical chemistry has been an indispensable area in the development of nanotechnology. Surface characterization instruments, electron microscopes and scanning probe microscopes enables scientists to visualize atomic structures with chemical characterizations. Analytical chemistry is pursuing the development of practical applications and commercial instruments rather than elucidating scientific fundamentals. Among active contemporary analytical chemistry research fields, micro total analysis system is considered as a great promise of revolutionary technology. In this approach, integrated and miniaturized analytical systems are being developed to control and analyze single cells and single molecules. This cutting edge technology has a promising potential of leading a new revolution in science as integrated circuits in computer developments.

1.7. ADVANTAGES AND LIMITATIONS

1.7.1. Advantages of instrumental methods

i. They are rapid, sensitive and require only a small amount of sample.

ii. They can easily handle even complex samples.

iii. They are quite reliable

1.7.2 Advantages of chemical methods

i. In them the method used in quite simple.

ii. They require only cheap equipment.

iii. These methods are based on absolute measurements.

iv. They can be carried out easily and don't require any specialized training.

1.7.3. Limitations of instrumental methods

   i. They require costly equipment.

   ii. They need a sizable space.

   iii. To handle the equipment a specialized training is required.

   iv. An initial or continuous calibration is needed.

   v. The concentration range is limited.

   vi. The sensitivity and accuracy is dependent on the instrument or the wet chemical method.

1.7.4. Limitations of chemical methods

   i. They consume a lot of time.

   ii. In them the accuracy decreases as the amount of species decreases in the sample.

   iii. They are not very specific.

   iv. They are not very versatile.

1.8. ERRORs

Error is not defined through the use of statistical techniques. But an amount of error is provided by the statistical methods for objectively evaluating the source of these errors by the analytical methods. Source of errors is sampling method, analyst technique and instrument response. Errors are classified into majorly two types (a) In-determine/Random errors: Uncertain nature of the measurement technique is the source of random error. This is present in every analysis method. Thermal, shot and flicker noises are the mainly source of this type of errors. These types of errors are usually small so can minimize by filtering methods (Hardware or Software) (b) Determine or Systematic errors: sources of determine errors are improper instrument calibration procedures, insufficient purity of reagents and improper operation of the measurement instrument. These errors are reduced by the applications of the statistical methods and may be identified and minimized by modifying the analytical procedure.

Error may be absolute terms as the difference between an analytical result x, and the known true value, μ:

$$D = \mu - x \qquad\qquad ...\ eq.\ 1.4$$

Above equation is used for only absolute errors.

Absolute error:

1. This represents difference between the result and the known true value.

2. It must be expressed in the same units as quantities.

3. This error has no significance when separated from the result or true value.

Example: An absolute error is 5.1µg/ml may be acceptable in an analysis of a sample containing 5.11 µg/ml lead but unacceptable in a sample containing 1.7 µg/ml lead.

Relative error (E_{rel})

It is used to determine the accuracy of measurement and is expressed as the % of the known true value.

$$E_{rel} = d/\mu \qquad\qquad \textit{... eq. 1.5}$$

It is used to determine the accuracy of results as well as to compare the accuracies of results expressed in different units.

1.9. SIGNIFICANT FIGURES

An analytical method to discover the source and magnitude of errors requires careful technique and processing of data and also appropriate statistical methods. Initial data is must be reported with a precision means that indicated by the no. of figures. Some operation and calculations involves in these data must preserve the correct no. of figures, so, that the results give a true indication of both the accuracy and precision of analysis.

Significant figures usually have economic importance, i.e. reported in a measurement and are more costly for the analysis. That means increased cost include more expensive instrumentation, higher quality reagents and increased expenditure of time and effort in the analysis. Once experimental observations have been recorded using the number of significant figures can must be taken to the results of mathematical operations correctly represents the precision of the original measurement. In simple arithmetic operations the precision of experimental measurements are not increase means the no. of significant figures are not reduced by simple operation. Calculators and computers are used to produce of original data and these are used for the present to ignore the rules for handling significant figures. Ex: Limit of error of a measurement is known, i.e. represented as: 3946 ± 11. If the error is not stated, the last digit of the measured value is assumed to be uncertain and therefore determines the no. of significant figures. So, the value 3946 contains four significant figures and may represent a value between 3945.5 and 3946.5.

1.10. ACCURACY AND PRECISION

In the fields of science, engineering, industry and statistics, the accuracy of a measurement system is the degree of closeness of measurements of a quantity to its actual (true) value. Accuracy indicates proximity of measurement results to the true value. The precision of a measurement system also called reproducibility or repeatability is the degree to which repeated measurements under unchanged conditions show the same results.

A measurement system can be accurate but not precise, precise but not accurate neither or both. For example, if an experiment contains a systematic error then increasing the sample size generally increases precision but does not improve accuracy. Eliminating the systematic error improves accuracy but does not change precision. A measurement system is called valid if it is both accurate and precise.

1.10.1. Target analogy for accuracy versus precision

The analogy used to explain the difference between accuracy and precision. In this analogy, repeated measurements are compared to arrows that are shot at a target. Accuracy describes the closeness of arrows to the bulls eye at the target center. Arrows that strike closer to the bull's eye are considered more accurate. The closer a system's measurements to the accepted value, the more accurate the system is considered to be.

To continue the analogy, if a large number of arrows are shot, precision would be the size of the arrow cluster. When only one arrow is shot, precision is the size of the cluster one would expect if this were repeated many times under the same conditions. When all arrows are grouped tightly together the cluster is considered precise since they all struck close to the same spot even if not necessarily near the bulls-eye. The measurements are precise though not necessarily accurate.

Ideally a measurement device is both accurate and precise with measurements all close to and tightly clustered around the known value. The accuracy and precision of a measurement process is usually established by repeatedly measuring some traceable reference standard. Such standards are defined in the International System of Units and maintained by national standards organizations such as the National Institute of Standards and Technology. This also applies when measurements are repeated and averaged.

In that case, the term standard error is properly applied: the precision of the average is equal to the known standard deviation of the process divided by the square root of the number of measurements averaged. Further, the central limit theorem shows that the probability distribution of the averaged measurements will be closer to a normal distribution than that of individual measurements. With regard to accuracy we can distinguish the difference between the mean of the measurements and the reference value, the bias. Establishing and correcting for bias is necessary for calibration and the combined effect of that and precision.

A common convention in science and engineering is to express accuracy and/or precision implicitly by means of significant figures. Here, when not explicitly stated, the margin of error is understood to be one-half the value of the last significant place. For instance, a recording of 843.6 m, or 843.0 m, or 800.0 m would imply a margin of 0.05 m while a recording of 8,436 m would imply a margin of error of 0.5 m (the last significant digits are the units).

A reading of 8,000 m, with trailing zeroes and no decimal point is ambiguous; the trailing zeroes may or may not be intended as significant figures. To avoid this ambiguity, the number could be represented in scientific notation: 8.0×10^3 m indicates that the first zero is significant (margin of 50 m) while 8.000×10^3 m indicates that all three zeroes are significant, giving a margin of 0.5 m. Similarly, it is possible to use a multiple of the basic measurement unit: 8.0 km is equivalent to 8.0×10^3 m. In fact, it indicates a margin of 0.05 km (50 m). However, reliance on this convention can lead to false precision errors when accepting data from sources that do not obey it. Looking at this in another way, a value of 8 would mean that the measurement has been made with a precision of 1 (the measuring instrument was able to measure only down to 1s place) whereas a value of 8.0 (though mathematically equal to 8) would mean that the value at the first decimal place was measured and was found to be zero. The measuring instrument was able to measure the first decimal place. The second value is more precise neither of the measured values may be accurate (the actual value could be 9.5 but measured inaccurately as 8 in both instances). Thus, accuracy can be said to be the correctness of a measurement while precision could be identified as the ability to resolve smaller differences.

Precision is sometimes stratified into Repeatability the variation arising when all efforts are made to keep conditions constant by using the same instrument and operator, and repeating during a short time period and Reproducibility - the variation arising using the same measurement process among different instruments and operators, and over longer time periods.

1.11. CONFIDENTIAL LIMIT

Data analysis can be said to connect with the study of populations and variation. Each measurement is containing as an individual value, then repetition of the measurement produces a cluster or aggregate of values know as the population. Statistics provides a means of determining the features of the parent population from a few repetitive individual measurements. Three major functions of statistics are used they are (1) To determine the properties of the aggregate populations. (2) To study the variations. (3) To reduce a large amount of data to a more easily comprehensible form.

The precision of an analytical method is usually indicated by its Confidence Limit i.e. the range of values about the mean that includes a specified % of the total observation. Precision is increase with decreasing value of the standard deviation. Other statistical methods that are used in analysis are (a) The Q-test for outlaying measurement (detects gross errors). (b) The t-test for significance in the difference between two means useful in comparing results from different instruments, methods or samples of the same composition. (c) The t-test for accuracy of results for comparing the results of several methods against know true value.

To individual tests for significance of results, control charts are useful monitoring tools in laboratories where large amount of data from a given method are generated over extended periods of time. Other statistical methods can't solve all the problems involved in error analysis; they do provide the analyst with a systematic, objective approach to these problems. Statistical analysis of data should determine when differences among results have exceeded reasonable limits and have become large enough of real variations in sample compositions or analytical techniques.

1.12. SIGNAL TO NOISE RATIO

As concentrations decrease to trace levels or as signal sources become weak, the problem of distinguishing signals from noise becomes increasingly difficult, resulting in decreased accuracy and precision in measurements. The ability of an instrument system to discriminate between signals and noise is usually expressed as a signal to noise ratio (S/N) where

$$S/N = \text{average signal amplitude/}$$
$$\text{average noise amplitude}$$

In the case of dc signals, an increase in the S/N ratio usually indicates a reduction in noise and thus a more desirable measurement. Once the physical or chemical quantity of interest is converted to an electrical signal, the S/N ratio cannot the increased by simple amplification alone, since each increase in the magnitude of the signal is accompanied by a corresponding increase in the value of the noise. Thus higher S/N ratios are usually obtained by electronic hardware devices (filters, lock in amplifiers) or software algorithms.

1.13. SENSITIVITY

The word sensitivity is often used in describing an analytical method. Unfortunately, it is occasionally used indicrimininately and incorrectly. The

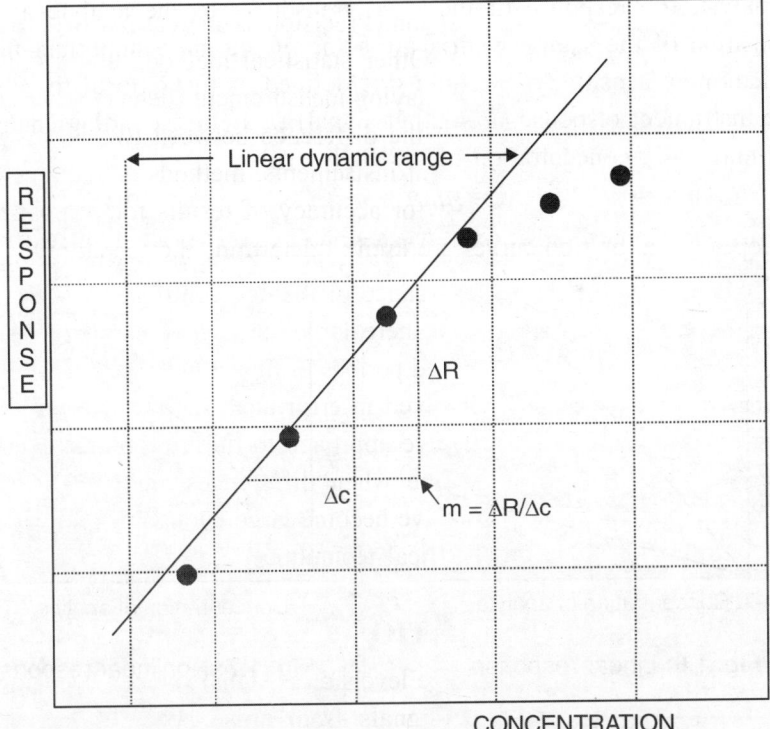

Fig. 1.5: Calibration curve of response vs. concentration

most often used definition of sensitivity is the calibration sensitivity or the change in the response signal per unit change in analyte concentration. The calibration sensitivity is thus the slope of the calibration curve shown in Fig. 1.5. The slope of the calibration curve is called the calibration sensitivity, m. The detection limit, DL, designated the lowest concentration that can be measured at a specified confidence level.

If the calibration curve is linear, the sensitivity is constant and independent of concentration. If nonlinear sensitivity is changes with concentration and is not single value. The calibration sensitivity does not indicate what concentration differences can be detected. Noise in the response signals must be taken into account to be quantitative about what differences can be detected. For this reason, the term analytical sensitivity is sometimes use. The analytical sensitivity is the ratio of the calibration curve slope to the standard deviation of the analytical signal at a given analyte concentration. The analytical sensitivity is usually a strong function of concentrations.

A number of parameters, including the *S/N* ratio, affect the sensitivity of a particular instrumental method. Physical and chemical properties of the analyte, the response of the input transducer to the analyte and the composition of the sample matrix are some of the more important factors that determine sensitivity. Sensitivity is defined as the ratio of the change in the instrument response (*Io*, output signal) to a corresponding change in the stimulus (*C*, concentration of the analyte):

$$S = dIo/Dc \qquad \qquad \text{... eq. 1.6}$$

Slopes of calibration curves are used to determine the sensitivity values.

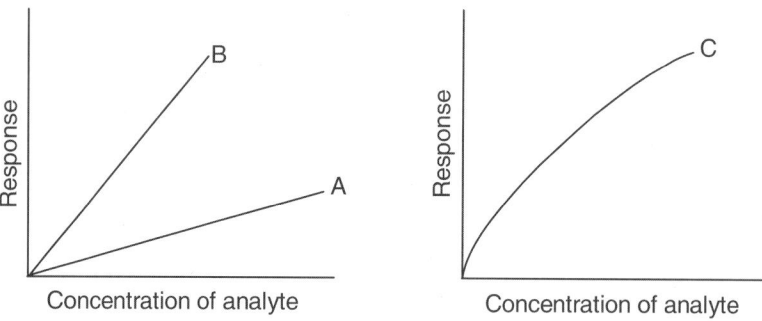

Fig. 1.6: Linear response   **Fig. 1.7:** Nonlinear response

Fig. 1.6 shows a linear response (constant sensitivity) over the entire range of measured concentrations for both substance *A* and *B*. From the slopes of the cures, the sensitivity of the method is much greater for substance *B* than for substance *A*. The nonlinear response in Fig. 1.7 indicated a constantly changing value for sensitivity as a function of concentration. Measurements of substance C become less sensitive with increasing concentration. Sensitivity may also be expressed as the concentration of analyte required to cause a given instrument response. For example In atomic absorption spectroscopy, sensitivity is expressed as concentration in micrograms per milliliter of analyte that produces an absorbance of 0.0043 absorbance unit (1.0% absorption). When comparing different techniques or instrument, one should be alert to the procedures used by the practitioners to arrive at sensitivity values.

1.14. DETECTION LIMIT

Detection limit is commonly understood to be the smallest concentration we can measure with a particular technique. In fact it is the point at which we can make a decision whether the element or compound is present or not. To be able to measure it we need at least three times the detection limit. Three times the detection limit is often called the limit of determination. A different term used for detection limits is "Limits of Detection" with the abbreviation *LoD*.

The minimum single result with a stated probability can be distinguished from a suitable blank value. The limit defines the point at which the analysis becomes possible and this may be different from the lower limit of the determinable analytical range. In chemical analysis, the minimum amount of a particular component that can be determined by a single measurement with a stated confidence level. *LOD* or detection limit (*DL*), is the lowest amount of analyte in a sample that can be detected, but not necessarily quantities as an exact value. The *LOD* may be expressed as:

$$LOD = 3.3 * SD/S \hspace{3cm} ... eq. \ 1.7$$

Where:

SD = the standard deviation of the response

S = the slope of the calibration curve

As the concentration of the analyte approaches zero the signal disappears into the noise and the detection limit is exceeded. The detection limit is most generally defined as the concentration of analyte that give a signal, x, significantly different from the blank or background signal x_B. This definition leaves the analyst with considerable freedom to define the phrase significantly different. When working with analytes in trace amounts the analyst is confronted with two problems:

  i. Reporting an analyte present when in fact it is absent and

ii. Reporting an analyte absent when it is present.

The analytical chemistry has defined this difference to be an analyzed concentration that produces a signal two times the standard deviation of the blank signal. Current guidelines define the Detection Limit as

$$x - x_B = 3s_B \hspace{4cm} ...\ eq.\ 1.8$$

Where x is the signal with minimum detectable analyte concentration, x_B is the signal of the blank and s_B is the standard deviation of the blank readings.

1.15. SUMMARY

Like all fields of chemistry, analytical chemistry is too broad and too active a discipline for us to define completely. Analytical chemistry is often described as the area of chemistry responsible for characterizing the composition of matter both qualitatively and quantitatively. An analytical chemistry as the application of chemical knowledge ignores the unique perspective that analytical chemists bring to the study of chemistry.

Qualitative Analysis is that branch of analytical chemistry which treats of the determination of the elements present in a chemical but does not deal with the quantities. Qualitative inorganic analysis seeks to establish the presence of a given element or inorganic compound in a sample, the presence of a given functional group or organic compound in a sample, it establishes the chemical identity of the species in the sample and an integral part of the separation step.

Quantitative analysis seeks to establish the amount of a given element or compound in a sample, determines the relative amount of the species or analytes in numerical terms and identity of the analytes.

Gravimetric methods determine the mass of the analyte or some compound chemically related to it.

In volumetric methods, the volume of a solution containing sufficient reagent to react completely with the analyte is measured. Electro-analytical methods involve the measurement of such electrical properties as potential, current, resistance and quantity of electrical charge.

Law of Universal Gravitation - Every object in the Universe attracts every other object with a force directed along the line of centers for the two objects that is proportional to the product of their masses and inversely proportional to the square of the separation between the two objects.

$$Fg = Gm_1m_2/r^2$$

Precipitation is accomplished by combining a selected ion(s) in solution with a suitable counter ion in sufficient concentrations to exceed the solubility of the resulting compound and produce a supersaturated solution. Precipitation at higher temperature slows nucleation and crystal growth because of the increased thermal motion of the particles in solution

Solvent: The nature of the solvent affects the solubility of an ionic solid (precipitate) in the solvent. The polarity of water can be reduced by the addition of other miscible solvents such as alcohols thereby reducing the solubility of precipitates.

Spectroscopy is a technique that uses the interaction of energy with a sample to perform an analysis. The data that is obtained from spectroscopy is called a spectrum.

Spectroscopy measures the interaction of the molecules with electromagnetic radiation, applications such as atomic absorption spectroscopy, atomic emission spectroscopy, ultraviolet-visible spectroscopy, X-ray fluorescence spectroscopy, infrared spectroscopy, Raman spectroscopy, nuclear magnetic resonance spectroscopy, photoemission spectroscopy and so on.

Crystallography is a technique that characterizes the chemical structure of materials at the atomic level by analyzing the diffraction patterns of usually x-rays that have been deflected by atoms in the material. From the raw data the relative placement of atoms in space may be determined.

A hyphenated separation technique refers to a combination of two or more techniques to detect and separate chemicals from solutions. Hyphenated techniques are widely used in chemistry and biochemistry.

Analytical chemistry has played critical roles in the understanding of basic science to a variety of practical applications, such as biomedical applications, environmental monitoring, quality control of industrial manufacturing and forensic science.

The ability of an instrument system to discriminate between signals and noise is usually expressed as a signal to noise ratio (S/N) where S/N = average signal amplitude/average noise amplitude.

Detection limit is commonly understood to be the smallest concentration we can measure with a particular technique. The limit defines the point at which the analysis becomes possible and this may be different from the lower limit of the determinable analytical range. *LOD* or detection limit (*DL*), is the lowest amount of analyte in a sample that can be detected, but not necessarily quantities as an exact value. The *LOD* may be expressed as:

$$LOD = 3.3 * SD/S.$$

Sensitivity is the calibration sensitivity or the change in the response signal per unit change in analyte concentration. If the calibration curve is linear the sensitivity is constant and independent of concentration. If nonlinear sensitivity is changes with concentration and is not a single value. Sensitivity is defined as the ratio of the change in the instrument response (*Io*, output signal) to a corresponding change in the stimulus (*C*, concentration of the analyte):

$$S = dIo/Dc.$$

Chapter 2
MICROSCOPY

2.0. INTRODUCTION

One of the tools essential for studying living organisms that are too small to be seen with the naked eye, is the microscope. The Microscope was invented by either Zacharias Janssen of the Netherlands or Galielo Galielei of Italy by the end of 15th century. In 1976, Antony van Leeuwenhoek, for the first time used crude, simple microscope consisting of a biconcave lens enclosed in two metal plates to describe bacterial and protozoa. These simple microscopes magnified objects 200 to 300 times. The inventors and early discoverers of the compound microscope, which consisted of a two lens system are still in dispute but it is true that Robert Hooke had used a compound microscope in 16th century. In a compound microscope the second lens magnifies the image from the first lens and consequently the size of the observed image depends on the magnification of both lenses.

Theoretically, the size of an image can be increased by additional lenses. The magnification is a process of enlarging the image of an object is only useful and the enlarged images are clearly visible. Effective magnification depends on the resolving power of a microscope. The resolving power is the ability to distinguish two adjacent objects as separate and distinct images rather than as a single blurred image. The lens system of the human eye has a resolving power of approximately 0.2 mm which means that the eye cannot see anything that is smaller than 0.2 mm. Proper illuminations is essential for efficient magnification and resolution of a microscope. The light from the illuminating source is reflected into the condenser by the mirror or passed directly through the condenser. Artificial light from a tungsten lamp is most commonly used in microscopy. The amount of light entering the condenser is controlled by the iris diaphragm. Excessive illumination should be avoided because it obscures the specimen (lack of contrast).

2.1. PRINCIPLE

Understanding of basic principles such as magnification, resolution, numerical aperture, illumination and focusing are essential to make use the microscope efficiently to overcome the small troubles and to translate the observations.

2.1.1. Magnification

Magnification or enlargement can be defined as the number of times the specimen is enlarged when observed through a particular microscope. It is the cumulative effect brought about by two lenses system, the ocular lens (eye piece) and the objective lens. Linear magnifications obtained of different combinations of objective lens and ocular lens (table 2.1). The magnification of a microscope depends on the type of objective lens used with the ocular. The total magnification of the object is calculated by multiplying the magnification of the objective lens by the magnification of the ocular lens. The useful magnification of a microscope is that which makes the smallest visible object clearly resolvable.

Table 2.1: Linear magnifications of different combinations of Objective lens and Ocular lens.

Magnification	Objective lens	Ocular lens	Total Magnification
Scanning	4X	10X	40X
Low power	10X	10X	100X
High power	45X	10X	450X
Oil immersion	100X	10X	1000X

2.1.2. Resolving Power or Resolution

The resolving power is the ability of a lens to show two adjacent objects as discrete units. Increased magnification without proper resolving power gives a blurred picture without finer details or clarity. The resolving power of a lens is dependent on the wavelength of the light used (resolution increases with decrease in wavelength of the light), characteristics of the lens and the numerical aperture.

$$\text{Resolving power} = \frac{\text{Wave length of light}}{2(\text{Numerical aperture})} \qquad \textit{... eq. 2.1}$$

Microscope resolving power is dependent on the wavelength of the beam used for illumination and on the optical quality of the lenses. Shorter wavelengths give better resolution. Resolving power (RP) is determined by the following formula:

$$\text{Resolving power (RP)} = \frac{\lambda}{2xN.A.} \qquad \dots eq.\ 2.2$$

Where

λ = Wavelength of light used which is usually set at 550 nm.

NA = Numerical aperture

The resolving power similar to magnification and can be increased by increasing the numerical aperture.

2.1.3. Numerical aperture (NA)

The numerical aperture (NA) can be defined as function of the diameter of the objective lens in relation to its focal length. Numerical aperture can be expressed by the formula:

Numerical aperture (NA) = $\eta \sin\theta$ *... eq. 2.3*

Where

   η = The refractive index of the medium

   θ = The trigonometric sine of one-half the angle

Numerical aperture (NA) is fixed on the barrel of objectives and specifies the maximum obtainable resolving power. With dry objectives the value of is one since one is the refractive index of air. When immersion oil is employed as the medium, η is 1.56 and θ is 58° then,

   Numerical aperture (NA) = sin

$$= 1.56 \times \sin 58°$$

$$= 1.56 \times 0.85 = 1.33$$

For the low power objective the Numerical aperture (NA) is 0.25, for the high power objective the Numerical aperture (NA) 0.65 and for the oil-immersion objective the Numerical aperture (NA) 1.25.

Numerical aperture is doubled by the use of sub stage condenser that illuminates the specimen with rays of light which pass through the specimen indirectly as well as directly. Resolving power of a lens can be expressed as follows:

$$RP = \frac{\lambda}{2(NA)} \qquad \dots eq.\ 2.4$$

Thus Resolving power (RP) for the low-power, high-power and oil-immersion objectives can be calculates as follows:

$$\text{RP for the low-power objective} = \frac{550\text{nm}}{2 \times 0.25} = 1100\text{nm} = 1.1\mu\text{m}$$

$$\text{RP for the high-power objective} = \frac{550\text{nm}}{2 \times 0.65} = 423.07\text{nm} = 0.42\mu\text{m}$$

$$\text{RP for the oil-immersion objective} = \frac{550\text{nm}}{2 \times 1.25} = 220\text{nm} = 0.22\mu\text{m}$$

Thus the objects smaller than and/or placed close to each other cannot be seen clearly with low-power or oil-immersion objectives, if they are 1.1μm and 0.22μm respectively.

2.1.4. Refractive Index (RI)

Refractive Index (RI) can be defined as the bending power of light passing through one medium to another medium. The RI of air is lower than that of glass and as light rays pass from the glass slide into air, they are bent or refracted so that they do not pass into the objective lens. This could cause a loss of light, which would reduce the numerical aperture and diminish the resolving power of the objective lens. Loss of refracted light can be compensated by interposing mineral oil, which has the same R.I. as glass, between the slide and the objective lens. In this way, decreased light refraction occurs and more light rays enter directly into the objective lens, producing a clear image with high resolution.

2.1.5. Illumination

High possible magnification and resolution can be achieved only through proper illumination. A variety of light sources including day light are used since the intensity of day light of an uncontrolled variable and an artificial light from a tungsten lamp is the most commonly used light source in microscopy. For any light a condenser should be used for proper illumination.

2.1.6. Working distance

It is the distance between the objective lens and slide. The working length decreases as the magnification increases. Similarly, the angle aperture of the objective lens increases.

2.2. TYPES

Depending upon the lens systems, microscopes are categorized into two types (1) Simple Microscope (2) Compound Microscope. A simple microscope consists of a single lens system while a compound microscope consists of two or more lens system. Depending upon the source of illumination, the microscope can be classified into two types (Table 2.1) (1) Light Microscopes - Where the specimen is illuminated by visible light or ultraviolet rays. Ex.: Bright filed, Dark field Microscope, Ultraviolet Microscope, Phase-contrast Microscope and Fluorescent Microscope (2) Electron Microscope - Where the image is formed on a fluorescent screen by electron beams instead of light rays and focused by magnets instead of lenses.

2.2.1. Bright field microscope

Bright field microscopes are the classical microscopes in which the sample is illuminated from the back and the image is formed due to the absorbing properties of the imaged objects. In bright field imaging all light from the specimen and its surroundings is collected by the objective to form an image against a bright background. All light microscopes are capable of bright field imaging.

2.2.1.1. Instrumentation

The Bright field microscope consists of a sturdy metal body or stand composed of a base and an arm to which the remaining parts are attached. A good quality microscope has a built-in illuminator, adjustable condenser with aperture diaphragm (contrast) control, mechanical stage and binocular eyepiece tube. A light source either a mirror or an electric illuminator is located in the base. Two focusing knobs, the fine and coarse adjustment knobs are located on the arm and can move either the stage or the nosepiece to focus the image (Fig. 2.1). The coarse and fine focus knobs are used to sharpen the image of the specimen. They are frequently unaware of adjustments to the condenser that can affect resolution and contrast.

Some condensers are fixed in position; others are focusable so that the quality of light can be adjusted. Usually the best position for a focusable condenser is as close to the stage as possible. The bright field condenser usually contains an aperture diaphragm, a device that controls the diameter of the light beam coming up through the condenser, so that when the diaphragm is stopped down (nearly closed) the light comes straight up through the center of the condenser lens and contrast is high (Fig. 2.1).

Table 2: Comparison of five types of Microscopes

S. No.	Name of the Microscope	Maximum Magnification	Appearance of Specimen	Common Uses	Advantages	Disadvantages
1.	Bright-field Microscopy	1000X	Specimen stained or unstained; bacterium generally stained and appear colour of stain	Observing stained specimens; counting microscopes	Easy to use; readily available; relatively inexpensive; allows staining reactions to be interpreted	Lack of contrast; inability to resolve very thin bacteria and viruses; introduction of artifacts during staining procedures
2.	Dark-field Microscopy	1000X	Generally unstained; appears bright or lighted in an otherwise dark field.	Viewing unstained cells not observable with bright field microscope	Allow living microbes to be viewed Ex: Spirochetes	Staining reactions cannot be used for examining stained specimens.
3.	Fluorescence Microscopy	1000X	Bright and coloured; colour of the fluorescent dye	Detecting specific infectious agents in tissue, detecting immunological reactions	Rapid identification of infectious microorganisms	Specimens that naturally fluoresce or which are stained with a fluorescent dye are only observed.
4.	Phase-Contrast Microscopy	1000X	Structures varying degrees of darkness	Observing living unstained cell; revealing intracellular structures	Enhances subcellular anatomy; allows observation of motility, phagocytes and biological activities	Inability to evaluate staining reactions
5.	Electron Microscopy	1,000.000X	Viewed on fluorescent screen	Examination of viruses and ultra structure of cells; diagnosis certain virus diseases and cancers; detecting certain giant molecules.		

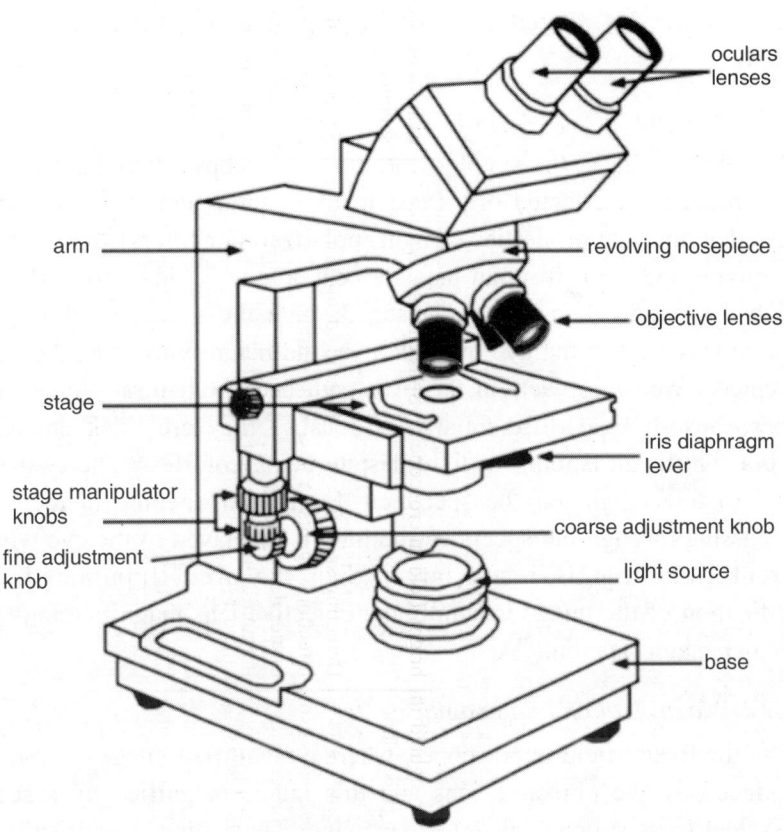

oculars
lenses

arm

revolving nosepiece

objective lenses

stage

iris diaphragm
lever

stage manipulator
knobs

coarse adjustment knob

fine adjustment
knob

light source

base

Fig. 2.1: Components of the Bright Field Microscope.

When the diaphragm is wide open the image is brighter and contrast is low. The stage is positioned about halfway up the arm and holds the microscope slides by either simple slide clips or a mechanical stage clip. A mechanical stage allows the operator to move a slide around smoothly during viewing by use of stage control knobs. The sub-stage condenser is mounted within or beneath the stage and focuses a cone of light on the slide. Its position often is fixing in simpler microscopes.

The curved upper part of the arm holds the body assembly to which a nosepiece and one or more eyepieces or ocular lenses are attached. More advanced microscopes have eyepieces for both eyes and are called binocular microscopes. The body assembly itself contains a series of mirrors and prisms so that the barrel holding the eyepiece may be tilted for ease in viewing. The nosepiece holds three to five objective lenses of differing magnifying power and can be rotated to position any objective beneath the body assembly.

2.2.1.2. Principle

Bright field imaging is the simplest form of microscopy where light is either passed through or reflected off, a specimen. Illumination is not altered by devices that alter the properties of light (polarizer's or filters). With a bright field microscope, light from an incandescent source is aimed toward a lens beneath the stage called the condenser, through the specimen, through an objective lens, and to the eye through a second magnifying lens, the ocular or eyepiece. We see objects in the light path because natural pigmentation or stains absorb light differentially or because they are thick enough to absorb a significant amount of light despite being colorless. The condenser is used to focus light on the specimen through an opening in the stage. After passing through the specimen the light is displayed to the eye with an apparent field that is much larger than the area illuminated. The magnification of the image is simply the objective lens magnification times the ocular magnification.

2.2.1.2.1. Magnification and imaging

Most of the Bright field microscopes, where a magnified image of an object is produced by the objective lens and this image magnified by a second lens system (ocular or eyepiece) for viewing. Thus, final magnification of the microscope is dependent on the magnifying power of the objective times the magnifying power of the ocular. Ocular magnification ranges are typically 8X-12X though 10X oculars are most common.

Each objective lens consists of six or more pieces of glass that combine to produce a clear image of an object. The six or more lenses in the objective lens are needed to provide corrections that produce image clarity. The interaction of light with the glass in a lens produce aberrations that result in a loss in image quality because light waves will be bent or refracted, differently in different portions of a lens and different colors of light will be refracted to different extents by the glass. The oculars in most microscopes are designed to work optimally with the objective lenses from the same manufacturer. The ocular designed to provide a corrected virtual image when viewed by eye is not suitable for the generation of photographic or video images through the microscope.

2.2.1.2.2. Illumination

The best way to illuminate the specimen involves the use of another lens system, known as a condenser. Two apertures in the illumination system allow, regulating the diameter of the illumination beam by closing or opening iris diaphragms. The first diaphragm is housed within the bright field condenser known as the condenser diaphragm, allow increasing contrast and the second diaphragms known as the field aperture diaphragm, does not affect resolution as dramatically and is regularly adjusted for optimal illumination.

2.2.1.2.3. Mount the specimen on the stage

High magnification objective lenses can't focus through a thick glass slide; they must be brought close to the specimen. The stage may be equipped with simple clips or with some type of slide holder. The slide may require manual positioning or there may be a mechanical stage that allows precise positioning without touching the slide.

2.2.1.3. Applications

1. In biological applications, bright field observation is widely used for stained or naturally pigmented or highly contrasted specimens mounted on a glass microscope slide.

2. It is widely used in pathology to view fixed tissue sections or cell films or smears.

3. Light is reflected from opaque samples and this is exploited in industrial environments where bright field imaging is used for wafer inspection and liquid crystal board inspection.

4. Bright field microscopy is best suited to viewing stained or naturally pigmented specimens such as stained prepared slides of tissue sections or living photosynthetic organisms.

5. It is useless for living specimens of bacteria and inferior for non-photosynthetic protists or metazoans, or unstained cell suspensions or tissue sections.

6. A *Paramecium* should show up fairly well in a bright field microscope, although it will not be easy to see cilia or most organelles.

2.2.1.3. Care of microscope

1. Everything on a good quality microscope is unbelievably expensive, so be careful.

2. Hold a microscope firmly by the stand only. Never grab it by the eyepiece holder, for example, hold the plug (not the cable) when unplugging the illuminator, since bulbs are expensive and have a limited life turn the illuminator off when you are done.

3. Always make sure the stage and lenses are clean before putting away the microscope.

4. Never use a paper towel, a Kim wipe, shirt, or any material other than good quality lens tissue or a cotton swab (must be 100% natural cotton) to clean an optical surface. Organic solvents may separate or damage the lens elements or coatings.

5. Cover the instrument with a dust jacket when not in use.

6. Focus smoothly don't try to speed through the focusing process or force anything.

2.2.2. Dark field microscope

Microscopes are used to magnify objects. Through magnification, an image is made to appear larger than the original object. The magnification of an object can be calculated roughly by multiplying the magnification of the objective lens times and the magnification of the ocular lens. Objects are magnified to be able to see small details. There is no limit to the magnification that can be achieved however, there is a magnification beyond which detail does not become clearer. The result is called empty magnification when objects are made bigger but their details do not become clearer. Therefore, not only magnification but resolution is important to the quality of the information in an image.

Light microscopy is an important investigative tool for biology. The structure of many biological specimens is of low contrast that cannot be revealed by the bright field microscope. Standard bright field microscopy relies upon light from the lamp source being gathered by the sub stage condenser and shaped into a cone whose apex is focused at the plane of the specimen. Specimens are seen because of their ability to change the speed and the path of the light passing through them, this ability is dependent upon the refractive index and the opacity of the specimen. To see a specimen in a bright field microscope, the light rays passing through it must be changed sufficiently to be able to interfere with each other which produce contrast (differences in light intensities) and, thereby build an image. If the specimen has a refractive index too similar to the surrounding medium between the microscope stage and the objective lens, it will not be seen. To visualize biological materials well, the materials must have this inherent contrast caused by the proper refractive indices or be artificially stained. These limitations require instructors to find naturally high contrast materials or to enhance contrast by staining them. Adequately visualizing transparent living materials or thin unstained specimens is not possible with a bright field microscope.

A simple, inexpensive modification that changes a bright field microscope into a dark field microscope allowing low contrast samples to be examined. Dark field microscopy is a specialized illumination technique that capitalizes on oblique illumination to enhance contrast in specimens that are not imaged well under normal bright field illumination conditions. The direct light has been blocked by an opaque stop in the sub-stage condenser, light passing through the specimen from oblique angles at all azimuths is diffracted, refracted and reflected into the microscope objective lens to form a bright image of the specimen superimposed onto a dark background.

Dark field microscopy relies on a different illumination system rather than illuminating the sample with a filled cone of light, the condenser is designed to form a hollow cone of light. The light at the apex of the cone is focused at the plane of the specimen; as this light moves through the specimen plane it spreads again into a hollow cone. The objective lens sits in the dark hollow of this cone; although the light travels around and past the objective lens, no rays enter. The entire field appears dark when there is no sample on the microscope stage; thus the name dark field microscopy.

2.2.2.1. Principle

When a sample is on the stage, the light at the apex of the cone strikes it. The image is made only by those rays scattered by the sample and captured in the objective lens. The image appears bright against the dark background. This situation can be compared to the glittery appearance of dust particles in a dark room illuminated by strong shafts of light coming in through a side window. The dust particles are very small, but are easily seen when they scatter the light rays. This is the working principle of dark field microscopy and explains how the image of low contrast material is created: an object will be seen against a dark background if it scatters light which is captured with the proper device such as an objective lens.

The highest quality dark field microscopes are equipped with specialized costly condensers constructed only for dark field application. This dark field effect can be achieved in a bright field microscope by the addition of a simple stop. The stop is a piece of opaque material placed below the sub stage condenser; it blocks out the center of the beam of light coming from the base of the microscope and forms the hollow cone of light needed for dark field illumination. To view a specimen in dark field, an opaque disc is placed underneath the condenser lens, so that only light that is scattered by objects on the slide can reach the eye. Instead of coming up through the specimen, the light is reflected by particles on the slide, everything is visible regardless of color, usually bright white against a dark background.

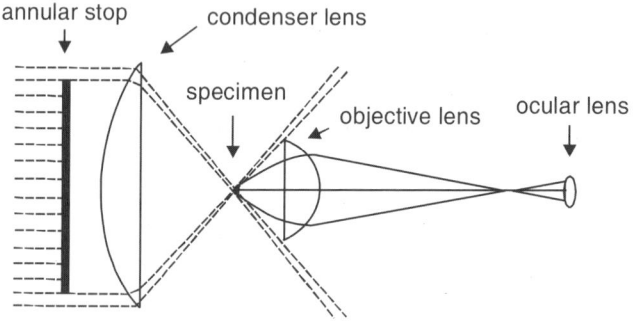

Fig. 2.2: The setup of the Dark Field Microscopy light path

Dark field microscopy is a method which also creates contrast between the object and the surrounding field. As the name implies, the background is dark and the object is bright. An annular stop is also used for dark field, but the stop is now outside the field of view. Only light coming from the outside of the beam passes through the object and it cannot be seen directly. Only when light from the stop is deflected and deviated by the object can it be seen. This method also produces a great deal of glare and therefore the specimen often appears as a bright shape rather than as a bright object of which much detail can be determined. Fig. 2.2 shows the setup of the dark field light path. Do need a higher intensity light, since seeing only reflected light. At low magnification (up to 100x) any decent optical instrument can be set up so that light is reflected toward the viewer rather than passing through the object directly toward the viewer.

2.2.2.2. Applications

1. Suspensions of cells and samples of pond water look spectacular in dark field.

2. While specimens may look washed out and lack detail in bright

field, Metazoans, Cell suspensions, Algae, and other Microscopic Organisms are clearly distinguished and their details show up well.

3. At 100x can see bacteria, even distinguish some structure of rods, curved rods, spirals, or cocci and movement.

4. Non-motile bacteria look like vibrating bright dots against a dark background.

5. Motile bacteria can be seen moving in a definite direction, sometimes remarkably fast.

6. In pond water samples may find *Spirillum volutans* a very large (up to 0.5 mm) motile spiral bacterium.

7. It also is used to identify certain bacteria like the thin and distinctively shaped *Treponema pallidum* the causative agent of syphilis.

8. It can reveal considerable internal structure in larger eukaryotic microorganisms.

2.2.3. Fluorescence microscope

2.2.3.1. Introduction

A fluorescence microscope is basically a conventional light microscope with added features and components that extend its capabilities. (1) A conventional microscope uses light to illuminate the sample and produce a magnified image of the sample. (2) A fluorescence microscope uses a much higher intensity light to illuminate the sample. This light excites fluorescence species in the sample, which then emit light of a longer wavelength. A fluorescent microscope also produces a magnified image of the sample, but the image is based on the second light source, the light emanating from the fluorescent species, rather than from the light originally used to illuminate and excite the sample.

British scientist Sir George G. Stokes first described fluorescence in 1852 and observed that the mineral fluorspar emitted red light when it was illuminated by ultraviolet excitation. Stokes noted that fluorescence emission always occurred at a longer wavelength than that of the excitation light. Early investigations in the 19th century showed that many specimens including minerals, crystals, resins, crude drugs, butter, chlorophyll, vitamins, and inorganic compounds fluoresce when irradiated with ultraviolet light. However, it was not until the 1930s that the use of fluorochromes was initiated in biological investigations to stain tissue components, bacteria and other pathogens. Several of these stains were highly specific and stimulated the development of the fluorescence microscope. Fluorescent Microscope process consists of coating a microbe with fluorescent dye such as fluorescent and illuminating it with ultraviolet light. As the electrons in the dye are excited, they move to high energy levels, and then quickly drop back giving off the excess energy as visible light.

The technique of fluorescence microscope has become an essential tool in biology and biomedical sciences as well as in material science due to attributes that are not readily available in other contrast modes with traditional optical microscopy. The application of an array of fluorochromes has made it possible to identify cells and sub-microscopic cellular components with a high degree of specificity amid non-fluorescing material. In fact, the fluorescence microscope is capable of revealing the presence of a single molecule, through the use of multiple fluorescence labeling, different probes can simultaneously identify several target molecules. Although the fluorescence microscope cannot provide spatial resolution below the diffraction limit of specific specimen features, the detection of fluorescing molecules below such limits is readily achieved. Fluorescence microscope is a rapid expanding technique, both in the medical and biological sciences. The technique has made it possible to identify cells and cellular components with a high degree of specificity. For example, certain antibodies and disease conditions or impurities in inorganic material can be studied with the fluorescence microscopy.

2.2.3.2. Principle

The basic function of a fluorescence microscope is to irradiate the specimen with a desired and specific band of wavelengths and then to separate the much weaker emitted fluorescence from the excitation light. In a properly configured microscope, only the emission light should reach the eye or detector so that the resulting fluorescent structures are superimposed with high contrast against a very dark or black background. The limits of detection are generally governed by the darkness of the background and the excitation light is typically several hundred thousand to a million times brighter than the emitted fluorescence.

The basic task of the fluorescence microscope is to let excitation light radiate the specimen and then sort out the much weaker emitted light to make up the image. The microscope has a filter that only lets through radiation with the desired wavelength that matches fluorescing material. The radiation collides with the atoms in specimen and electrons are excited to a higher energy level. When they relax to a lower level, they emit light; to become visible the emitted light is separated from the much brighter excitation light in a second filter (Fig. 2.3). Here, the fact that the emitted light is of lower energy and has a longer wavelength. The fluorescing areas can be observed in the microscope and shine out against a dark background with high contrast.

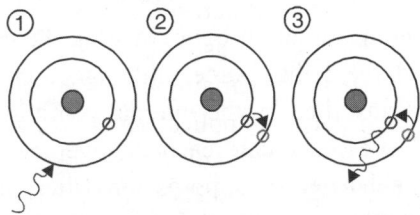

Fig. 2.3: Excitation and Emission processes (1) Energy is absorbed by the atom which becomes excited. (2) The electron jumps to a higher energy level. (3) Soon, the electron drops back to the ground state, emitting a photon or a packet of light the atom is fluorescing.

The selected wavelengths after passing through the excitation filter, reach the dichromatic beam splitting mirror which is a specialized interference filter that efficiently reflects shorter wavelength light and efficiently passes longer wavelength light. The dichromatic beam splitter is tilted at a 45°

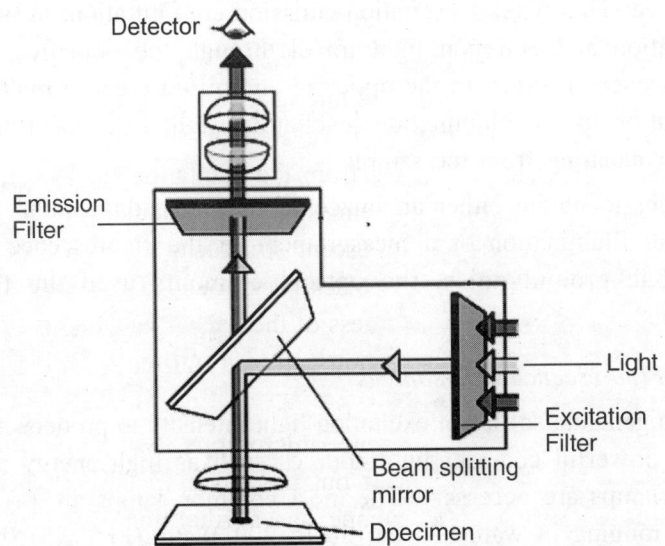

Fig. 2.4: The Light pathway of the Fluorescence Microscope

angle with respect to the incoming excitation light and reflects this illumination at a 90° angle directly through the objective optical system and onto the specimen. Fluorescence emission produced by the illuminated specimen is gathered by the objective, now serving in its usual image forming function. Because the emitted light consists of longer wavelengths than the excitation illumination, it is able to pass through the dichromatic mirror and upward to the observation tubes or electronic detector (Fig. 2.4).

Most of the scattered excitation light reaching the dichromatic mirror is reflected back toward the light source, although a minute quantity often passes through and is absorbed by the internal coating of the mirror block. Before the emitted fluorescence can reach the eyepiece or detector, it must first pass through the barrier or suppression filter, this filter blocks or suppresses any residual excitation light and passes the desired longer emission wavelengths.

2.2.3.3. Instrumentation

Basic Requirements of Fluorescence Microscope are Optics; nearly all fluorescence microscopes use the objective lens to perform two functions (1) Focus the illumination (excitation) light on the sample: In order to excite fluorescent species in a sample, the optics of a fluorescent microscope must focus the illumination (excitation) light on the sample to a greater extent than is achieved using the simple condenser lens system found in the illumination light path of a conventional microscope. (2) Collect the emitted fluorescence: This type of excitation-emission configuration, in which both the excitation and emission light travel through the objective, is called epifluorescence. The key to the optics in an epifluorescence microscope is the separation of the illumination (excitation) light from the fluorescence emission emanating from the sample.

In order to obtain either an image of the emission without excessive background illumination or a measurement of the fluorescence emission without background noise, the optical elements used the following components.

2.2.3.3.1. Fluorescence light sources

In order to generate sufficient excitation light intensity to produce detectable emission, powerful compact light sources such as high energy short arc-discharge lamps are necessary. The most common lamps are (a) Mercury burners - ranging in wattage from 50 to 200 Watts (Fig. 2.5) (b) Xenon burners – ranging in wattages from 75 to 150 Watts (Fig. 2.6). These light sources are usually powered by an external direct current supply, furnishing enough startup power to ignite the burner through ionization of the gaseous vapor and to keep it burning with a minimum of flicker. The arc-discharge lamp external power supply is usually equipped with a timer to track the number of hours. Arc lamps lose efficiency and are more likely to shatter if used beyond their rated lifetime (200-300 hours). The mercury burners do not provide even intensity across the spectrum from ultraviolet to infrared, and much of the intensity of the lamp is expended in the near ultraviolet. Prominent peaks of intensity occur at 313, 334, 365, 406, 435, 546 and 578 nm.

Fig. 2.5: Mercury HBO Arc-Discharge Fluorescence lamp

Fig. 2.6: Xenon HBO Arc-Discharge Fluorescence lamp

2.2.3.3.2. Dichroic mirror

In a fluorescence microscope, a dichroic mirror is used to separate the excitation and emission light paths. Within the objective, the excitation and emission shares the same optics. Fig. 2.7 shows the dichroic mirror's position in an inverted fluorescence microscope. In a fluorescence microscope, the dichroic mirror separates the light paths. (1) The excitation light reflects off the surface of the dichroic mirror into the objective. (2) The fluorescence emission passes through the dichroic to the eyepiece or detection system.

The dichroic mirror's special reflective properties allow it to separate the two light paths. Each dichroic mirror has a set wavelength value called the transition wavelength value which is the wavelength of 50% transmission. The mirror reflects wavelengths of light below the transition wavelength value and transmits wavelengths above this value, this property accounts for the name given to this mirror dichroic (two color). The dichroic mirror is a key element of the fluorescence microscope, but it is not able to perform all of the required optical functions on its own. Typically, about 90% of the light at wavelengths below the transition wavelength value is reflected and about 90% of the light at wavelengths above this value is transmitted by the dichroic mirror.

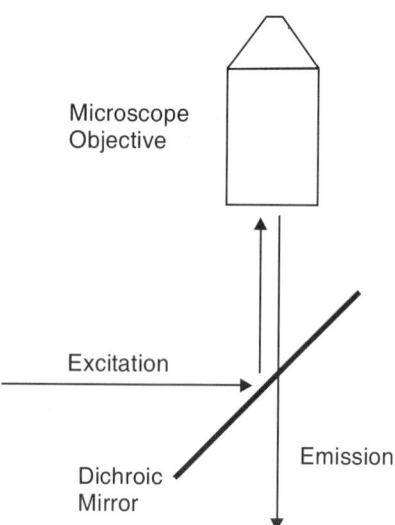

Microscope
Objective

Excitation

Emission

Dichroic
Mirror

Fig. 2.7: Dichroic mirror's position in fluorescence microscope.

When the excitation light illuminates the sample, a small amount of excitation light is reflected off the optical elements within the objective and some excitation light is scattered back into the objective by the sample. Some of this excitation light is transmitted through the dichroic mirror along with the longer wavelength light emitted by the sample. This contaminating light would otherwise reach the detection system.

2.2.3.3.3. Excitation and emission filters

Two filters are used along with the dichroic mirror (1) Excitation filter - In order to select the excitation wavelength and placed in the excitation path just prior to the dichroic mirror (Fig. 2.8a,b). (2) Emission filter - In order to more specifically select the emission wavelength of the light emitted from the sample and to remove traces of excitation light, an emission filter is placed beneath the dichroic mirror (Fig. 2.8a,b). In this position, the filter functions to both select the emission wavelength and to eliminate any trace of the wavelengths used for excitation. These filters are usually a special type of filter referred to as an interference filter because of the way in which it blocks the out of band transmission. Interference filters exhibit an extremely low transmission outside of their characteristic band pass. Thus, they are very efficient in selecting the desired excitation and emission wavelengths.

2.2.3.3.4. Filter cube

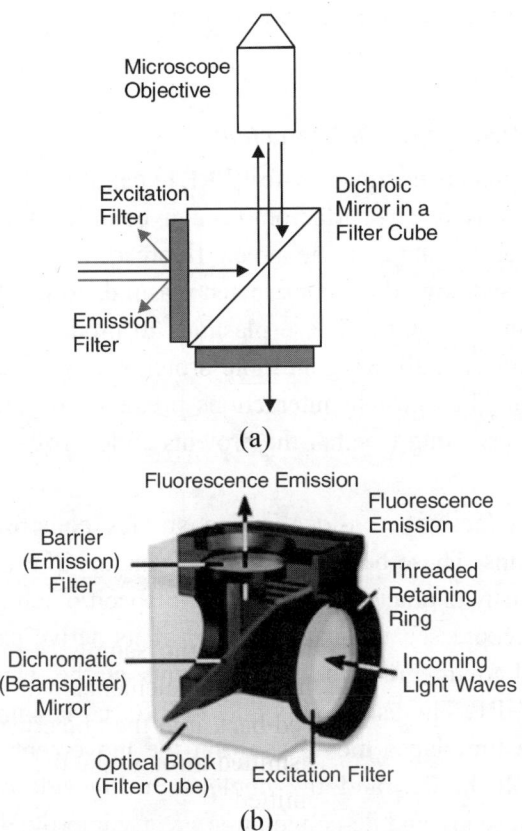

(a)

(b)

Fig. 2.8 a and b: Light path through the filter cube in a fluorescence microscope

The dichroic mirror is mounted on an optical block commonly referred to as a filter cube. The excitation and emission filters are usually affixed to the filter cube. This cube provides a convenient means to change the dichroic mirror without direct handling of either the mirror or filters. The narrow red line emanating from the objective to the filter cube represents the scattered and reflected emission light that must be removed by these optical elements (Fig. 2.8a and b).

It is often the case that a specific combination of excitation filter, emission filter and dichroic mirror are needed to visualize and/or quantities the fluorescence emission from a particular fluorescent species. In newer models of fluorescence microscopes, manufacturers have provided a means to change these optical elements in a convenient manner by arranging a set of four or more filter cubes in a circular or linear turret under the objective.

With a turret arrangement, a specific filter cube can be selected in a manner similar to that of selecting a specific objective.

2.2.3.4. Applications

1. Measuring Protein-Protein Interactions

Fluorescent resonance energy transfer (FRET) can directly measure protein-protein interactions *in vivo*. Macromolecular complexes can be studied *in vivo* using color variants of the green fluorescent protein (GFP) to tag proteins. Wach and coworkers constructed plasmids to label proteins at their carboxy-terminus with GFP. These plasmids to produce the CFP and YFP color variants of GFP, allowing multiple proteins to be labeled concurrently in a single cell. Thus protein interactions predicted by other methods can be tested by determining whether the proteins co-localize to the same region of the cell.

In order to facilitate co-localization studies constructed a group of benchmark strains. These benchmark strains contain CFP or YFP fusions to proteins previously characterized to reside in specific sub-cellular domains. Each fusion is expressed under the control of its native promoter by using a PCR method to integrate the GFP variants at the 3'prime end of the targeted yeast ORF. The use of CFP and YFP for co-labeling studies in live cells, present a time-lapse movie showing the movement of microtubules, the spindle pole bodies and the nucleus during cell division. Optical sectioning microscopy and de-convolution greatly improve image resolution. Optical sectioning also makes 3-D rendering possible. The sub-cellular localization of the hydrophobic TGBp3 protein of *Poa semilatent virus* (PSLV, genus *Hordeivirus*) was studied in transgenic plants using fluorescent microscopy to detect green fluorescent protein (GFP)-tagged protein and immune-detection with monoclonal antibodies (mAbs) rose against the GFP-based fusion expressed in *E. coli*.

2. Detecting single molecules

Under ideal conditions, it is often possible to detect the fluorescence emission from a single molecule, provided that the optical background and detector noise are sufficiently low. A single fluorescein molecule could emit as many as 300,000 photons before it is destroyed by photo-bleaching. Assuming a 20% collection and detection efficiency, about 60,000 photons would be detected.

Using avalanche photodiode or electron multiplying CCD detectors have been able to monitor the behavior of single molecules for many seconds and even minutes. The major problem is adequate suppression of the optical background noise. Because many of the materials utilized in construction of microscope lenses and filters display some level of auto-fluorescence, efforts were initially directed toward the manufacture of very low fluorescence components. However, it soon became evident that fluorescence microscopy techniques utilizing total internal reflection (TIR) provided the desired combination of low background and high excitation light flux. Total internal reflection fluorescence microscopy (TIRFM) takes advantage of the evanescent wave that is developed when light is totally internally reflected at the interface between two media having dissimilar refractive indices. In this technique, a beam of light usually an expanded laser beam is directed through a prism of high refractive index such as glass or sapphire, which abuts a lower refractive index medium of glass or aqueous solution. If the light is directed into the prism at higher than the critical angle, the beam will be totally internally reflected at the interface (Fig. 2.9).

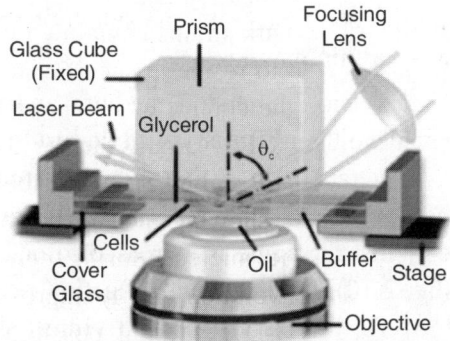

Fig. 2.9: The principle employing an external light source in Total Internal Reflection Fluorescence Microscopy (TIRFM).

The reflection phenomenon develops an evanescent wave at the interface by the generation of an electromagnetic field that permeates about 200 nanometers or less into the lower refractive index space. The light intensity in the evanescent wave is sufficiently high to excite the fluorophores within it, but because of its shallow depth, the volume excited is very small. The result is an extremely low level background because so little of the specimen is exposed to the excitation light.

Total internal reflection fluorescence microscopy can also be conducted through a modification of the epi-illumination approached utilized in wide field techniques (Fig. 2.10). This method requires a very high numerical aperture objective at least 1.4, but preferably 1.45 to 1.6 and partial

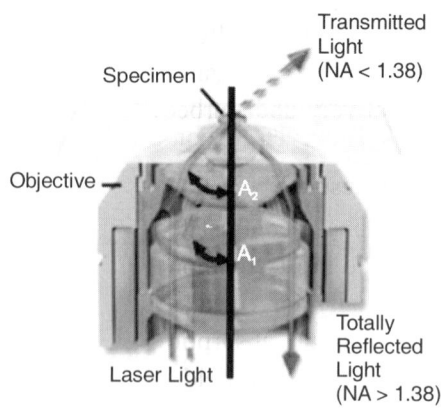

Fig. 2.10: Epi-illumination approach in Total Internal
Reflection Fluorescence Microscopy (TIRFM)

illumination of the microscope field from one side by a small sport or more
uniform illumination by a thin annulus.

High refractive index lens immersion medium and microscope cover
glass are required to achieve the illumination angle resulting in total internal
reflection. Fig. 2.10, light rays exiting the objective front lens element at
an angle less than the critical angle (A(1)) (fig. 2.10) are transmitted away
from the microscope. When the angle is increased to or beyond the critical
angle (angle A (2) Fig. 2.10), total internal reflection results. The result is
a very powerful tool for the study of individual fluorophores and
fluorescently labeled molecules. The advantages resulting from the study of
the properties of single molecules are only beginning to be appreciated.
Thus, the current range of optical microscopy now extends from the single
molecule to the entire animal.

3.   The fluorescence microscope has become an essential tool in medical
     microbiology and microbial ecology.

4.   Bacterial pathogens can be identified after staining them with
     fluorochromoes using immunofluorescence procedures. Example:
     Mycobacterium tuberculosis.

5.   In ecological studies the fluorescence microscope is used to observe
     microorganisms stained with fluorochrome labeled probes or
     fluorochromoes that bind specific cell constituents.

6.   To visualize photosynthetic microbes, as their pigments naturally
     fluoresce when excited by light of specific wavelengths.

7. It is even possible to distinguish live bacteria from dead bacterial by the color they fluoresce after treatment with a special mixture of stains. Thus, the microorganisms can be viewed and directly counted in a relatively undisturbed ecological niche.

2.2.4. Phase contrast microscope

Prior to the invention of phase contrast technique, transmitted bright field illumination was one of the most commonly utilized observation modes in optical microscopy, especially for fixed, stained specimens or other types of samples having high natural absorption of visible light. Collectively, specimens readily imaged with bright field illumination are termed amplitude objects or specimens because the amplitude or intensity of the illuminating wave fronts is reduced when light passes through the specimen. The addition of phase contrast optical accessories to a standard bright field microscope can be employed as a technique to render a contrast enhancing effect in transparent specimens that is reminiscent of optical staining. Light waves that are diffracted and shifted in phase by the specimen termed a phase object can be transformed by phase contrast into amplitude differences that are observable in the eyepieces (Fig. 2.11).

Phase contrast is a method used in microscopy and developed in the early 20th century by Dutch physicist Frits Zernike. Zernike discovered that if speed up the direct light path, it can cause destructive interference patterns in the viewed image. These patterns make details in the image appear darker against a light background. To cause these interference patterns, Zernike developed a system of rings located both in the objective lens and in the condenser system. When aligned properly, a light wave emitted from the illuminator arrive at your eye 1/2 wavelength out of phase, the image of the specimen then becomes greatly enhanced. In effect, the phase contrast technique employs an optical mechanism to translate minute variations in phase into corresponding changes in amplitude, which can be visualized as differences in image contrast. One of the major advantages of phase contrast microscopy is that living cells can be examined in their natural state without previously being killed, fixed and stained. As a result, the dynamics of ongoing biological processes can be observed and recorded in high contrast with sharp clarity of minute specimen detail.

Phase is only useful on specimens that are colorless and transparent and usually difficult to distinguish from their surroundings, these specimens are called phase objects. Examples of phase objects include cell parts in Protozoan's, Bacteria, Sperm tails and other types of unstained cells. This phase contrast technique proved to be such advancement in microscopy that Zernike was awarded the Nobel Prize (physics) in 1953. Phase contrast microscope is a contrast enhancing optical technique that can be utilized to produce high contrast images of transparent specimens such as living cells (culture), microorganisms, thin tissue slices, lithographic patterns, fibers, latex dispersions, glass fragments and sub-cellular particles including nuclei and other organelles.

2.2.4.1. Principle

Phase contrast microscope imparts contrast to unstained biological material by transforming phase differences of light caused by differences in refractive index between cellular components into differences in amplitude of light, i.e., light and dark areas, which can be observed. As light rays pass through areas within the tissue of different optical path (refractive index and geometric path length) they may be retarded in phase by up to 1/4Ü but will remain unchanged in amplitude. Since the eye cannot discern phase differences, a mechanism for transforming phase changes into amplitude changes is required. Highly refractive structures bend light to a much greater angle than do structures of low refractive index. The same properties that cause the light to bend also delay the passage of light by a quarter of a wavelength or so. In a light microscope in bright field mode, light from highly refractive structures bends farther away from the center of the lens

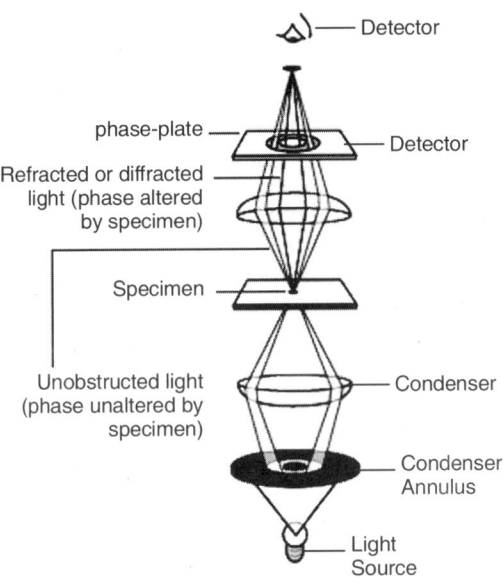

Fig. 2.11: The light pathway of the Phase Contrast Microscope

than light from less refractive structures and arrives about a quarter of a wavelength out of phase.

Light from most objects passes through the center of the lens as well as to the periphery. Now if the light from an object to the edges of the objective lens is retarded a half wavelength and the light to the center are not retarded at all, then the light rays are out of phase by a half wavelength. They cancel each other when the objective lens brings the image into focus. A reduction in brightness of the object is observed. The degree of reduction in brightness depends on the refractive index of the object (Fig. 2.10).

2.2.4.2. Instrumentation

The most important concept underlying the design of a phase contrast microscope is the segregation of surround and diffracted wave fronts emerging from the specimen, which are projected onto different locations in the objective rear focal plane (the diffraction plane at the objective rear aperture). In addition, the amplitude of the surround (undeviated) light must be reduced and the phase advanced or retarded by a quarter wavelength in order to maximize differences in intensity between the specimen and background in the image plane. The mechanism for generating relative phase retardation is a two step process, with the diffracted waves being retarded in phase by a quarter wavelengths at the specimen, while the surround waves are advanced or retarded in phase by a phase plate positioned in or very near the objective rear focal plane. Only two specialized accessories are required to convert a bright field microscope for phase contrast observation. (1) Annular diaphragm (2) Optically conjugates to an internal phase plate residing in the objective rear focal plane.

The condenser annulus (Fig. 2.12) is typically constructed as an opaque flat black (light absorbing) plate with a transparent annular ring, which is positioned in the front focal plane (aperture) of the condenser, so the specimen can be illuminated by defocused, parallel light wave-fronts emanating from the ring. The microscope condenser images the annular diaphragm at infinity, while the objective produces an image at the rear focal plane. It should be noted that many texts describe the illumination emergent from the condenser of a phase contrast microscope as a hollow cone of light with a dark center. This concept is useful for describing the configuration, but is not strictly accurate. The condenser annulus either replaces or resides close to the adjustable iris diaphragm in the front aperture of the condenser.

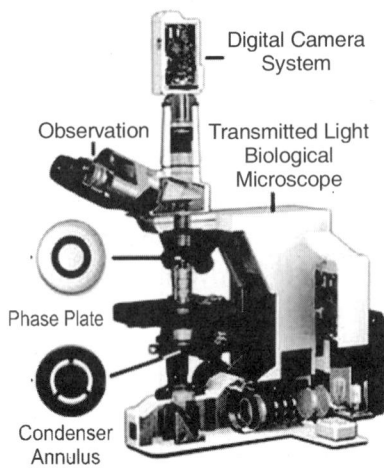

Fig. 2.12: Phase Contrast Microscope.

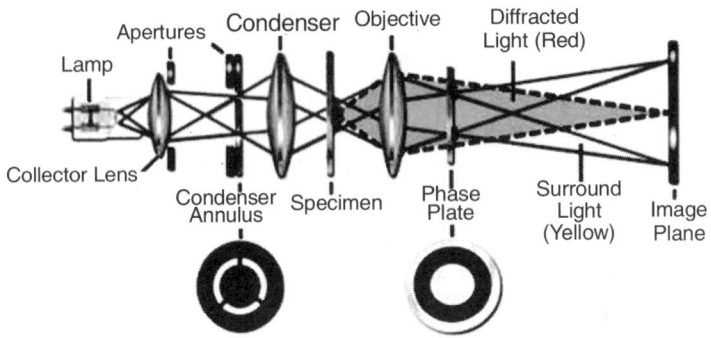

Fig. 2.13: The Phase Contrast Microscope Optical Train.

When utilizing phase contrast microscope, a condenser with a phase annulus and an aperture diaphragm, check to ensure that the iris diaphragm is opened wider than the periphery of the phase annulus. Phase contrast is also insensitive to polarization and birefringence effects, which is a major advantage when examining living cells growing in plastic tissue culture vessels. A phase plate is mounted in or near the objective rear focal plane in order to selectively alter the phase and amplitude of the surround or undeviated light passing through the specimen (Fig. 2.13). In some phase contrast objectives, the thin phase plate contains a ring etched into the glass that has reduced thickness in order to differentially advance the phase of the surround wave by a quarter-wavelength.

In order to modify the phase and amplitude of the spatially separated surround and diffracted wave fronts in phase contrast optical systems, a number of phase plate configurations have been introduced. Because the phase plate is positioned in or very near the objective rear focal plane (the diffraction plane) all light passing through the microscope must travel through this component. The portion of the phase plate upon which the condenser annulus is focused is termed the conjugate area, while the remaining regions are collectively referred to as the complementary area. The conjugate area contains the material responsible for altering the phase of the surround (undiffracted) light by either plus or minus 90° with respect to that of the diffracted wave fronts. In general, the phase plate conjugate area is wider (by about 25%) than the region defined by the image of the condenser annulus in order to minimize the amount of surround light that spreads into the complementary area.

2.2.4.3. Applications

1. Cilia and flagella are nearly invisible in bright field but show up in sharp contrast in phase contrast.

2. Amoebae look like vague outlines in bright field, but show a great deal of detail in phase.

3. The cells are human Glial brain tissue grown in monolayer culture bathed with a nutrient medium containing amino acids, vitamins, mineral salts, and fetal calf serum. In bright field illumination (Fig. 2.14a), the cells appear semi-transparent with only highly refractive regions, such as the membrane, nucleus and unattached cells (rounded or spherical), being visible. When observed using phase contrast optical accessories, the same field of view reveals

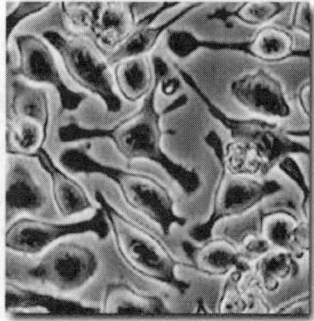

Fig. 2.14: Living cells in (a) Bright filed microscope
(b) Phase contrast microscope.

significantly more structural detail (Fig. 2.14b). Cellular attachments become discernable, as does much of the internal structure. In addition, the contrast range is dramatically improved.

4. Most living microscopic organisms are much more obvious in phase contrast (Fig. 2.15).

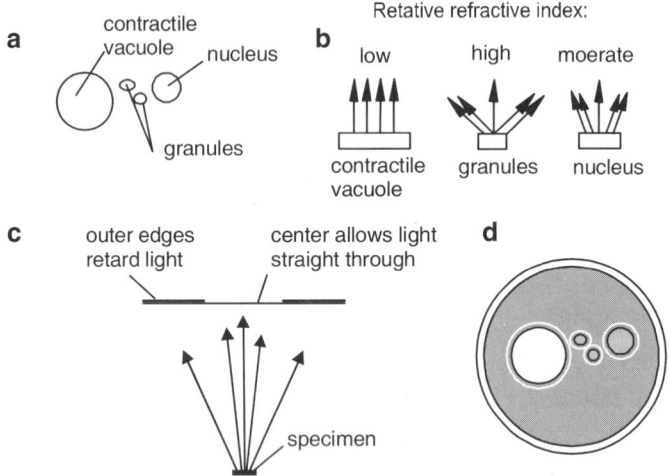

Fig. 2.15: Observation of organelles under Phase Contrast Microscope.

(a) Organelles are nearly invisible in bright field although they have different refractive indexes.

(b) Light is bent and retarded more by objects with a high refractive index.

(c) In phase contrast a phase plate is placed in the light path. Barely refracted light passes through the center of the plate and is not retarded. Highly refracted light passes through the plate farther from center and is held back another one quarter wavelength.

(d) The microscope field shows a darker background (the cell cytoplasm has a higher refractive index than the contractile vacuole) with the organelles in sharp contrast.

2.2.5. Confocal microscopy

About 8-9 years ago, two investigators at Cambridge, Brad Amos and John White were attempting to look at the mitotic divisions in the first few divisions in embryos of *C. elegans*. They were doing anti-tubulin immunofluorescence and were trying to determine the cleavage planes of the cells, but were frustrated in their attempt in that the majority of the fluorescence they observed was out of focus no matter how much they adjusted the focus. They looked at the technique called confocal imaging which was first proposed by Nipkow who made the first stage scanning confocal microscope in 1957. His microscope was commercially unfeasible because the technology needed to produce useful images was not available at the time. In 1986-87, a confocal microscope with the capabilities of producing very useful images could be built by combining the technologies of the laser, the computer and microelectronics. Amos and White built the first prototype incorporating the technologies and obtained much better in-focus confocal images of the *C. elegans* embryos.

Confocal microscopy offers several advantages over conventional optical microscopy, including controllable depth of field, the elimination of image degrading out-of-focus information and the ability to collect serial optical sections from thick specimens. The key to the confocal approach is the use of spatial filtering to eliminate out-of-focus light or flare in specimens that are thicker than the plane of focus. There has been a tremendous explosion in the popularity of confocal microscopy in recent years, due in part to the relative ease with which extremely high quality images can be obtained from specimens prepared for conventional optical microscopy and in its great number of applications in many areas of current research interest.

2.2.5.1. Principle

The principle of confocal imaging advanced by Marvin Minsky in 1957 is employed in all modern confocal microscopes. In a conventional wide field microscope, the entire specimen is bathed in light from a Mercury or Xenon source, and the image can be viewed directly by eye or projected onto an image capture device or photographic film. In contrast, the method of image formation in a confocal microscope is fundamentally different. Illumination is achieved by scanning one or more focused beams of light, usually from a laser or arc-discharge source, across the specimen. This point of illumination is brought to focus in the specimen by the objective lens, and laterally scanned using some form of scanning device under computer control. The sequences of points of light from the specimen are detected by a photomultiplier tube (PMT) through a pinhole or a slit and the output from the PMT is built into an image and displayed by the computer. Although unstained specimens can be viewed using light reflected back from the specimen, they usually are labeled with one or more fluorescent probes.

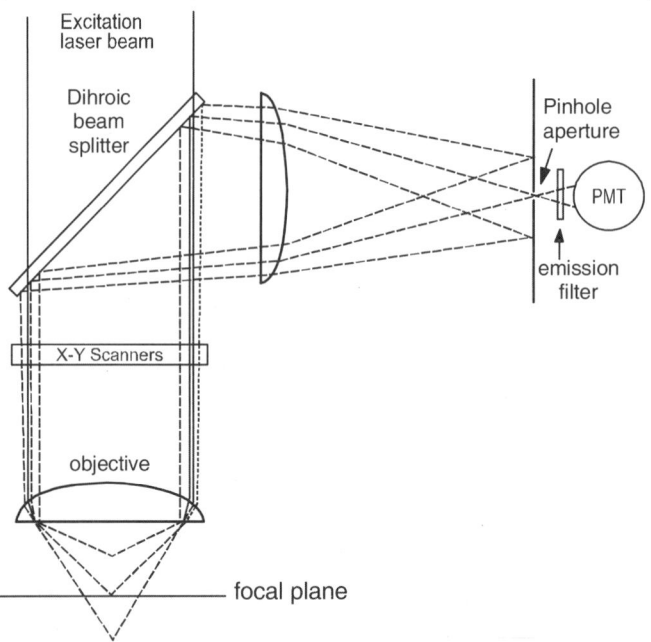

Fig. 2.16: A diagram of the Confocal Microscope principle.

'Confocal' is defined as having the same focus, means in the microscope is that the final image has the same focus as or the focus corresponds to the point of focus in the object and the object and its image are confocal. The most important feature of a confocal microscope is the capability of isolating and collecting a plane of focus from within a sample, thus eliminating the out of focus haze normally seen with a fluorescent sample. Normally when an object is imaged in the fluorescence microscope, the signal produced is from the full thickness of the specimen which does not allow most of it to be in focus to the observer. The confocal microscope eliminates this out-of-focus information by means of a confocal pinhole situated in front of the image plane which acts as a spatial filter and allows only the in-focus portion of the light to be imaged. Light from above and below the plane of focus of the object is eliminated from the final image. A diagram of the confocal principle is shown Fig. 2.16.

Compared to a normal fluorescence microscope, the amount of light that is seen in the final image is greatly reduced by the pinhole, sometimes up to 90-95%. To compensate for this loss of light somewhat, two components have been incorporated into modern confocal microscopes. (1) First, lasers are used as light sources instead of the conventional mercury arc lamps because they produce extremely bright light at very specific wavelengths for fluorochrome excitation. (2) Second, highly sensitive photomultiplier-detectors (PMTs) were employed as imaging devices to pick up the reduced signal. The signal for detection in the original design of modern confocal microscopes is created by scanning a focused laser beam across a square or rectangular field. A system of motorized scanner mirrors sequentially scans a horizontal beam across the specimen.

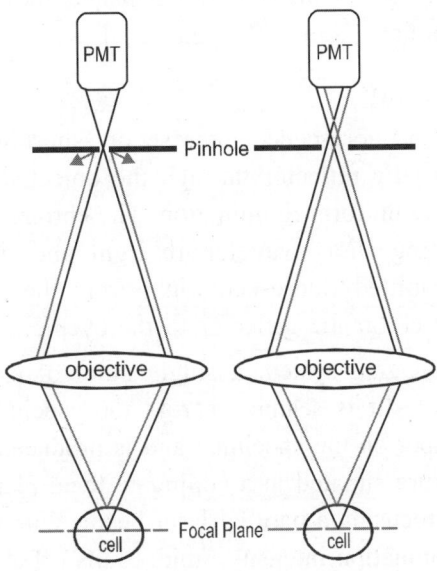

Fig. 2.17: The light path way of Confocal Microscope

The confocal microscope has a stepper motor attached to the fine focus, enabling the collection of a series of images through a three dimensional object. These images can then be used for a two or three dimensional reconstruction. Double and triple labels can be collected with a confocal microscope. Since these images are collected from an optical plane within the sample, precise co-localizations can be performed.

A plane of focus within a specimen is defined by the optics of the microscope. In a fluorescent microscope a small part of a sample may be in focus but you look at the entire object. With the confocal microscope, the z-resolution or optical sectioning thickness depends on a number of factors: the wavelength of the excitation or emission light, pinhole size, numerical aperture of the objective lens, refractive index of components in the light path and the alignment of the instrument.

Figure 2.17 depicts the effect of the pinhole or iris diaphragm on the thickness of the optical plane that is collected. The pinhole and focal plane in the sample are at conjugate planes of focus. The small pinhole opening in the Fig. 2.17 on the left enables data collection from a thin optical plane within the specimen. Points that are out of the plane of focus will have a different secondary focal plane thus most of the data is deflected. Although some of the out-of-focus light enters the photomultiplier tube (PMT) in the Fig. 2.17 on the left, the intensity is too dim to be visualized. All of the data at the plane of focus is collected; in this manner the confocal microscope can collect only the data from within the focal plane.

2.2.5.2. Confocal imaging

In a conventional epifluorescence microscope, short wavelength light is reflected by a chromatic reflector through the objective and the whole of the specimen in fairly uniform illumination. The chromatic reflector has the property of reflecting short wavelength light and transmitting longer wavelength light. Emitted fluorescent light from the specimen is passes straight through the chromatic reflector to the eyepiece.

In a confocal imaging system a single point of excitation light or a group of points or a slit is scanned across the specimen. The point is a diffraction limited spot on the specimen and is produced either by imaging an illuminated aperture situated in a conjugate focal plane to the specimen or more usually by focusing a parallel laser beam. With only a single point illuminated, the illumination intensity rapidly falls off above and below the plane of focus as the beam converges and diverges, thus reducing excitation of fluorescence for interfering objects situated out of the focal plane being examined. Fluorescent light (signal) passes back through the dichroic reflector and then passes through a pinhole aperture situated in a conjugate focal plane to the specimen. Any light emanating from regions away from the vicinity of the illuminated point will be blocked by the aperture, thus providing yet further attenuation of out-of-focus interference. Light passing through the image pinhole is detected by a photo detector. Usually a computer is used to control the sequential scanning of the sample and to assemble the image for display onto a video monitor. Most confocal microscopes are implemented as imaging systems that couple to a conventional microscope.

2.2.5.3. Confocal Scanning Laser Microscope (CSLM)

A laser is used to provide the excitation light in order to get very high intensities. The laser light reflects off a dichroic mirror. From there, the laser hits two mirrors which are mounted on motors these mirrors scan the laser across the sample. Dye in the sample fluoresces and the emitted light (green) gets descanned by the same mirrors that are used to scan the excitation light (blue) from the laser. The emitted light passes through the dichroic and is focused onto the pinhole. The light that passes through the pinhole is measured by a detector (photomultiplier tube) (Fig. 2.18). So, there never is a complete image of the sample at any given instant, only one point of the sample is observed. The detector is attached to a computer

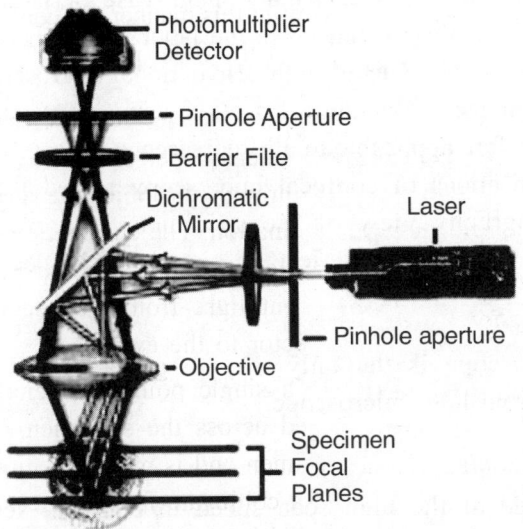

Fig. 2.18: Schematic diagram of a Confocal Scanning Laser Microscopy (CSLM).

which builds up the image, one pixel at a time. This can be done perhaps 3 times a second, for a 512×512 pixel image. The limitation is in the scanning mirrors.

The lasers commonly employed in laser scanning confocal microscopy are high intensity monochromatic light sources, which are useful as tools for a variety of techniques including optical trapping, lifetime imaging studies, photo bleaching recovery and total internal reflection fluorescence. In addition, lasers are also the most common light source for scanning confocal fluorescence microscope and have been utilized, although less frequently in conventional wide field fluorescence investigations. Confocal microscope uses a special Acoustic Optical Deflector (AOD) in place of one of the mirrors in order to speed up the scanning. This uses a high frequency sound wave in a special crystal to create a diffraction grating which deflects the laser light. By varying the frequency of the sound wave, the AOD changes the angle of the diffracted light, helping scan the sample quickly allowing us to take 512x480 pixel images 30 times per second.

2.2.5.4. Applications

Confocal microscope will always give a better image than can be obtained with a standard epifluorescence microscope. All this improvement essentially comes from the rejection of out-of-focus interference. The improvement can vary between marginal in the case of very flat specimens like chromosome squashes, to spectacular, in the case of large, whole-mount specimens such as embryos. Indeed in the latter case it is often impossible to distinguish any interior detail with conventional microscopy yet obtain a perfectly clear image of an optical section using confocal imaging. The technique is therefore applicable to all fluorescence microscopy applications. The major applications of confocal microscopy rooted from this feature including tomography view of the specimen (microscopic CT), 3-D reconstruction from z-section series and resolve thick specimen.

1. 3-D reconstruction

Confocal Microscope is the only way to get a true Z-stack for 3-D reconstruction from light microscope.

2. Physiological study

Taking advantage of the high scan speed in confocal scanner and fast sampling speed on the PMTs, confocal microscope can be used to monitor the highly dynamic intra cellular events such as calcium release, concentration change of calcium and other ions like K, Na, Mg, Ze, P^H change in the cytoplasm of a cell. Taking advantage of the high scan speed of line scan in confocal scanner and fast sampling speed on the PMTs, confocal microscope can be used to monitor the highly dynamic intra cellular events such as calcium release, concentration change of calcium and other ions like K, Na, Mg, Ze and pH change in the cytoplasm of a cell.

Confocal microscope can use multiple PMTs as detector simultaneously detecting signal from different emission channel is achieved without any time delay. Up to six detecting channels can be fitted to the single system. This is another advantage of confocal microscopy vs. digital camera based system. With UV laser installed, the ratio image of calcium can be easily acquired without need of extra device. In digital camera system, special configuration is needed for ratio image. Taking advantage of the high intensity of laser, fast laser AOTF control and accurate scan register of confocal system, multiple region of interesting (ROI) scan can be configured on the same field with different excitation light on and off for overall background image and different ROIs. This makes point bleaching experiment, fluorescence recovery after photo bleaching (FRAP) and fluorescent resonance energy transfer (FRET) very easy to be implemented.

3. Thick specimen

Confocal microscopy is its power to resolve thick specimen to its depth discrimination property. It can be used for tissue block, small organ, embryo, etc. The images are taken from the same field of the plant of 60 μm thick by different means. The first one is taken by CCD camera in the wild field microscope configuration. The second one is taken by confocal microscopy with pinhole size at one Airy unit. The difference on the results is striking. It is worth to mention that the difference is not just the sharpness or crisp of the images, but also the ability of resolving the internal structures. Many fine structures clearly visible in confocal image are obscured or totally invisible in the image taken under wild field microscope.

4. Tomography view of specimen

The depth discrimination feature of confocal microscope enables user to look specimen along certain axis, usually Z-axis, slice by slice at a chosen distance interval to reveal the structural differences or fluorescence distribution differences inside the specimen. Somehow like a Röntgenist examines patients with CT: computerized X-ray tomography, the tomography of a cell or thick tissue can be obtained in the similar way. That is why people sometimes refer this as Microscopic CT, although not accurate as a full description for all the function of confocal microscopy, it is very suitable for this application. This function is very useful especially when you think the spatial distribution of the structures make sense in addition to its positive staining, even if you are not interested to get a Z-series for 3D reconstruction at all.

2.2.6 Electron microscope

2.2.6.1. Introduction

The light microscope has been important instrument for studying microorganisms. Even very best light microscopes have a resolution limit of about 0.2 μm, which greatly compromises their usefulness for detailed studies of many organisms for example Viruses are too small to be seen with light microscope. Prokaryotes can be observed because they are usually only 1 μm to 2 μm in diameter, just their general shape and major morphological features are visible. The detailed internal structure of larger microorganisms also cannot be effectively studied by light microscope. These limitations arise from the nature of visible light waves not from any inadequacy of the light microscope itself.

Electron microscopes are using a beam of electrons to illuminate and create magnified images of specimen. The resolution of a light microscope increases with a decrease in the wavelength of the light, it uses for illumination. Electrons replace light as the illuminating beam. They can be focused, much as light is in a light microscope, but their wavelength is around 0.005 nm, approximately 100,000 times shorter than that of visible light. The electron microscope have a practical resolution roughly 1,000 times better than the light microscope, point closer than 0.5 nm can be distinguished and the useful magnification is well over 100,000X.

Electron microscopes have much greater resolution. Electron Microscopes were developed due to the limitations of light microscope which are limited by light to 500x or 1000x magnification and a resolution of 0.2 micrometers. In the early 1930's this theoretical limit had been reached and there was a scientific desire to see the fine details of the interior structures of organic cells such as nucleus, mitochondria etc. This required 10,000x plus magnification which was just not possible using Light Microscopes. Electron Microscopes (EMs) function exactly as their optical counterparts except that they use a focused beam of electrons instead of light to image the specimen and gain information as to its structure and composition. The following basic steps are involved in all electron microscopes.

1. A stream of electrons is formed by the electron source and accelerated toward the specimen using a positive electrical potential.

2. This stream is confined and focused using metal apertures and magnetic lenses, and this monochromatic beam is focused onto the sample.

3. Interactions occur inside the irradiated sample affecting the electron beam. These interactions and effects are detected and transformed into an image.

The electron beam comes from a filament, made of various types of materials; the most common type material is Tungsten hairpin gun. This filament is a loop of tungsten which functions as the cathode. A voltage is applied to the loop causing it to heat up. The anode which is positive with respect to the filament forms powerful attractive forces for electrons this causes electrons to accelerate toward the anode (Fig. 2.19). Some accelerate right by the anode and on down the column to the sample. Other examples of filaments are Lanthanum hexaboride filaments and Field emission guns.

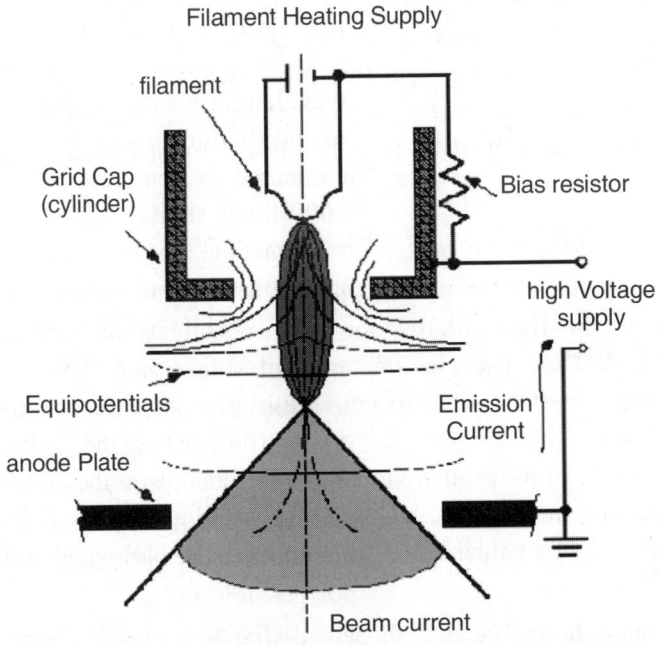

Fig. 2.19: Components of Electron Microscope

Electron Microscopes are scientific instruments that use a beam of highly energetic electrons to examine objects on a very fine scale. This examination can yield the following information.

a. **Topography:** The surface features of an object or how it looks, its texture; direct relation between these features and materials properties like hardness and reflectivity.

b. **Morphology:** The shape and size of the particles making up the object; direct relation between these structures and materials properties such as ductility, strength, reactivity etc.

c. **Composition:** The elements and compounds that the object is composed of and the relative amounts of them; direct relationship between composition and materials properties such as melting point, reactivity, hardness etc.

2.2.6.2. Types

Electron microscopes are classified into two types (1) Transmission Electron Microscope (TEM) and (2) Scanning Electron Microscope (SEM). The Transmission Electron Microscope (TEM) was the first type of Electron Microscope to be developed and is patterned exactly on the Light Transmission Microscope except that a focused beam of electrons is used instead of light to see through the specimen. Transmission Electron Microscope was developed by Max Knoll and Ernst Ruska in Germany in 1931.

2.2.6.2.1. Transmission Electron Microscope (TEM)

The transmission electron microscope (TEM) operates on the same basic principles as the light microscope but uses electrons instead of light (Table 2.2). A TEM use electron as light source and their much lower wavelength makes it possible to get a resolution of thousand times better than with a light microscope. We can see objects to the order of a few angstroms (10^{-10} m) for example to study small details in the cell or different materials down to near atomic levels. The possibility for high magnifications has made the TEM a valuable tool in both medical, biological and materials research.

   i. **Morphology:** The size, shape and arrangement of the particles which make up the specimen as well as their relationship to each other on the scale of atomic diameters.

   ii. **Crystallographic information:** The arrangement of atoms in the

Table 2.2: Characteristics of Light microscope and Transmission electron microscope (TEM)

Feature	Light microscope	TEM
Highest practical magnification	About 1,000 – 1,500	Over 100,000
Best Resolution	0.2µm	0.5 nm
Radiation source	Visible light	Electron beam
Medium of travel	Air	High vacuum
Type of lens	Glass	Electromagnet
Source of contrast	Differential light absorption	Scattering of electrons
Focusing mechanism	Adjust lens position mechanically	Adjust current to the magnetic lens
Method of changing magnification	Switch the objective lens or eyepiece	Adjust current to the magnetic lens
Specimen mount	Glass slide	Metal grid (copper)

specimen and their degree of order, detection of atomic-scale defects in areas a few nanometers in diameter.

iii. **Compositional information:** The elements and compounds the sample is composed of and their relative ratios, in areas a few nanometers in diameter.

2.2.6.2.1.1 Principle

A TEM works much like a slide projector. A projector shines a beam of light through (transmits) the slide as the light passes through it is affected by the structures and objects on the slide. These effects result in only certain parts of the light beam being transmitted through certain parts of the slide. This transmitted beam is then projected onto the viewing screen, forming an enlarged image of the slide. TEM work the same way except that they shine a beam of electrons like the light through the specimen like the slide.

A modern Transmission Electron Microscope (TEM) is complex and sophisticated but the basic principles behind its operation can be readily understood. A heated tungsten filament in the electron gun generates a beam of electrons that is then focused on the specimen by the condenser. Since electrons cannot pass through a glass lens, doughnut shaped electromagnets called magnetic lenses are used to focus the beam. The column containing the lenses and specimen must be under high vacuum to obtain a clear image because electrons are deflected by collisions with air molecules. The specimen scatters some electrons but those that pass through are used to form an enlarged image of the specimen on a fluorescent screen. A denser region in the specimen scatters more electrons and therefore appears darker in the image since fewer electrons strike that area of the screen; these regions are said to be electron dense, in contrast, electron-transparent regions are brighter.

The TEM has distinctive feature that place harsh restrictions on the nature of the samples that can be viewed by which those samples must be prepared. Since electrons are deflected by air molecules and are easily absorbed and scattered by solid matter, only thin slices (20 – 100 nm) of a microbial specimen can be viewed in the average TEM. Such a thin slice cannot be cut unless the specimen has support of some kind; the necessary support is provided by plastic. After fixation with chemicals like glutaraldehyde or osmium tetroxide to stabilize cell structure, the specimen is dehydrated with organic solvents like Acetone and Ethanol. Complete dehydration is essential because most plastics used for embedding are not water soluble. Next the specimen is soaked in unpolymerized, liquid epoxy plastic until it is completely permeated and then the plastic is hardened to form a solid block. Thin sections are cut from this block with a glass or diamond knife using a special instrument called an ultramicrotome.

As with bright-field microscopy, cells usually must be stained before they can be seen clearly. The probability of electron scattering is determined by the density (atomic number) of the specimen atoms. Biological molecules are composed primarily of atoms with low atomic numbers (H, C, N and O) and electron scattering is fairly constant throughout the unstained cell. Therefore specimens are prepared for observation by soaking thin sections with solutions of heavy metal salts like lead citrate and uranyl acetate. The lead and uranium ions bind to cell structure and make them more electron opaque thus increasing contrast in the material. Heavy osmium atoms from the osmium tetroxide fixative also stain cells and increase their contrast. The stained thin sections are then mounted on tiny copper grids and viewed.

Two other important techniques used for preparing the specimens for transmission electron microscope (1) **Negative staining:** In negative staining, the specimen is spread out in a thin film with either phsopho-tungstic acid or uranyl acetate. Just as in negative staining for light microscopy, heavy metals do not penetrate the specimen but render the background dark, whereas the specimen appears bright in photographs. Negative staining is an excellent way to study the structure of viruses, bacteria gas vacuoles and other similar objects. (2) **Shadowing staining :** In shadowing, a specimen is coated with a thin film of platinum or other heavy metal by evaporation at an angle of about 45° from horizontal so that the metal strikes the microorganism on only one side. In one commonly used imaging method, the area coated with metal appears dark in photographs, whereas the uncoated side and the shadow region created by the object are light. This technique is particularly useful in studying virus morphology, prokaryotic flagella and DNA.

2.2.6.2.1.2 Instrumentation

A light source at the top of the microscope emits the electrons that travel through vacuum in the column of the microscope. Instead of glass lenses focusing the light in the light microscope, the TEM uses electromagnetic lenses to focus the electrons into a very thin beam (Fig. 2.22). The electron beam then travels through the specimen. Depending on the density of the material present, some of the electrons are scattered and disappear from the beam. At the bottom of the microscope the un-scattered electrons hit a fluorescent screen, which gives rise to a shadow image of the specimen with its different parts displayed in varied darkness according to their density. The image can be studied directly by the operator or photographed with a camera (Fig. 2.20 & 2.21).

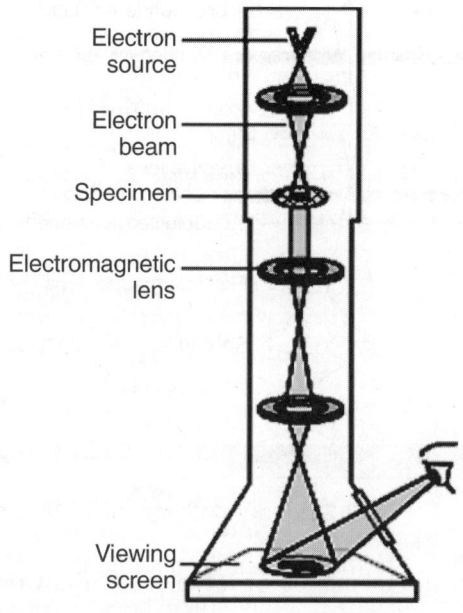

Electron source

Electron beam

Specimen

Electromagnetic lens

Viewing screen

Fig. 2.20: Components of TEM

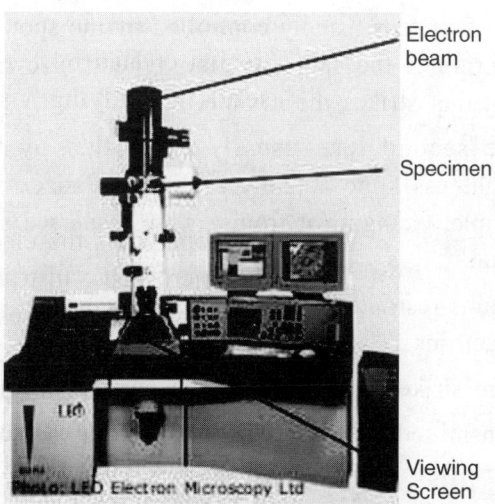

Electron beam

Specimen

Viewing Screen

Fig. 2.21: TEM instrument

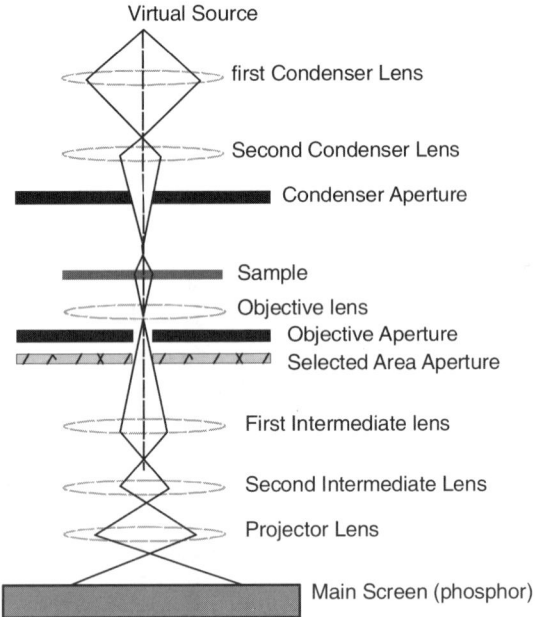

Fig. 2.22: Light pathway of TEM

A more technical explanation of a typical TEMs working is as follows.

1. The Virtual source at the top represents the electron gun, producing a stream of monochromatic electrons.

2. This stream is focused to a small, thin, by the use of condenser lenses i and ii.

   i. The first lens usually controlled by the spot size knob largely determines the spot size; the general size range of the final spot that strikes the sample.

   ii. The second lens usually controlled by the intensity or brightness knob actually changes the size of the spot on the sample; changing it from a wide dispersed spot to a pinpoint beam.

3. The beam is restricted by the condenser aperture, knocking out high angle electrons.

4. The beam strikes the specimen and parts of it are transmitted.

5. This transmitted portion is focused by the objective lens into an image.

6. Optional objective and selected area metal apertures can restrict

the beam; the objective aperture enhancing contrast by blocking out high-angle diffracted electrons, the selected area aperture enabling the user to examine the periodic diffraction of electrons by ordered arrangements of atoms in the sample.

7. The image is passed down the column through the intermediate and projector lenses, being enlarged all the way.

8. The image strikes the phosphor image screen and light is generated, allowing the user to see the image. The darker areas of the image represent those areas of the sample that fewer electrons were transmitted through (they are thicker or denser). The lighter areas of the image represent those areas of the sample that more electrons were transmitted through (they are thinner or less dense).

2.2.6.2.1.3. Applications

The Transmission Electron Microscope (TEM) allows the user to determine the internal structure of materials either of biological or non-biological origin.

Materials for TEM must be specially prepared to thicknesses which allow electrons to transmit through the sample, much like light is transmitted through materials in conventional optical microscopy. Because the wavelength of electrons is much smaller than that of light, the optimal resolution attainable for TEM images is many orders of magnitude better than that from a light microscope. Thus, TEMs can reveal the finest details of internal structure in some cases as small as individual atoms. Magnifications of 350,000 times can be routinely obtained for many materials, whilst in special circumstances; atoms can be imaged at magnifications greater than 15 million times.

For biological samples, cell structure and morphology is commonly determined whilst the localization of antigens or other specific components within cells is readily undertaken using specialized preparative techniques. For non-biological materials, phase determination as well as defect and precipitate orientation are typical outcomes of conventional TEM experiments. Micro structural characterization of non-biological materials, including unit cell periodicities can be readily determined using various combinations of imaging and electron diffraction techniques. Images obtained from a TEM are two dimensional sections of the material under study but applications which require three dimensional reconstructions can be accommodate by these techniques.

The energy of the electrons in the TEM determines the relative degree of penetration of electrons in a specific sample or alternatively, influence the thickness of material from which useful information may be obtained. Thus a 400 kV TEM not only provides the highest resolution allows for the observation of relatively thick samples less than 0.2 micrometers when compared with the more conventional 100 kV or 200 kV instruments. Because of the high spatial resolution obtained, TEMs are often employed to determine the detailed crystallography of fine grained or rare materials. Thus, for the physical and biological sciences, TEM is a complementary tool to conventional crystallographic methods such as X-ray diffraction.

2.2.6.2.2. Scanning Electron Microscope (SEM)

The first Scanning Electron Microscope (SEM) debuted in 1942 with the first commercial instruments around 1965. Its late development was due to the electronics involved in scanning the beam of electrons across the sample. Scanning electron microscope (SEM) is used for inspecting topographies of specimens at very high magnifications. SEM magnifications can go to more than 300,000 X but most semiconductor manufacturing applications require magnifications of less than 3,000 X only. SEM inspection is often used in the analysis of die/package cracks and fracture surfaces, bond failures, and physical defects on the die or package surface.

2.2.6.2.2.1. Principle

The transmission electron microscope forms an image from radiation that has passed through a specimen. The Scanning Electron Microscope (SEM) works in a different manner. It produces an image from electrons released from atoms on an object's surface. The SEM has been used to examine the surfaces of microorganisms in great detail; many SEMs have a resolution of 7 nm or less.

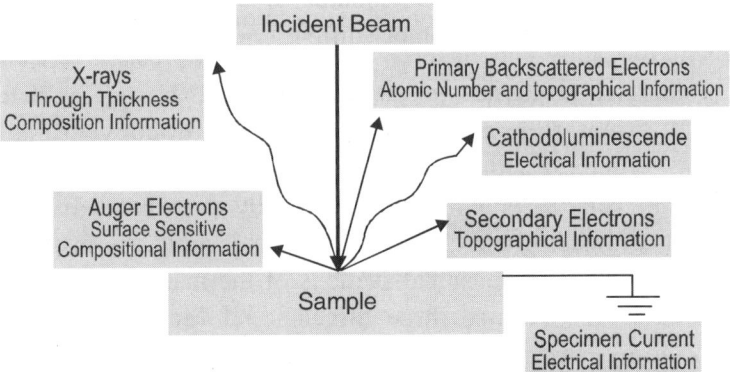

Fig. 2.23: SEM Setup: Electron or specimen interactions, when the electron beam strikes the sample, both photon and electron signals are emitted

During SEM inspection, a beam of electrons is focused on a spot volume of the specimen, resulting in the transfer of energy to the spot (fig.-2.23). These bombarding electrons, also referred to as primary electrons, dislodge electrons from the specimen itself. The dislodged electrons also known as secondary electrons are attracted and collected by a positively biased grid or detector and then translated into a signal. To produce the SEM image, the electron beam is swept across the area being inspected, producing many such signals. These signals are then amplified, analyzed and translated into images of the topography being inspected. Finally, the image is shown on a CRT.

The energy of the primary electrons determines the quantity of secondary electrons collected during inspection. The emission of secondary electrons from the specimen increases as the energy of the primary electron beam increases, until a certain limit is reached (Fig. 2.23). Beyond this limit, the collected secondary electrons diminish as the energy of the primary beam is increased, because the primary beam is already activating electrons deep below the surface of the specimen. Electrons coming from such depths usually recombine before reaching the surface for emission.

Aside from secondary electrons, the primary electron beam results in the emission of backscattered or reflected electrons from the specimen (Fig. 2.23). Backscattered electrons possess more energy than secondary electrons, and have a definite direction. As such, they cannot be collected by a secondary electron detector, unless the detector is directly in their path of travel. All emissions above 50 eV are considered to be backscattered electrons. Backscattered electron imaging is useful in distinguishing one material from another, since the yield of the collected backscattered electrons increases monotonically with the specimen's atomic number. Backscatter imaging can distinguish elements with atomic number differences of at least 3, i.e., materials with atomic number differences of at least 3 would appear with good contrast on the image. For example, inspecting the remaining Au on an Al bond pad after its Au ball bond has lifted off would be easier using backscatter imaging, since the Au islets would stand out from the Al background.

Specimen preparation for SEM is relatively easy and in some cases air dried material can be examined directly. Most often, however, microorganisms must first be fixed, dehydrated and dried to preserve surface structure and prevent collapse of the cells when they are exposed to the SEM's high vacuum. Before viewing, dried samples are mounted and coated with a thin layer of metal to prevent the buildup of an electrical charge on the surface and to give a better image. To create an image, the SEM scans a narrow, tapered electron beam back and forth over the specimen. When the beam strikes a particular area, surface atoms discharge a tiny shower of electrons called secondary electrons and these are trapped by a special detector. Secondary electrons entering the detector strike a scintillator causing it to emit light flashes that a photomultiplier converts to an electrical current and amplifies. The signal is sent to a cathode ray tube and produces an image like a television picture, which can be viewed or photographed.

The number of secondary electrons reaching the detector depends on the nature of the specimen's surface. When the electron beam strikes a raised area, a large number of secondary electrons enter the detector; in contrast fewer electrons escape a depression in the surface and reach the detector. Thus raised are appear lighter on the screen and depressions are darker, a realistic three dimensional image of the microorganism's surface results. The actual *in-situ* location of microorganisms in ecological niches such as the human skin and the lining of the gut also can be examined.

2.2.6.2.2.2. Instrumentation

An SEM is comprised of a large number of basic systems, all of which need to work in order for the basic machine to function.

2.2.6.2.2.2.1. Control console

Most of the parts of the microscope are controlled through the control console. Electron beam control (electron lenses, beam deflections, electron gun controls), image processing (linear and nonlinear signal amplification, static frame stores, image display), scan generation and synchronization, and vacuum system controls are all principally controlled through the console. This essentially makes up all the bits 'n pieces that constitute the SEM. One potential advantage of this approach is to automate some operations such as focusing and the like through an image analysis feedback loop in the computer. The disadvantage is usually the inability to operate multiple controls simultaneously, greatly slowing and impairing operator control.

2.2.6.2.2.2.2. Vacuum system

The actual moving guts of the scope are all either in or on the vacuum system or a part of it. These include the electron gun and lenses and all the active parts of beam manipulation controls, signal collection devices and of course the vacuum pumps and valves.

2.2.6.2.2.2.3. Electron guns

Every electron microscope has an electron gun this is where the electrons come from to make it an electron microscope. A thermionic electron gun consists essentially of a heated wire or compound from which electrons are given enough thermal energy to overcome the work function of the source, combined with an electric potential to give the newly free electrons a direction and velocity. Common materials that are used for these sources are Tungsten because it has a very high melting temperature, so more thermal energy can be made available; LaB_6; and Ce_6 because they have both a low work function and a high melting temperature. Thermal field emitters enhance the pure field emission effect by giving some thermal energy to the electrons in the metal so that the required tunneling distance is shorter for successful escape from the surface. A Schottky emitter is a thermal field emitter that has been further enhanced by doping the surface of the emitter to reduce the work function.

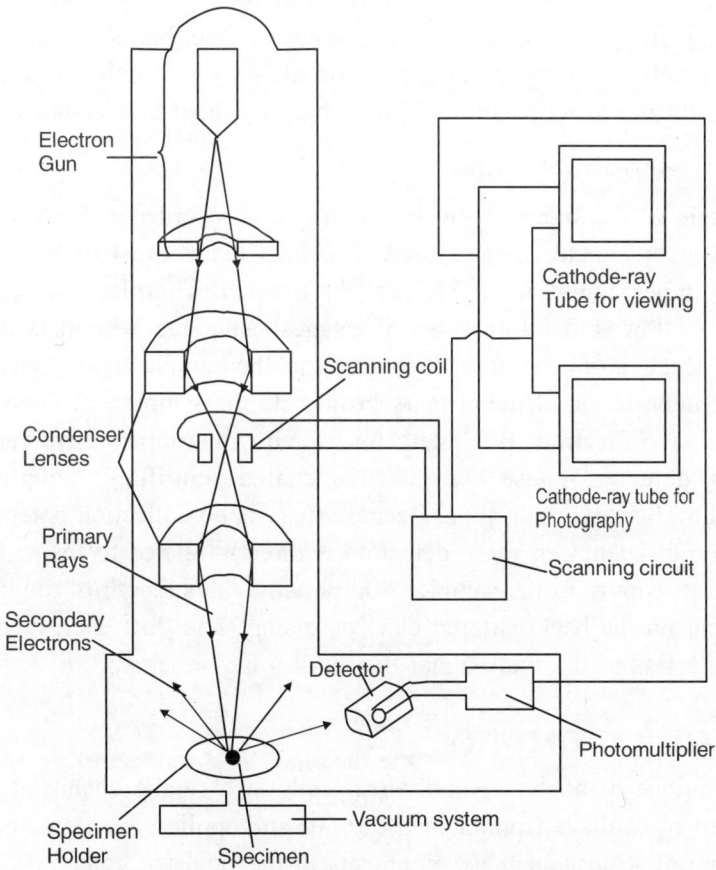

Fig. 2.24: Scanning Electron Microscope (SEM)

2.2.6.2.2.2.4. Thermal field emitters

Thermal field emitters are enhance the pure field emission effect by giving some thermal energy to the electrons in the metal, so that the required tunneling distance is shorter for successful escape from the surface.

2.2.6.2.2.2.5. Vacuum Systems

There are four types of vacuum pumps that are at least somewhat commonly employed in SEMs for basic operational mechanisms and operational parameters of each, then put each into their place in a typical SEM. (1) Roughing pump/mechanical pump: This is a pretty basic design that takes a large initial volume, squeezes it down into a smaller volume and expels it at the high pressure side. (2) Diffusion pump: This uses a high velocity stream of gas molecules to kinetically trap' random gas molecules from the vacuum system that blunder into the stream. (3) Turbo pump: A turbo pump is essentially a whole heck of a lot of axial compressors on really tight tolerances moving at truly frightening speeds. (4) Ion pump: This is an entrapment pump, so there is only an inlet but no outlet pressure. An ion pump works by ionizing any and all stray molecules or atoms that fall into it, then reacting these with or burying them in a chunk of metal.

2.2.6.2.2.2.6. Detectors

The nature of the signals collected by an SEM in order to form images is all dependent on the detector used to collect them. Most of the common detector types found in SEMs are (1) Everhart-Thornley detector (E-T detector) allowed the formation of images using the secondary electron signal, which is much more dependent on the sample topography at the point of intersection of the primary beam with the sample. (2) The simplest Backscattered Electron (BSE) detector is simply a stripped down version of the E-T detector. These consist of a coated scintillator coupled to a photomultiplier by a light pipe. Because there is no collection potential, the collection efficiency of these detectors is directly related to the scintillator size and proximity to the sample. Additionally, the strength of the signal is dependent on the backscattered electron energy, so a low-energy BSE will contribute less to the total signal than will a high energy BSE.

2.3. FLOW CYTOMETRY

The technique of analyzing individual cells in a fluidic channel was first described by Wallace Coulter in the 1950s and applied to automated blood cell counting. Subsequent developments in the fields of computer science, laser technology, monoclonal antibody production, cytochemistry and fluorochrome chemistry led to the development of the flow cytometer two decades later. Because the first commercial flows cytometers were large, complex, expensive, and difficult to operate and maintain they were primarily used in the research laboratory. However, the enormous value of the flow cytometer in the medical and biologic sciences was quickly appreciated, and its cost and complexity gradually decreased as its analytic capability increased. Flow cytometry is a technique of quantitative single cell analysis. The flow cytometer was developed in the 1970's and rapidly became an essential instrument for the biologic sciences. The major clinical application of flow cytometry is diagnosis of hematologic malignancy but a wide variety of other applications exist such as reticulocyte enumeration and cell function analysis.

2.3.1. Principle

Prepared single cell or particle suspensions are necessary for flow cytometric analysis. Various immunoflurescent dyes or antibodies can be attached to the antigen or protein of interest. The suspension of cells or particles is aspirated into a flow cell where, surrounded by a narrow fluid stream they pass one at a time through a focused laser beam. The light is either scattered or absorbed when it strikes a cell. Absorbed light of the appropriate wavelength may be reemitted as fluorescence if the cell contains a naturally fluorescent substance or one or more fluorochrome labeled antibodies are attached to surface or internal cell structures. Light scatter is dependent on the internal structure of the cell and its size and shape.

Flow cytometry uses the principles of light scattering, light excitation and emission of fluorochrome molecules to generate specific multi-parameter data from particles and cells in the size range of 0.5um to 40um diameter. Cells are hydro-dynamically focused in a sheath of PBS before intercepting an optimally focused light source (Fig. 2.25). Lasers are most often used as a light source in flow cytometry. The sample is injected into the center of a sheath flow. The combined flow is reduced in diameter forcing the cell into the center of the stream. This the laser one cell at a time, this schematic of the flow chamber in relation to the laser beams in the sensing area (Fig. 2.25).

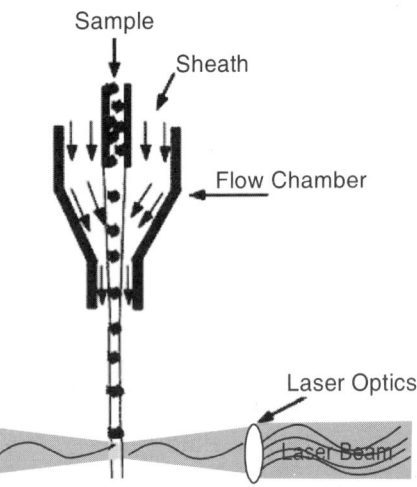

Fig. 2.25: Flow cytometers use the principle of hydrodynamic focusing for presenting cells to a laser or any other light excitation source.

2.3.2. *Analysis and sorting*

Flow cytometry analysis of a single cell suspension yields multi-parameter data corresponding to Forward Light Scatter (FLS), 90° Light Scatter (90LS)

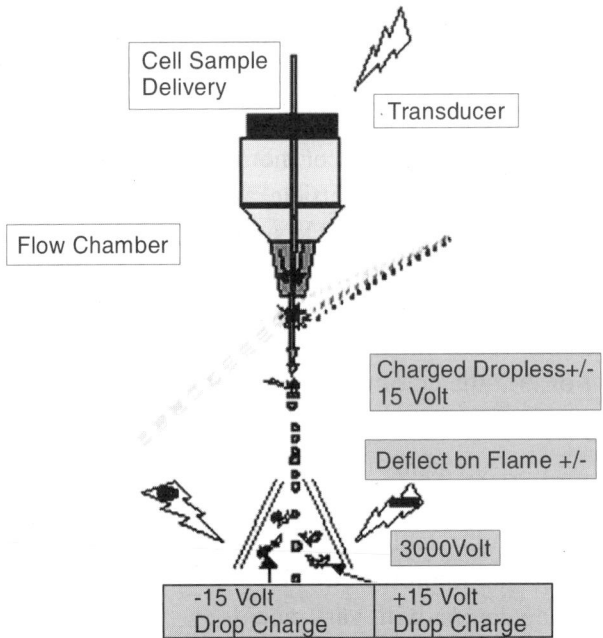

Fig. 2.26: Analysis and Storing.

and FL1-FL4. This information allows researchers to identify and characterize various subpopulations of cells (Fig. 2.26). The process of separating cells using flow cytometry multi-parameter data is referred to as sorting. They are capable of acquiring multi-parameter flow cytometry data but they can not separate or purify cells. Sorting is a specialized process that requires sophisticated electronic components not incorporated into most bench-top instruments.

A major application of flow cytometry is to separate cells according to subtype or epitope expression for further biological studies. This process is called cell sorting or FACSTM analysis. After the sample is hydro dynamically focused, each particle is probed with a beam of light. The scatter and fluorescence signal is compared to the sort criteria set on the instrument. If the particle matches the selection criteria, the fluid stream is charged as it exits the nozzle of the fluidics system.

Electrostatic charging actually occurs at a precise moment called the break-off point, which describes the instant the droplet containing the particle of interest separates from the stream. To prevent the break-off point happening at random distances from the nozzle and to maintain consistent droplet sizes, the nozzle is vibrated at high frequency. The droplets eventually pass through a strong electrostatic field, and are deflected left or right based on their charge. The speed of flow sorting depends on several factors including particle size and the rate of droplet formation. A typical nozzle is between 50 – 70 μm in diameter and depending on the jet velocity from it, can produce 30,000–100,000 droplets per second, which is ideal for accurate sorting. Higher jets velocities risk the nozzle becoming blocked and will also decrease the purity of the preparation.

Each particle passes through one or more beams of light. Light scattering or fluorescence emission if the particle is labeled with a fluorochrome provides information about the particle's properties. The laser and the arc lamp are the most commonly used light sources in modern flow cytometry. Lasers produce a single wavelength of light (a laser line) at one or more discreet frequencies (coherent light). Arc lamps tend to be less expensive than lasers and exploit the color emissions of an ignited gas within a sealed tube.

Fluorescence measurements taken at different wavelengths can provide quantitative and qualitative data about fluorochrome-labeled cell surface receptors or intracellular molecules such as DNA and cytokines. Flow cytometers use separate fluorescence (FL-) channels to detect light emitted. The number of detectors will vary according to the machine and its manufacturer. Detectors are either silicon photodiodes or photomultiplier tubes (PMTs). Silicon photodiodes are usually used to measure forward scatter when the signal is strong. PMTs are more sensitive instruments and are ideal for scatter and fluorescence readings.

The specificity of detection is controlled by optical filters, which block certain wavelengths while transmitting (passing) others. There are three major filter types (1) Long pass filters allow through light above a cut-off wavelength. (2) Short pass permit light below a cut-off wavelength. (3) Band pass transmit light within a specified narrow range of wavelengths. All these filters block light by absorption. When a filter is placed at a 45° angle to the oncoming light it becomes a dichroic filter or mirror. This type of filter performs two functions (i) First, to pass specified wavelengths in the forward direction (ii) Second, to deflect blocked light at a 90° angle. To detect multiple signals simultaneously, the precise choice and order of optical filters will be an important consideration.

2.3.3 Signal processing

When light hits a photodetector a small current (microamperes) is generated. Its associated voltage has amplitude proportional to the total number of light photons received by the detector. This voltage is then amplified by a series of linear or logarithmic amplifiers, and by analog to digital convertors (ADCs), into electrical signals large enough (5–10 volts) to be plotted graphically. Log amplification is normally used for fluorescence studies because it expands weak signals and compresses strong signals, resulting in a distribution that is easy to display on a histogram. Linear scaling is preferable where there is not such a broad range of signals Ex.: in DNA analysis. The measurement from each detector is referred to as a 'parameter' Ex.: forward scatter, side scatter or fluorescence.

2.3.4. System and components

A flow cytometer is made up of three main systems (1) The **fluidics system** transports particles in a stream to the laser beam for interrogation. (2) The **optics system** consists of lasers to illuminate the particles in the sample stream and optical filters to direct the resulting light signals to the appropriate detectors. (3) The **electronics system** converts the detected light signals into electronic signals that can be processed by the computer. For some instruments equipped with a sorting feature, the electronics system is also capable of initiating sorting decisions to charge and deflect particles.

A tunable transducer permits the breaking of the fluid sheath into individual droplets. These individual droplets will encapsulate single cells. Electric charge delays for charging individual droplets. Deflection plates for deflecting individually charged droplets into collection tubes. Software settings for defining sorting criteria these include regions defining populations to be sorted.

Flow cytometry is a technology that simultaneously measures and then analyzes multiple physical characteristics of single particles, usually cells, as they flow in a fluid stream through a beam of light. The properties measured include a particle's relative size, relative granularity or internal complexity, and relative fluorescence intensity. In the flow cytometer, particles are carried to the laser intercept in a fluid stream. Any suspended particle or cell from 0.2–150 micrometers in size is suitable for analysis.

Cells from solid tissue must be disaggregated before analysis. The portion of the fluid stream where particles are located is called the sample core. When particles pass through the laser intercept, they scatter laser light. Any fluorescent molecules present on the particle fluoresce.

2.3.5. Applications

The identification and quantization of cellular antigens with fluoro-chrome labeled monoclonal antibodies (immunophenotyping) is one of the most important applications of the flow cytometer. Immuno-phenotypic analysis is critical to the initial diagnosis and classification of the acute leukemias, chronic lymphoproliferative diseases and ma-lignant lymphomas since treatment strategy often depends upon an-tigenic parameters. In addition, immunophenotypic analysis provides prognostic information not available by other techniques provides a sensitive means to monitor the progress of patients after chemother-apy or bone marrow transplantation and often permits the detection of minimal residual disease. Flow cytometric analysis of apoptosis, multidrug resistance, leukemia-specific chimeric proteins, cytokine receptors and other parameters may provide additional diagnostic or prognostic information in the near future.

Leukemias represent abnormal proliferations of hematopoietic cells that are arrested at a discrete stage of differentiation. Leukemias are classified into acute and chronic forms based on a constellation of clinical and laboratory findings. The acute leukemias are classified into two subclasses; the lymphoblastic (ALL) type and myeloid (AML) type based on morphologic, cytochemical and immunophenotypic features.

The flow cytometer works, Fluorochrome-labeled monoclonal antibody solutions are added to a cell suspension from a peripheral blood, bone marrow aspirate or lymph node. The tubes are incubated at room temperature for a short period of time. The labeled cell suspensions are passed through the flow cell of a flow cytometer. Many flow cytometers are automated, but some models require the operator to process the tubes individually.

More than 10,000 cells from each tube are typically analyzed to produce statistically valid information. Each cell passes individually through the highly focused laser beam of the flow cytometer, a process termed single cell analysis. The fluorochrome of each labeled monoclonal antibody attached to the cell is excited by the laser light and emits light of a certain wavelength. The cells also scatter light at multiple angles. Photo detectors placed a forward angle and at right angles to the axis of the laser beam collect the emitted or scattered light. Forward and right angle scatter signals, and as many as five fluorochrome signals can be detected from each cell (multiparametric analysis). The signals from each photodiode are digitized and passed to a computer for storage, display, and analysis. Typically, all data recorded from each cell is stored for possible later recall for further analysis.

A variety of histograms for visual display can be generated automatically or at the discretion of the operator. List mode data can also be transferred to a separate computer for analysis. Presently, most commercial flow cytometers utilize a standardized file format for list mode storage, and a variety of computer programs are commercially available for data analysis and display.

2.4. SUMMARY

The Microscope was invented by either Zacharias Janssen of the Netherlands or Galielo Galielei of Italy by the end of 15th century. In 1976, Antony van Leeuwenhoek for the first time used crude simple microscope consisting of a biconcave lens enclosed in two metal plates to describe bacterial and protozoa

Magnification or enlargement can be defined as the number of times the specimen is enlarged when observed through a particular microscope. It is the cumulative effect brought about by two lenses system, the ocular lens (an eye piece) and the objective lens.

The resolving power is the ability of a lens to show two adjacent objects as discrete units. Increased magnification without proper resolving power gives a blurred picture without finer details or clarity. The resolving power of a lens is dependent on the wavelength of the light used (resolution increases with decrease in wavelength of the light), characteristics of the lens and the numerical aperture.

The numerical aperture (NA) can be defined as function of the diameter of the objective lens in relation to its focal length. Numerical aperture can be expressed by the formula: Numerical aperture (NA) = $\eta \sin \theta$.

Refractive Index (RI) can be defined as the bending power of light passing through one medium to another medium. The RI of air is lower than that of glass and as light rays pass from the glass slide into air, they are bent or refracted so that they do not pass into the objective lens.

Bright field imaging is the simplest form of microscopy where light is either passed through or reflected off, a specimen. Illumination is not altered by devices that alter the properties of light (polarizer's or filters). In bright field imaging all light from the specimen and its surroundings is collected by the objective to form an image against a bright background.

Dark field microscopy is a specialized illumination technique that capitalizes on oblique illumination to enhance contrast in specimens that are not imaged well under normal bright field illumination conditions. The direct light has been blocked by an opaque stop in the sub-stage condenser, light passing through the specimen from oblique angles at all azimuths is diffracted, refracted and reflected into the microscope objective lens to form a bright image of the specimen superimposed onto a dark background.

A fluorescent microscope produces a magnified image of the sample, but the image is based on the second light source, the light emanating from the fluorescent species, rather than from the light originally used to illuminate and excite the sample. **Fluorescent Microscope** process consists of coating a microbe with **fluorescent** dye such as fluorescent and illuminating it with ultraviolet light. As the electrons in the dye are excited, they move to high energy levels, and then quickly drop back giving off the excess energy as visible light.

Dichroic mirror: In a fluorescence microscope, a dichroic mirror is used to separate the excitation and emission light paths. The dichroic mirror is a key element of the fluorescence microscope, but it is not able to perform all of the required optical functions on its own. Typically, about 90% of the light at wavelengths below the transition wavelength value is reflected and about 90% of the light at wavelengths above this value is transmitted by the dichroic mirror.

Phase contrast is a method used in microscopy and developed in the early 20th century by Dutch physicist Frits Zernike. Phase is only useful on specimens that are colorless and transparent and usually difficult to distinguish from their surroundings, these specimens are called phase objects. Phase contrast microscope is a contrast enhancing optical technique that can be utilized to produce high contrast images of transparent specimens such as living cells (culture), microorganisms, thin tissue slices, lithographic patterns, fibers, latex dispersions, glass fragments and sub-cellular particles including nuclei and other organelles.

'Confocal' is defined as having the same focus, means in the microscope is that the final image has the same focus as or the focus corresponds to the point of focus in the object and the object and its image are confocal. Confocal microscopy offers several advantages over conventional optical microscopy, including controllable depth of field, the elimination of image degrading out-of-focus information, and the ability to collect serial optical sections from thick specimens. The most important feature of a confocal microscope is the capability of isolating and collecting a plane of focus from within a sample, thus eliminating the out of focus haze normally seen with a fluorescent sample.

The lasers commonly employed in laser scanning confocal microscopy are high intensity monochromatic light sources, which are useful as tools for a variety of techniques including optical trapping, lifetime imaging studies, photo bleaching recovery and total internal reflection fluorescence. In addition, lasers are also the most common light source for scanning confocal fluorescence microscope and have been utilized, although less frequently in conventional wide field fluorescence investigations.

Electron microscopes have much greater resolution. Electron Microscopes were developed due to the limitations of light microscope which are limited by light to 500x or 1000x magnification and a resolution of 0.2 micrometers.

Electron microscopes are classified into two types (1) Transmission Electron Microscope (TEM) and (2) Scanning Electron Microscope (SEM). A TEM use electron as light source and their much lower wavelength makes it possible to get a resolution of thousand times better than with a light microscope. The Transmission Electron Microscope (TEM) allows the user to determine the internal structure of materials either of biological or non-biological origin.

The first Scanning Electron Microscope (SEM) debuted in 1942 with the first commercial instruments around 1965. The Scanning Electron Microscope (SEM) works in a different manner. It produces an image from electrons released from atoms on an object's surface. The SEM has been used to examine the surfaces of microorganisms in great detail; many SEMs have a resolution of 7 nm or less.

Flow cytometry is a technique of quantitative single cell analysis. The flow cytometer was developed in the 1970's and rapidly became an essential instrument for the biologic sciences. The major clinical application of flow cytometry is diagnosis of hematologic malignancy but a wide variety of other applications exist such as reticulocyte enumeration and cell function analysis.

Chapter 3
CENTRIFUGATION

3.0 INTRODUCTION

Centrifugation is one of the most important and widely applied research techniques in biochemistry, cellular and molecular biology and in medicine. Current research and clinical applications rely on isolation of cells, sub-cellular organelles, and macromolecules, often in high yields. A centrifuge uses centrifugal force (g-force) to isolate suspended particles from their surrounding medium on either a batch or a continuous-flow basis. Applications for centrifugation are many and may include sedimentation of cells and viruses, separation of sub-cellular organelles and isolation of macromolecules such as DNA, RNA, proteins or lipids.

A centrifuge is a device for separating particles from a solution according to their size, shape, density, viscosity of the medium and rotor speed. In biology, the particles are usually cells, sub cellular organelles, viruses large molecules such as proteins and nucleic acids. Centrifugation separates on the basis of the particle size and density difference between the liquid and solid phases. Sedimentation of material in a centrifugal field may be described by the following equation.

$$v = \frac{d^2(\rho_s - \rho_l)\omega^2 r F_s}{18\eta\theta} \qquad \textit{... eq. 3.1}$$

Where

v = The rate of sedimentation
d = The particle diameter
ρ_s = The particle density
ρ_l = The solution density
ω = The angular velocity in radians s^{-1}

$r =$ The radius of rotation

$\eta =$ The dynamic viscosity

$F_s =$ A correction factor for particle interaction during hindered settling and

$\theta =$ A shape factor (=1 for spherical particles).

F_s depends on the volume fraction of the solids present; approximately equaling 1, 0.5, 0.1 and 0.05 for 1%, 3%, 12% and 20% solids volume fraction respectively. Only material which reaches a surface during the flow through continuous centrifuges will be removed from the centrifuge feedstock, the efficiency depending on the residence time within the centrifuge and the distance necessary for sedimentation (D). This residence time will equal to the volumetric throughput (ϕ) divided by the volume of the centrifuge (V).

The maximum throughput of a centrifuge for efficient use is given by

$$\phi = \frac{d^2 (\rho_s - \rho_l)\omega^2 rVF_s}{18\eta\theta D} \qquad \dots \ eq.\ 3.2$$

The efficiency of the process is depend on the solids volume fraction, the effective clarifying surface (V/D) and the acceleration factor ($\omega^2 r/g$, where g is the gravitational constant, 981 cm s^{-2}; a rotor of radius 25 cm spinning at 1 rev s^{-1} has an acceleration factor of approximately 1 g). Low acceleration factors of about 1500 g may be used for harvesting cells whereas much higher acceleration factors are needed to collect enzyme efficiently. The product of these factors ($\omega^2 rV/gD$) is called the sigma factor (Σ) and is used to compare centrifuges and to assist scale up.

3.1 PRINCIPLE

Centrifugation is based on Stoke's Law. The particle sedimentation velocity increases with increasing diameter, Increasing difference in density between the two phases and decreasing viscosity of the continuous phase. If raw milk were allowed to stand, the fat globules would begin to rise to the surface in phenomena called creaming. Raw milk in a rotating container also has centrifugal forces acting on it. This allows rapid separation of milk fat from the skim milk portion and removal of solid impurities from the milk. The idea here is pretty straight forward and mechanical. If wants more dense materials to be separated from the less dense materials, need a force that differentiates between particles of different density. Think about a swimming pool with a rock and a piece of styrofoam. The rock is denser than water and thus it sinks. The styrofoam is less dense than water and thus it floats. Density is of course mass per unit volume, so, if have a bag full of rocks and styrofoam and want to separate one from the other, just dump the mixture into some water under the influence of the earth's gravity. The rocks will displace the water because they have greater mass for a given volume and gravity will pull them through the water. On the other hand the water will displace the styrofoam because a certain volume of water weighs more than the same volume of styrofoam.

The greater difference in density the particles move faster. If there is no difference in density (isopyknic conditions) the particles stay steady. To take advantage of even tiny differences in density to separate various particles in a solution, gravity can be replaced with the much more powerful centrifugal force provided by a centrifuge (Fig. 3.1 a and b).

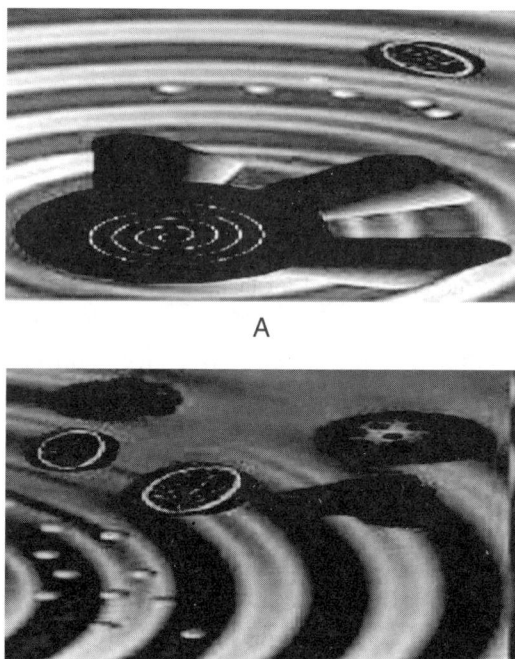

A

B

Fig. 3.1 a & b: Centrifugal force provided by a centrifuge

Many particles or cells in a liquid suspension will eventually settle at the bottom of a container due to gravity ($1 \times g$) in a given time. However the length of time required for such separations is impractical. Other particles extremely small in size will not separate at all in solution unless subjected to high centrifugal force. When a suspension is rotated at a certain speed or revolutions per minute (RPM), centrifugal force causes the particles to move radially away from the axis of rotation. The force on the particles (compared to gravity) is called Relative Centrifugal Force (RCF). For example an RCF of 500Xg indicates that the centrifugal force applied is 500 times greater than Earths gravitational force. Table 3.1 illustrates common centrifuge classes and their applications.

Table 3.1: Classes of centrifuges and their applications

Particulars	Centrifuge classes		
	Low speed	High-speed	Ultra/micro-ultra
Maximum RCF ($\times 10^3$)	7	100	800/900
Pelleting applications Bacteria	Yes	Yes	(Yes)
Animal and plant cells	Yes	Yes	(Yes)
Nuclei	Yes	Yes	(Yes)
Precipitates	Some	Most	(Yes)
Membrane fractions	Some	Saome	Yes
Ribosomes or Polysomes	—	—	Yes
Macromolecules	—	—	Yes
Viruses	—	Most	Yes

( ) = Can be done but not usually used for this purpose

3.1.1. Sedimentation

The rate of sedimentation is dependent upon the applied centrifugal field (*G*) being directed radially outwards; this is determined by the square of the angular velocity of the rotor (ω, in radians s^{-1}) and the radial distance (*r*, in centimeters) of the particle from the axis of rotation, according to the equation:

$$G = \omega^2 r \qquad \text{... } eq.\ 3.3$$

Since one revolution of the rotor is equal to 2π radians its angular velocity in radians s^{-1} can be readily expressed in terms of revolutions per minute (rev min^{-1}) the common way of expressing rotor speed being:

$$\omega = \frac{2\pi\ \text{rev min}^{-1}}{3600} \qquad \text{... } eq.\ 3.4$$

The centrifugal field (G) in terms of rev min^{-1} is then

$$G = \frac{4\pi 2 (\text{rev min}^{-1}) 2r}{3600} \qquad \ldots eq.\ 3.5$$

And is generally expressed as a multiple of the Earth's gravitational field ($g = 981$ cm s^{-2}) i.e the ratio of the weight of the particle in the centrifugal field to the weight of the same particle when acted on by gravity alone and is then referred to as the relative centrifugal field (RCF) or more commonly as the number times g:

Hence, $\qquad$ RCF $= \dfrac{r\pi 2 (\text{rev min}^{-1}) 2r}{3600 \times 981} \qquad \ldots eq.\ 3.6$

Which may be shortened to give:

$$\text{RCF} = (1.118 \times 10^{-5}) (\text{rev min}^{-1})^2 r \qquad \ldots eq.\ 3.7$$

When conditions for the centrifugal separation of particles are reported therefore rotor speed, radial dimensions and time of operation of the rotor must all be quoted. Since biochemical experiments are usually conducted with particles dissolved or suspended in solution the rate of sedimentation of a particle is dependent not only upon the applied centrifugal field but also upon the mass of the particle which may be expressed as the product of its volume and density, the density and viscosity of the medium in which it is sedimenting and the extent to which its shape deviates from spherical.

When particle sediments must displace some of the solution in which it is suspended, resulting in an apparent up thrust on the particle equal to the weight of liquid displaced. If a particle is assumed to be spherical and of known volume and density the later being corrected for the buoyancy due to the density of the medium then the net outward force (F) when centrifuge at an angular velocity of radians s^{-1} is given by:

$$F = \frac{4}{3}\pi r_p^3 \left(p_p - p_m \right) \omega^2 r \qquad \ldots eq.\ 3.8$$

Where $\dfrac{4}{3}\pi r_p^3$ is the volume of a sphere of radius

p_p = The density of the particle

p_m = The density of the suspending medium and

r $\;$ = The distance of the particle form the centre of rotation

3.1.1.1 *Problems*

1. An ultracentrifuge is operated at a speed of 58000 rev min^{-1}.

i. Calculate the angular velocity (in radians per second) and the centrifugal field at a point equivalent to 6.2 cm from the centre of rotation.

ii. How many 'times g' is this equivalent to?

Answer:

i. The angular velocity may be calculated using the following equation:

$$\omega = \frac{2\pi \, \text{rev} \, \text{min}^{-1}}{3600} = \frac{2 \times 3.14 \times 58000}{60} = 6070.7 \text{ radians } s^{-1}$$

The centrifugal field (G) at a point 6.2 cm from the centre of rotation may be calculated the equation

$$G = \omega^2 r$$

$$G = (6070.7)^2 \times 6.2 = 2.285 \times 10^8 \text{cm s}^{-2}$$

ii. The Earth's gravitational field (g) = 980cm s^{-2}

$$\text{Relative centrifugal field (RCF)} = \frac{\omega 2r}{980}$$

And, since

$$G = \omega^2 r,$$

$$\text{RCF} = G/980 = 2.285 \times 10^8/980 = 2333163 \times g$$

2. Calculate the relative centrifugal field (RCF) exerted at the top and bottom of a centrifuge tube being centrifuged in a fixed-angle rotor, assuming that the rotor dimensions for the minimum radius (r_{min}) at the top of the tube is 4.8 cm and for the maximum radius (r_{max}) at the bottom of the tube is 9 cm, and that the rotor is spinning at a speed of 12000 rev min^{-1}.

Answer: Using the equation

$$\text{RCF} = (1.118 \times 10^{-5})(\text{rev min}^{-1})^2 r$$

$$\text{RCF}_{\text{top}} = (1.118 \times 10^5)(12000)^2 \times 4.8$$

$$= 77728 \times g$$

And $\quad \text{RCF}_{\text{bottom}} = (1.118 \times 10^5)(12000)^2 \times 9$

$$= 14489 \times g$$

As can be seen the centrifugal field exerted at the top and bottom of the centrifuge tube differs by nearly two-fold.

3. A protein was subjected to ultra centrifugal analysis at a rotor speed of 59,780 Rev min^{-1} and a temperature of 20°C. Using the technique of sedimentation Velocity, measurements were taken of the radial position of the sedimenting protein boundary as a function of centrifugation time ...

Time (Min)	Radial position (r in cm)
8	6.089
16	6.179
24	6.270
32	6.362
40	6.454
48	6.549

i. From the information provided plot a graph of log r (centimeters) time (seconds) and hence calculate the sedimentation coefficient of the protein.

ii. In subsequent analyses, also carried out at 20°C the protein was found to have an average diffusion coefficient of 4.0 × 10^{-11}m^{2}s^{-1} and a partial specific volume of 0.734 × 10^{-3}m^{3}kg^{-1}.

The density of water at 20°C is 998 kgm^{-3}.

The gradient of the plot of log r (centimeters) against time (seconds) = s^2/2.3

Where $\omega = \dfrac{2\pi \text{ rev min}^{-1}}{60}$

R = 8.314 j K^{-1} mol^{-1}; 0°C = 273 K

Answer:

i. Calculation of the sedimentation coefficient of the protein:

The sedimentation coefficient may be calculated using the equation

$$S = \frac{2.3(gradient)}{\omega 2}$$

Where $\omega = \dfrac{2\pi \text{ rev min}^{-1}}{60} = \dfrac{2\pi 59780}{60}$ =6260.15 rads s^{-1}

The gradient of the line is obtained by measuring the radial position of the sedimenting protein boundary (in centimeters) at different times (in seconds) and then plotting log r (centimeters) against time (seconds).

Time (min)	Time (S)	r(cm)	log r (cm)
8	480	6.089	0.7845
16	960	6.179	0.7909
24	1440	6.270	0.7973
32	1920	6.362	0.8036
40	2400	6.454	0.8098
48	2880	6.549	0.8161

From the graph obtained the gradient of the line (log r/time) may be calculated and found to be 1.32×10^{-5}

Hence

$$s = \frac{2.3 \times 1.32 \times 10^{-5}}{(6260.15)2} = 7.75 \times 10^{-13} s$$

ii. Calculation of the relative molecular mass of the protein:

The answer may be calculated using the Svedberg equation

Substituting the values given in the equation (in SI units) into the equation the answer may be calculated to be

$$M_r = \frac{RTs}{D(1 - \nabla p)} = \frac{8.314 \times 293 \times 7.75 \times 10^{-13}}{4 \times 10^{-11} (1 - 0.734 \times 10^{-3} \times 998)}$$

$$= 176.46 \text{ kg mol}^{-1}$$

The relative molecular mass of a molecule is a relative quantity and is defines as the ratio of the mass of a molecule relative to 1/12 of the mass of the carbon isotope ^{12}C; it is thus dimensionless. The answer obtained from the calculation (i.e. 176.46 kg mol^{-1}) can be converted to a relative molecular mass by dividing by 1g mol^{-1} (the equivalent of multiplying the answer by 1000 and canceling units) to give a relative molecular mass of 176 460.

Molecular mass expressed in Daltons. The molecular mass is the mass of one molecule of a substance expressed in Daltons (Da) or atomic mass units (1dalton = 1 atomic mass unit equals one-twelfth of the mass of one atom of ^{12}C). This may be obtained by multiplying the answer given in the calculation by 1000 to give an answer of 176 460.

3.1.2 Sedimentation Velocity

Sedimentation Velocity provides hydrodynamic information about molecules in solution. The experimentally determined parameters include the sedimentation coefficient, s, the diffusion constant D and in some cases the molecular mass M. If the molecular weight is known the sedimentation coefficient can be used to obtain an estimate of the molecular shape of the molecule in solution. Sedimentation velocity is an analytical ultracentrifugation (AUC) method that measures the rate at which molecules move in response to centrifugal force generated in a centrifuge. This sedimentation rate provides information about both the molecular mass and the shape of molecules. In some cases this technique can also measure diffusion coefficients and molecular mass.

The data is gathered at high rotor speeds where the sedimentation transport dominates the diffusion. One measures a molecule's rate of transport from the top to the bottom of the cell. A sedimentation velocity experiment requires approximately 500µg of sample protein in a volume of 1 ml.

Sedimentation Equilibrium provides thermodynamic information about molecules in solution. The experimentally determined parameters include the molecular mass M, the solution assembly state and thermodynamic parameters for associating systems most notably the equilibrium constant K from which the free energy of the association reaction can be calculated. Measurements collected at different temperatures can yield other thermo-dynamic parameters including the enthalpy sometimes the heat capacity of the reaction. The data are collected at slower rotor speeds so that the sedimentation and diffusion forces can balance. At sedimentation equilibrium, these forces are equal in magnitude but opposite in direction and the molecule is exponentially distributed across the cell. There is no longer any net transport of molecules in the system. All shape factors are thus cancelled out yielding the molecular weight. A typical sedimentation equilibrium experiment requires 300µl of sample at an OD of 0.8 at the wavelength of interest usually 280 nm. This is generally on the order of 200 – 400µg of total protein but varies greatly depending on protein's molar extinction coefficient.

Sedimentation velocity is particularly valuable for:

1. Verifying whether a sample is entirely homogeneous in mass
2. Detecting aggregates in protein samples and quantifying the amount of aggregate.
3. Comparing the conformations for samples from different lots, manufacturing processes or expression systems or comparing different mutants of the same protein.
4. Establishing whether the native state of a protein is a monomer, dimer, trimer etc.
5. Determining the overall shape of non-glycosylated protein molecules in solution.
6. Measuring the distribution of sizes in samples which contain a very broad range of sizes.
7. Detecting changes in protein conformation for example partial unfolding or transitions to molten globule states.
8. Studying the formation and stoichiometry of tight complexes between proteins for example receptor-ligand or antigen-antibody complexes.

A major advantage of this method over sedimentation equilibrium is that experiments usually require only 3–5 hours, as opposed to the several days typical of sedimentation equilibrium. Thus sedimentation velocity can be used with samples that are too labile for sedimentation equilibrium. The major drawback relative to sedimentation equilibrium applies to interacting systems proteins that reversibly self-associate or protein-protein complexes, where the non-equilibrium nature of the measurement can lead to significant changes in species distributions over the course of an experiment. Further, for interacting systems it is generally more difficult and less accurate to derive binding constants (K_d's) from sedimentation velocity data.

An important strength of sedimentation velocity is its ability to study samples over a fairly wide range of pH and ionic strength conditions and often directly in formulation buffers and at temperatures from 4° to 40°C. The amount of protein required depends on the application but each sample is usually ~0.45ml at typical protein concentrations of 0.1 mg/ml. Protein concentration can range as low as ~10 micrograms/ml or as high as ~10 mg/ml in some cases. Up to 3 samples can be run at one time.

3.1.3. Rotors

Rotors can be broadly classified into three common categories, that type of rotor has strengths and limitations depending on the type of separation.

(1) **Swinging bucket rotors:** In swinging bucket rotors, the sample tubes are loaded into individual buckets that hang vertically while the rotor is at rest. When the rotor begins to rotate the buckets swing out to a horizontal position (Fig. 3.2). This rotor is particularly useful when samples are to be resolved in density gradients. The longer path length permits better separation of individual particle types from a mixture. However, this rotor is relatively inefficient for pelleting (Table 3.2). Also, care must be taken to avoid point loads caused by spinning CsCl or other dense gradient materials that can precipitate.

(2) **Fixed angle rotors:** In fixed-angle rotors, the sample tubes are held fixed at the angle of the rotor cavity. When the rotor begins to rotate, the solution in the tubes reorients (Fig. 3.2). This rotor type is most commonly used for pelleting applications. Examples include pelleting bacteria, yeast, and other mammalian cells. It is also useful for isopycnic separations of macromolecules such as nucleic acids (Table 3.2).

(3) **Vertical rotors:** In vertical rotors, sample tubes are held in vertical position during rotation (Fig. 3.2). This type of rotor is not suitable

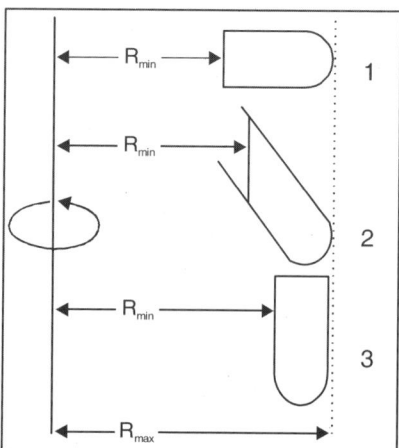

Fig.-3.2: Rotor types (1) swinging bucket rotors
(2) fixed-angle rotors (3) vertical rotors

for pelleting applications but is most efficient for isopycnic (density) separations due to the short path length. Applications include plasmid DNA, RNA and lipoprotein isolations (Table 3.2). Other rotors include continuous flow and elutriation rotors.

Table 3.2: Types of rotors and their applications

S.No.	Type of rotor	Pelleting	Rate-zonal Sedimentation	Isopycnic
1.	Fixed-angle	Excellent	Limited	Variable*
2.	Swinging-Bucket	Inefficient	Good	Good**
3.	Vertical	Not suitable	Good	Excellent
4.	Zonal	Not suitable	Excellent	Good

*Good for macromolecules, poor for cells and organelles
**Good for cells and organelles, caution needed if used with CsCl

3.1.4 Centrifuge tubes

Centrifuges tubes are designed prevent sample leakage or loss, ensures chemical compatibility and allows easy sample recovery (Table 3.3). In selection of a centrifuge tube some major factors are involved (a) Clarity (b) Chemical resistance (c) Sealing mechanism.

Table 3.3: Chemical compatibility of popular tube materials

S.No	Tube Plastic type	Clarity	Chemical Resistance
1.	Polypropylene (PP)	Opaque	Good
2.	Polyallomer (PA)	Opaque	Good
3.	Polycarbonate (PC)	Clear	Poor
4.	Polyethylene terephtalate (PET)	Clear	Poor

3.2. TYPES

The major distinguishing features between centrifuge types are speed and capacity. (1) Low speed, up to about 5000 rpm (2) High speed machines of up to about 25,000 rpm (3) Ultracentrifuges which will turn at up to 100,000 rpm. In a typical biochemistry laboratory will find three different centrifuges.

(a) **Microfuge centrifuges** - These are made for spinning 1 to 2 ml plastic centrifuge tubes at speeds up to 12 or 13 thousand rounds per minute. They have very small, light rotors in them (the rotor is the part of the centrifuge that contains the holes for the sample tubes) which speed up and slow down rapidly. These centrifuges are very convenient for low to medium speed centrifugation of small quantities of material.

(b) **Large super-speed centrifuge** - These are common size centrifuges. These have speeds up to about 20,000 rpm and can take tubes of various sizes, depending on the rotors.

(c) **Ultracentrifuge** - Most biochemistry laboratories have access to an ultracentrifuge. Speeds up to 70,000 rpm are available on typical modern versions. Again the size of tube and the maximum speed vary from rotor to rotor but tube sizes up to about 60 ml are available.

The two most common types of centrifugation are (1) **Analytical centrifugation:** Analytical centrifugation involves measuring the physical properties of the sedimenting particles such as sedimentation coefficient or molecular weight. Optimal methods are used in analytical centrifugation. Molecules are observed by optical system during centrifugation to allow observation of macromolecules in solution as they move in gravitational field. The samples are centrifuged in cells (tubes with quartz windows) having windows that lie paralleled to the plane of rotation of the rotor head. As the rotor turns, the images of the cell (proteins) are projected by an optical system on to film or a computer. The concentration of the solution at various points in the cell is determined by absorption of a light of the appropriate wavelength (Beer's law is followed). This can be accomplished either by measuring the degree of blackening of a photographic film or by the pen deflection of the recorder of the scanning system or fed into a computer. (2) **Preparative centrifugation:** The other form of centrifugation is called preparative and the objective is to isolate specific particles which can be reused. There are many type of preparative centrifugation such as rate zonal, differential, isopycnic centrifugation.

3.2.1. Differential centrifugation

Separation is achieved primarily based on the size of the particles in differential centrifugation. This type of separation is commonly used in simple pelleting and in obtaining partially pure preparation of subcellular organelles and macromolecules. For the study of subcellular organelles, tissue or cells are first disrupted to release their internal contents; this crude disrupted cell mixture is referred to as a homogenate. During centrifugation of a cell homogenate larger particles sediment faster than smaller ones and this provides the basis for obtaining crude organelle fractions by differential centrifugation. A cell homogenate can be centrifuged at a series of progressively higher g-forces and times to generate pellets of partially purified organelles.

When a cell homogenate is centrifuged at $1000 \times g$ for 10 minutes, unbroken cells and heavy nuclei pellet to the bottom of the tube. The supernatant can be further centrifuged at $10,000 \times g$ for 20 minutes to pellet sub-cellular organelles of intermediate velocities such as mitochondria, lysosomes and microbodies. Some of these sedimenting organelles can obtain in partial purity and are typically contaminated with other particles. Repeated washing of the pellets by re-suspending in isotonic solvents and re-pelleting may result in removal of contaminants that are smaller in size (Fig. 3.3 and 3.4). Obtaining partially purified organelles by differential centrifugation serves as the preliminary step for further purification using other types of centrifugal separation (density gradient separation).

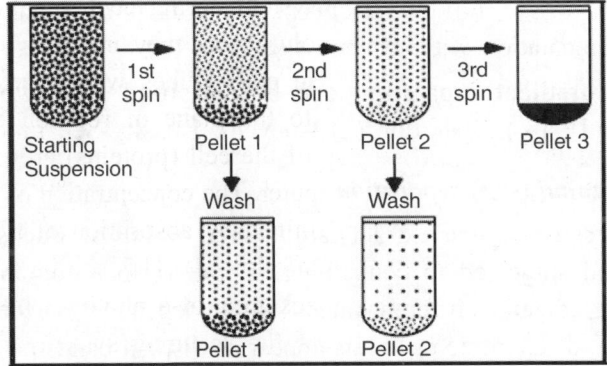

Fig. 3.3: Differential Centrifugation-1

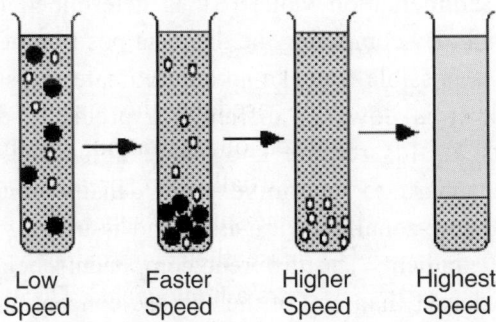

Fig. 3.4 Differential centrifugation-2

3.2.2. Density gradient centrifugation

Density gradient centrifugation is a technique that allows the separation of cells, organelles and macromolecules depending on their size, shape and density. A density gradient is created in a centrifuge tube by layering solutions of varying densities with the dense end at the bottom of the tube. Cells and large molecules are usually separated on a shallow gradient of sucrose or other inert carbohydrates even at relatively low centrifugation speeds while macromolecules such as proteins and nucleic acids are separated at higher centrifugation using ultracentrifuges.

Density gradients are used in many different operations:

i. To separate particles of different densities (isopycnography, which is short for equilibrium density gradient centrifugation)

ii. To separate particles of different sizes (sedimentation centrifugation)

iii. Isolation of diamond dust (isopycnography).

iv. Isolation of bovine X-sperm from Y-sperm (dairy industry) (sedimentation without centrifugation).

Density gradient separation can be classified into two categories (1) Rate-zonal (size) separation (2) Isopycnic/Density Separation.

3.2.2.1 Rate zonal (size) separation

When mixtures of cellular extracts are layered on top of a density gradient in a tube and subjected to centrifugation, the various components move through the gradient at different rates that are dependent on their sizes and shapes. These different components appear as distinct bands or zones in the gradient with large components migrating farthest in the tube in a given period of time.

The rate with a fraction moves the fixed distance in the gradient tube is dependant of its sedimentation value (S), is determined by the size and shape of that fraction. By comparing the different position of the components in the gradient, it is possible to make an approximate measurement of their molecular weight. It is however difficult to precisely determine these molecular weights as this require knowledge about the shape of these molecules which is hard to determine. This density gradient separation technique is called rate-zonal centrifugation and is usually performed with a shallow sucrose gradient. The different components being separated by this technique are denser than any of the sucrose concentrations used in the gradient. Samples are centrifuged just long enough to separate the components of interest. Longer centrifugation would allow all components to form a pellet at the bottom of the tube. One of the most important applications of this technique over the past decades was the separation of transfer RNA (4S) from ribosomal RNA that forms three different classes with distinct sedimentation values 23S, 16S and 5S. This help to facilitate the characterization of the protein synthesizing system.

Rate-zonal separation takes advantage of particle size and mass instead of particle density for sedimentation. Fig. 3.5 illustrates a rate-zonal separation process and the criteria for successful rate-zonal separation. Examples of common applications include separation of cellular organelles such as endosomes or separation of proteins, such as antibodies. For instance, Antibody classes all have very similar densities, but different masses. Thus, separation based on mass will separate the different classes, whereas separation based on density will not be able to resolve these antibody classes.

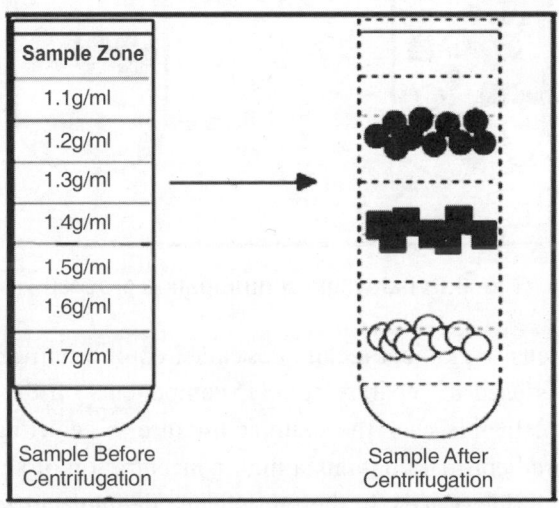

Fig. 3.5: Rate-Zonal centrifugation process.

Successful criteria for rate-zonal centrifugation depends on density of the sample solution must be less than that of the lowest density portion of the gradient and the path length of the gradient must be sufficient for the separation.

3.2.2.2. Isopycnic or density separation

In this type of separation, a particle of a particular density will sink during centrifugation until a position is reached where the density of the surrounding solution is exactly the same as the density of the particle. Once this quasi-equilibrium is reached, the length of centrifugation does not have any influence on the migration of the particle. A common example for this method is separation of nucleic acids in a CsCl gradient. Fig. 3.6 illustrates the isopycnic separation and criteria for successful separation.

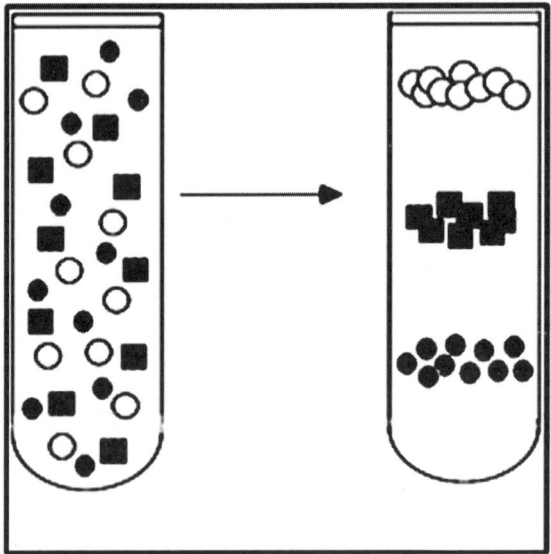

Fig. 3.6: Isopycnic centrifugation process

A second density gradient technique, called equilibrium density-gradient centrifugation is used to separate cellular components on the basis of their buoyant density. In this case the cellular mixture is centrifuged through a steep density gradient that contains a high concentration of sucrose or more often cesium chloride (CsCl). In these gradients, the molecules being studied have a density somewhere in between the highest and lowest densities of sucrose or CsCl generated in the gradient. The components of a sample begin to move down this gradient in the same way as they do in a rate-zonal density gradient. When a component of the mixture reaches a point where the density of the solution is equal to its own density, it stops moving further and forms a distinct band. The position of the band in the tube is characteristic of the buoyancy of that component. Buoyancy or buoyant density of a substance is its tendency to float in a medium, which in this case is the density gradient. Equilibrium density gradient centrifugation using CsCl was for decades, the method of choice in the purification of highly pure plasmid DNA. Meselson and Stahl, who developed this technique, were the first to use it in an experiment that provided evidence for the semi-conservative replication of DNA and confirmed the double helix structure of DNA proposed by Crick and Watson.

A variety of gradient media can be used for isopycnic separations and their biological applications are listed in Table 3.4.

Table 3.4: Applications of Density gradient media for Isopycnic separations.

S.No.	Gradient media	Cells	Viruses	Organelles	Nucleoproteins	Macro-molecules
1.	Sugars (Sucrose)	+	+++	+++	+	—
2.	Polysaccharides Ificoll)	++	++	++	—	—
3.	Colloidal Silica (Percoll)	+++	+	+++	—	—
4.	Iodinated media (Nycodenz)	++++	++	++++	+++	+
5.	Alkali metal salts (CsCl)	—	++	—	++	++++

++++ Excellent, +++ good, ++ good for some applications, + limited use, — unsatisfactory.

The successful criteria for equilibrium density-gradient centrifugation depend on density of the sample particle must fall within the limits of the gradient densities any gradient length is acceptable and run time must be sufficient for the particles to band at their isopycnic point. Excessive run times have no adverse effect.

3.2.3. Ultracentrifugation

Samples are dissolved or suspended in a suitable solvent or suspension medium. High gravitational force is applied by spinning the container at high speeds at a fixed centrifugal force and liquid viscosity. The sedimentation rate of a particle is proportional to its size (molecular weight) and the difference between its density and the density of the solution. By proper selection of suspension medium and gravitations force, sedimentation equilibrium can be established such that the sample is layered top to bottom in order of increasing density. The ultracentrifuge is a centrifuge optimized for spinning a rotor at very high speeds, capable of generating acceleration as high as 1,000,000g (9,800km/s^2). Ultracentrifugation is carried out at speed faster than 20,000rpm. Super speed ultracentrifugation is at speeds between 10,000 and 20,000rpm. Low speed centrifugation is at speed below 10,000rpm. There are two kinds of ultracentrifuges (1) Analytical ultracentrifuge (2) Preparative ultracentrifuge. Both classes of instruments find important uses in molecular biology, biochemistry and polymer science. Theodor Svedberg invented the analytical ultracentrifuge in 1923 and won the Nobel Prize in Chemistry in 1926 for his research on colloids and proteins using the ultracentrifuge. The vacuum ultracentrifuge was invented by Edward Greydon Pickels, the vacuum which allowed a reduction in friction generated at high speeds. Vacuum systems also enabled the maintenance of constant temperature.

3.2.3.1. Analytical ultracentrifuges

Analytical ultracentrifuges are capable of operating at speed approaching 70000rev min^{-1} (500000g) and consist of a motor, a rotor contained in a protective armoured chamber that is refrigerated and evacuated, and an optical system to enable the sedimenting material to be observed throughout the duration of centrifugation to determine concentration distributions in the sample at any time during centrifugation. Three types of optical system are available in the analytical ultracentrifuge (1) Light absorption system (2) The alternative Schlieren system (3) Rayleigh interferometric system.

The rotor is solid, with holes to hold the cells that contain the samples and is suspended on a wire coming from the drive shaft of a high speed motor that allows the rotor to find its own axis of rotation. The tip of the rotor contains a thermistor for measuring temperature. Several types of rotor are available, the simplest rotor model incorporates two cells which are the analytical cell and the counterpoise cell which counterbalances the analytical cell. Two holes are drilled through the counterpoise cell to facilitate the calibration of distances in the analytical cell. A wide variety of cells is available and has a capacity of between 0.4 and 1.0 cm^3. Analytical cells used with the ultraviolet light absorption optical system and the Schlieren optical system have a single 2° or 4° sector shape, to prevent convection and usually have a 12 mm optical path length centerpiece although centre pieces of 1.5 mm to 30 mm are available. Double sector cells, having two 2.5 sectors and an optical path length that can vary from 12 mm to 30 mm, are used with the interference optical system, the absorption optical system when used with the photoelectric scanner and the Schlieren optical system when a baseline is required. Double sector cells also facilitate measurements of differences in sedimentation coefficients and of diffusion coefficients.

The rotor chamber contains an upper condensing lens and a lower collimating lens together with a camera lens focuses light on to a photographic plate whereas the latter collimates the light so that the sample cell is illuminated by parallel light (Fig. 3.7).

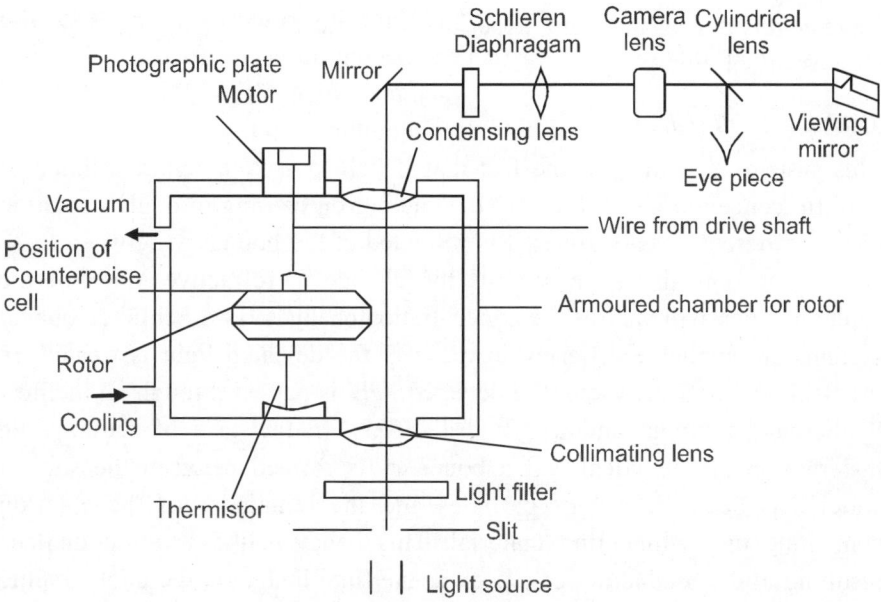

Fig. 3.7: An Analytical Ultracentrifuge system with a Schlieren Optical System

3.2.3.1.1. Ultraviolet light absorption system

The light of a suitable wavelength is passed through the moving analytical cell containing the solution under analysis. Example: Protein or nucleic acid, the intensity of the transmitted light recorded either on a photographic plate or by an automatic single or split be photoelectric scanning system.

The scanner system unlike the ultraviolet photographic method has the advantage of allowing direct visualization of the results during the course of the experiment and can provide a plot of the concentration of the sample at all points in the analytical cell at any particular time. Different wavelengths of light can be selected enabling the separate movement of single components in a mixture of substances to be monitored provided that they absorb light at different wavelengths.

Absorption optics is particularly sensitive for detection of macro-molecules containing strong chromophores. For example, taking advantage of the intense amide absorption in the far ultraviolet (UV) and the strong lamp output at 230 nm proteins can be characterized with good signal to noise ratio at concentrations as low as 10µg/ml. Similarly nucleic acids can be studied in the same concentration range by following their absorbance at 260 nm. For samples containing two or more components with different absorption spectra (protein and nucleic acids) data can be obtained at multiple wavelengths during the same experiment to selectively monitor the several species in solution.

3.2.3.1.2. Schlieren optical system

This system makes use of the fact that if light passes though a solution of uniform concentration it does not deviate but on passing through a solution having different density zones it is refracted at the boundary between these zones. The optical system records the change in refractive index of the solution which will vary as the concentration changes. Two Schlieren optical systems are available differing in the way the deviated light is treated. In the Schlieren optical system the deviated light is passed through an inclined Schlieren diaphragm and a cylindrical lens. In the case of sedimenting materials in an analytical cell a boundary is formed between the solvent which has been cleared of particles and the remainder of the solution containing the sedimenting material. This behaves like a refraction lens resulting in the production of a peak in the final image on the photographic plate which is used as the detector system. The peak is an exact record of the refractive index gradient and the area beneath it is proportional to the concentration of solute. As sedimentation proceeds the boundary the peak shifts and the rate at which the peak moves gives a measure of the rate at which the material is sedimenting. After a period of sedimentation the peak height diminishes and the width increases owing to radial dilution of the sample due to the sector shape of the cell.

3.2.3.1.3. Rayleigh interference optical system

In the bright field scanning optical system the deviated light is interrupted by a knife edge and the resultant image scanned with a photomultiplier, the resulting refractive index gradient appearing as a dark band against a bright background. The Schlieren Optical system plots the refractive index gradient against distance along the analytical cell which makes it useful for locating boundaries in sedimentation velocity measurements. For example the sedimentation equilibrium method used for relative molecular mass determinations the Schlieren system is not sufficiently sensitive to detect small concentration differences. Therefore made of the more sensitive Rayleigh interference optical system which employees a double sector cell in which one sector contains the solvent and the other the solution.

The optical system measures the difference in refractive index between the reference solvent and the solution by the displacement of interferences fringes caused by slits placed behind the two liquid columns each fringe tracing a curve of the refractive index gradient against distance in the cell. Since the position of the fringes is determined by solute concentration it is possible to measure the concentration of solute at any point along the cell.

The Rayleigh interference optical system is used to analyze macromolecules lacking intense chromophores Ex.: polysaccharides and samples that contain strongly absorbing buffer components Ex.: ATP/GTP. The data from each cell are acquired simultaneously on a charge couple device camera by the interference optical system, and the resulting rapid collection of large amounts of data is especially useful for certain types of sedimentation velocity experiments. Interference optics is also useful for sedimentation equilibrium experiments that require a higher radial resolution than is provided by the absorbance optical system. Because the refractive index increment is fairly constant for most proteins, combining absorbance and interference data provides a method to determine protein extinction coefficients.

3.2.3.1. Preparative ultracentrifuge

Preparative ultracentrifuges are available with a wide variety of rotors suitable for a great range of experiments. Most rotors are designed to hold tubes that contain the samples. Swinging bucket rotors allow the tubes to hang on hinges so the tubes reorient to the horizontal as the rotor initially accelerates. Fixed angle rotors are made of a single block of metal and hold the tubes in cavities bored at a predetermined angle. Zonal rotors are designed to contain a large volume of sample in a single central cavity rather than in tubes. Some zonal rotors are capable of dynamic loading and unloading of samples while the rotor is spinning at high speed.

Preparative rotors are used in biology for pelleting of fine particulate fractions such as cellular organelles (mitochondria, microsomes, ribosomes) and viruses. They can also be used for gradient separations, in which the tubes are filled from top to bottom with an increasing concentration of a dense substance in solution. Sucrose gradients are typically used for separation of cellular organelles. After the sample has spun at high speed for sufficient time to produce the separation, the rotor is allowed to come to a smooth stop and the gradient is gently pumped out of each tube to isolate the separated components.

3.3. SUMMARY

A centrifuge is a device for separating particles from a solution according to their size, shape, density, viscosity of the medium and rotor speed. Centrifugation separates on the basis of the particle size and density difference between the liquid and solid phases. Centrifugation is based on Stoke's Law.

Sedimentation Velocity provides hydrodynamic information about molecules in solution. The experimentally determined parameters include the sedimentation coefficient, s, the diffusion constant D and in some cases the molecular mass M. This sedimentation rate provides information about both the molecular mass and the shape of molecules. In some cases this technique can also measure diffusion coefficients and molecular mass.

Rotors: In swinging bucket rotors, the sample tubes are loaded into individual buckets that hang vertically while the rotor is at rest. This rotor is particularly useful when samples are to be resolved in density gradients. In fixed-angle rotors, the sample tubes are held fixed at the angle of the rotor cavity. When the rotor begins to rotate, the solution in the tubes reorients. This rotor type is most commonly used for pelleting applications. Examples include pelleting bacteria, yeast, and other mammalian cells. In vertical rotors, sample tubes are held in vertical position during rotation. This type of rotor is not suitable for pelleting applications but is most efficient for isopycnic (density) separations due to the short path length. Applications include plasmid DNA, RNA, and lipoprotein isolations. Other rotors include continuous flow and elutriation rotors.

Microfuge centrifuges: These are made for spinning 1 to 2 ml plastic centrifuge tubes at speeds up to 12 or 13 thousand rounds per minute. They have very small, light rotors in them (the rotor is the part of the centrifuge that contains the holes for the sample tubes) which speed up and slow down rapidly. These centrifuges are very convenient for low to medium speed centrifugation of small quantities of material

Large super-speed centrifuge: These are common size centrifuges. These have speeds up to about 20,000 rpm and can take tubes of various sizes, depending on the rotors.

Ultracentrifuge: Most biochemistry laboratories have access to an ultracentrifuge. Speeds up to 70,000 rpm are available on typical modern versions. The size of tube and the maximum speed vary from rotor to rotor but tube sizes up to about 60 ml are available.

Analytical centrifugation involves measuring the physical properties of the sedimenting particles such as sedimentation coefficient or molecular weight. Optimal methods are used in analytical centrifugation. Molecules are observed by optical system during centrifugation to allow observation of macromolecules in solution as they move in gravitational field.

Preparative centrifugation: The other form of centrifugation is called preparative and the objective is to isolate specific particles which can be reused. There are many type of preparative centrifugation such as rate zonal, differential, isopycnic centrifugation.

Separation is achieved primarily based on the size of the particles in differential centrifugation. This type of separation is commonly used in simple pelleting and in obtaining partially pure preparation of sub-cellular organelles and macromolecules. During centrifugation of a cell homogenate larger particles sediment faster than smaller ones and this provides the basis for obtaining crude organelle fractions by differential centrifugation. A cell homogenate can be centrifuged at a series of progressively higher g-forces and times to generate pellets of partially purified organelles.

Density gradient centrifugation is a technique that allows the separation of cells, organelles and macromolecules, depending on their size, shape and density. A density gradient is created in a centrifuge tube by layering solutions of varying densities with the dense end at the bottom of the tube.

Rate-zonal separation takes advantage of particle size and mass instead of particle density for sedimentation. Successful criteria for rate-zonal centrifugation depends on density of the sample solution must be less than that of the lowest density portion of the gradient, the path length of the gradient must be sufficient for the separation to occur and time If perform too long runs, particles may all pellet at the bottom of the tube.

Isopycnic or density separation a particle of a particular density will sink during centrifugation until a position is reached where the density of the surrounding solution is exactly the same as the density of the particle. Once this quasi-equilibrium is reached, the length of centrifugation does not have any influence on the migration of the particle. A common example for this method is separation of nucleic acids in a CsCl gradient.

The Schlieren Optical system plots the refractive index gradient against distance along the analytical cell which makes it useful for locating boundaries in sedimentation velocity measurements. For example the sedimentation equilibrium method used for relative molecular mass determinations the Schlieren system is not sufficiently sensitive to detect small concentration differences.

The Rayleigh interference optical system is used to analyze macromolecules lacking intense chromophores Ex.: polysaccharides and samples that contain strongly absorbing buffer components.

Chapter 4
SPECTROSCOPY–I

4.0. RADIATION

Radiation may be defined as energy traveling through space (Fig. 4.1). Radiation sources are found in a wide range of occupational settings. The most familiar form of electromagnetic (EM) radiation is sunshine, which provides light and heat. Sunshine consists primarily of radiation in infrared (IR), visible and ultraviolet (UV) frequencies. Radiation can be classified into two types (1) Non-ionizing radiation - Lower frequency radiation, consisting of ultraviolet (UV), infrared (IR), microwave (MW), Radio Frequency (RF), and extremely low frequency (ELF) (Fig. 4.2). Non-ionizing radiation is essential to life, but excessive exposures will cause tissue damage. (2) Ionizing radiation. All forms of ionizing radiation have sufficient energy to ionize atoms that may destabilize molecules within cells and lead to tissue damage. The two types of ionizing radiation are particulate (alpha, beta, neutrons) and the higher frequencies of Electromagnetic radition (X-rays, gamm rays) Fig. 4.2.

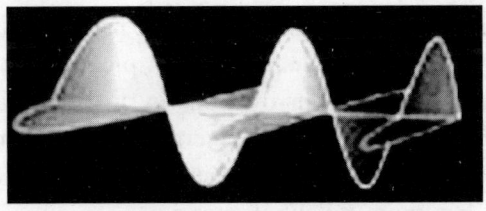

Fig. 4.1: Radiation waves

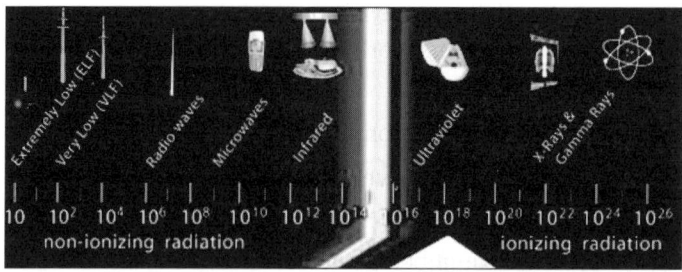

Fig. 4.2: Non-ionizing and ionizing radiation of electromagnetic radiation

4.1. ENERGY

Energy is defined as the ability to do work and it causes things to happen around us. During the day, the sun gives out light and heat energy. At night, street lamps use electrical energy to light our way. When a car drives by, it is being powered by gasoline, a type of stored energy. The foods we eat contain energy and use that energy to work and play. Energy powers our vehicles, trains, planes and rockets. Energy warms our homes, cooks our food, plays our music and gives us pictures on television. Energy from the sun gives us light during the day. It dries our clothes when they're hanging outside on a clothes line and it helps plants grow. Energy stored in plants is eaten by animals, giving them energy and predator animals eat their prey, which gives the predator animal energy. The forms of energy includes Electricity, Biomass Energy - energy from plants, Geothermal Energy, Fossil Fuels - Coal, Oil and Natural Gas, Hydro Power and Ocean Energy, Nuclear Energy and Solar Energy .

Energy makes everything happen and can be divided into two types (1) Stored energy is called potential energy (2) Moving energy is called kinetic energy. With a pencil, try this example to know the two types of energy. Put the pencil at the edge of the desk and push it off to the floor. The moving pencil uses kinetic energy. Now, pick up the pencil and put it back on the desk. You used your own energy to lift and move the pencil. Moving it higher than the floor adds energy to it. As it rests on the desk, the pencil has potential energy. The higher it is, the further it could fall that means the pencil has more potential energy.

4.2. ATOMIC STRUCTURE

The atoms of sub-atomic particles are protons, neutrons and electrons (Table 4.1). The nucleus is at the centre of the atom and contains the protons and neutrons. Protons and neutrons are collectively known as nucleons. Virtually all the mass of the atom is concentrated in the nucleus, because the electrons weigh so little. Working out the numbers of protons and neutrons:

Number of protons = Atomic number of the atom

The atomic number is also given the more descriptive name of proton number.

Number of protons + number of neutrons = Mass number the atom

Table 4.1: Relative mass and Relative charge of protons, neutrons and electrons

S.No	Sub-atomic particles	Relative mass	Relative charge
1.	Proton	1	+1
2.	Neutron	1	01
3.	Electron	1/1836	−1

The mass number is also called the nucleon number. For example the atomic number counts the number of protons (9); the mass number counts protons + neutrons (19). If there are 9 protons, there must be 10 neutrons for the total to add up to 19 (Fig. 4.3).

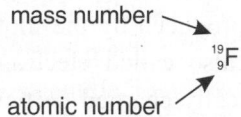

Fig. 4.3: The Mass number and Atomic number of Florin

The atomic number is tied to the position of the element in the periodic table and therefore the number of protons defines what sort of element you are talking about. So if an atom has 8 protons (atomic number = 8), it must be Oxygen. If an atom has 12 protons (atomic number = 12), it must be Magnesium. Similarly, every Chlorine atom (atomic number = 17) has 17 protons; every Uranium atom (atomic number = 92) has 92 protons.

The number of neutrons in an atom can vary within small limits for example there are three kinds of carbon atom ^{12}C, ^{13}C and ^{14}C. These different atoms of carbon are called isotopes. The fact that they have varying numbers of neutrons makes no difference whatsoever to the chemical reactions of the carbon. Isotopes are atoms which have the same atomic number but different mass numbers. They have the same number of protons but different numbers of neutrons. (Table 4.2).

Table 4.2: Protons, Neutrons and Mass number of different atoms of Carbon

S.No	Different atoms of Carbon	Protons	Neutrons	Mass number
1.	Carbon-12	6	6	12
2.	Carbon-13	6	7	13
3.	Carbon-14	6	8	14

Atoms are electrically neutral and the positiveness of the protons is balanced by the negativeness of the electrons (Table 4.1). It follows that in a neutral atom: Number of electrons = Number of protons. For example, if an oxygen atom (atomic number = 8) has 8 protons, it must also have 8 electrons; if a chlorine atom (atomic number = 17) has 17 protons, it must also have 17 electrons. The electrons are found at considerable distances from the nucleus in a series of levels called energy levels. Each energy level can only hold a certain number of electrons. The first level (nearest the nucleus) will only hold 2 electrons, the second holds 8, and the third also seems to be full when it has 8 electrons.

4.3. ELECTROMAGNETIC RADIATION

Electromagnetic waves are produced by the motion of electrically charged particles. These waves are also called electromagnetic radiation because they radiate from the electrically charged particles. As the name indicates, there are both electrical and magnetic components of the wave at right angles to each other. They travel through empty space as well as through air and other substances in the X direction at the speed of light, approximately 3.00×10^8 m/sec. in a vacuum.

Electromagnetic radiation can be described in terms of a stream of photons, which are mass less particles each traveling in a wave like pattern and moving at the speed of light. Each photon contains a certain amount or bundle of energy and all electromagnetic radiation consists of these photons. The only difference between the various types of electromagnetic radiation is the amount of energy found in the photons. Radio waves have photons with low energies, microwaves have a little more energy than radio waves, infrared has still more, then visible, ultraviolet, X-rays and the most energetic of all gamma-rays. Photons are particles of energy radiating from a source and characterized by an electromagnetic wave. The wavelength of the radiation λ can be visualized as the distance between maxima of either the electrical or magnetic component (Fig. 4.4).

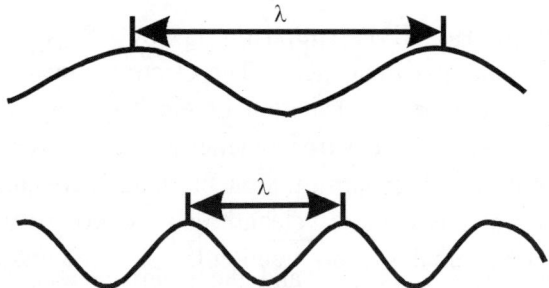

Fig. 4.4: The wavelength of the radiation (λ)

Associated with wavelength are frequency v and the number of waves that pass a fixed point such as P, in a unit of time. When a photon passes a particular region of space, the electric field in that region oscillates with the frequency v. Wavelength and frequency are related to the energy of a photon E by Plank's constant h 6.62×10^{-4} J sec and c the velocity of light in vacuum.

$$E = hv = \frac{hc}{\lambda} \qquad \text{... eq.4.1}$$

Only frequency is the truly characteristic of a particular radiation. Radiation is propagated through matter at velocities of less than c because of interactions between the electric vector and the bound electrons of the medium. When radiation of a particular wavelength enters matter its velocity decreases but its frequency remains constant. Wave numbers ∇ are sometimes used instead of frequency; they are calculated as follows:

$$\nabla = \frac{1}{\lambda} \qquad \text{... eq.4.2}$$

Wave number ∇ in units of /cm express, the number of waves that occur per centimeter; this number is directly proportional to the frequency but is should be emphasized that it is not a frequency:

$$V = c \quad \text{or} \quad \frac{v}{c} \qquad \text{... eq.4.3}$$

A beam that carries radiation with a very small wavelength spread approximating one discrete wavelength is said to be monochromatic. A polychromatic beam contains radiation with a wide distribution of wavelength. Both the amplitude and frequency of the electromagnetic radiation are important properties in spectrochemical measurements.

4.4. ELECTROMAGNETIC SPECTRUM

The electromagnetic (EM) spectrum is the range of all possible electromagnetic radiation frequencies. The electromagnetic spectrum of an object is the characteristic distribution of electromagnetic radiation from that particular object. The electromagnetic spectrum extends from below the frequencies used for modern radio at the long wavelength end through gamma radiation at the short wavelength end, covering wavelengths from thousands of kilometers down to a fraction the size of an atom. It is thought that the short wavelength limit is in the vicinity of the Planck length, and the long wavelength limit is the size of the universe itself, although in principle the spectrum is infinite and continuous.

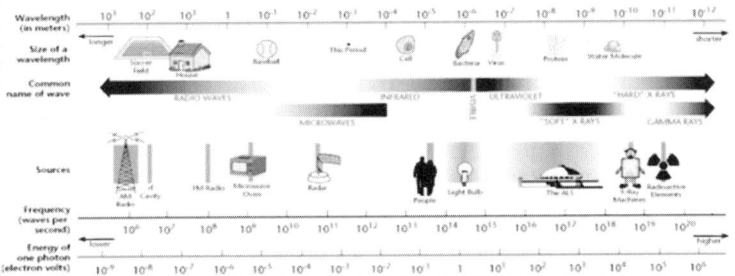

Fig. 4.5: The Electromagnetic spectrum

Different types of radiation in the EM spectrum, in order from lowest energy to highest (Fig. 4.5). While the classification scheme is generally accurate, in reality there is often some overlap between neighboring types of electromagnetic energy. For example, Super Low Frequency (SLF) radio waves at 60 Hz may be received and studied by astronomers or may be ducted along wires as electric power. Also, some low energy gamma rays actually have a longer wavelength than some high-energy X-rays. This is possible because gamma ray is the name given to the photons generated from nuclear decay or other nuclear and sub-nuclear processes, whereas X-rays on the other hand are generated by electronic transitions involving highly energetic inner electrons. Therefore the distinction between gamma ray and X-ray is related to the radiation source rather than the radiation wavelength. Generally, nuclear transitions are much more energetic than electronic transitions, so usually, gamma-rays are more energetic than X-rays. However, there are a few low-energy nuclear transitions that produce gamma rays that are less energetic than some of the higher energy X-rays.

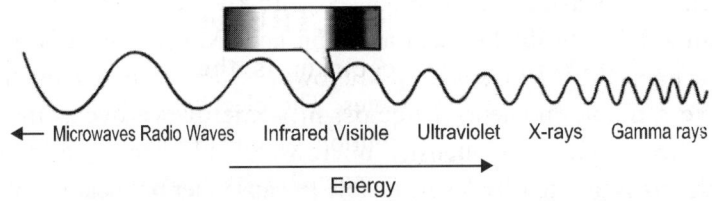

Fig. 4.6: Energy flow and Frequencies of electromagnetic radiation.

Energy and frequency both increases from left to right, while wavelength increases from right to left, making it inversely proportional to energy and frequency (Fig. 4.6). So, the highest energy light or radiation is gamma radiation, while the lowest energy is microwave. This is important because, as explained above, molecules only absorb light if it is at exactly the correct energy. In spectroscopic terms, UV cannot be used to investigate the vibrations of molecules because it is not correct energy, even though UV light has more energy. Similarly, even if we had the most powerful IR lamp in the world, we could not use it to investigate the electronic transitions that make UV spectroscopy investigates because IR light is not the correct wavelength or energy to investigate electronic transitions.

4.4.1. Radio frequency

Radio waves generally are utilized by antennas of appropriate size with wavelengths ranging from hundreds of meters to about one millimeter. They are used for transmission of data via modulation. Television, mobile phones, MRI, wireless networking and amateur radio all use radio waves. Radio waves can be made to carry information by varying a combination of the amplitude, frequency and phase of the wave within a frequency band and the use of the radio spectrum are regulated by many governments through frequency allocation. When Electromagnetic radiation impinges upon a conductor, it couples to the conductor, travels along it and induces an electric current on the surface of that conductor by exciting the electrons of the conducting material. Electromagnetic radiation may also cause certain molecules to absorb energy and thus to heat up, thus causing thermal effects and sometimes burns; this is exploited in microwave ovens.

4.4.2. Microwaves

The super high frequency (SHF) and extremely high frequency (EHF) of microwaves come next up the frequency scale. Microwaves are waves which are typically short enough to employ tubular metal waveguides of reasonable diameter. Microwave energy is produced with klystron and magnetron tubes, and with solid state diodes such as Gunn and IMiact ionization Avalanche Transit-Time (IMPATT) devices. Microwaves are absorbed by molecules that have a dipole moment in liquids. In a microwave oven, this effect is used to heat food. Low-intensity microwave radiation is used in Wi-Fi, although this is at intensity levels unable to cause thermal heating. Volumetric heating, as used by microwaves, transfer energy through the material electro-magnetically, not as a thermal heat flux. The benefit of this is a more uniform heating and reduced heating time; microwaves can heat material in less than 1% of the time of conventional heating methods.

4.4.3. Terahertz radiation

Terahertz radiation is a region of the spectrum between far infrared and microwaves. Recently, the range was rarely studied and few sources existed for microwave energy at the high end of the band sub-millimetre waves or so called terahertz waves, but applications such as imaging and communications are now appearing. Scientists are also looking to apply terahertz technology in the armed forces, where high frequency waves might be directed at enemy troops to incapacitate their electronic equipment.

4.4.4. Infrared radiation

The infrared part of the electromagnetic spectrum covers the range from roughly 300 GHz (1 mm) to 400 THz (750 nm). It can be divided into three parts:

   i. Far-infrared, from 300 GHz (1 mm) to 30 THz (10 μm). The lower part of this range may also be called microwaves. This radiation is typically absorbed by so called rotational modes in gas phase molecules by molecular motions in liquids and by phonons in solids. The water in the Earth's atmosphere absorbs so strongly in this range that it renders the atmosphere effectively opaque. However, there are certain wavelength ranges within the opaque range which allow partial transmission and can be used for astronomy. The wavelength range from approximately 200 μm up to a few mm is often referred to as sub-millimetre in astronomy, reserving far infrared for wavelengths below 200 μm.

   ii. Mid-infrared, from 30 to 120 THz (10 to 2.5 μm). Hot objects can radiate strongly in this range. It is absorbed by molecular vibrations, where the different atoms in a molecule vibrate around their equilibrium positions. This range is sometimes called the fingerprint region since the mid-infrared absorption spectrum of a compound is very specific for that compound.

   iii. Near-infrared, from 120 to 400 THz (2,500 to 750 nm). Physical processes that are relevant for this range are similar to those for visible light

4.4.5. Visible radiation (light)

Above infrared in frequency comes visible light. This is the range in which the sun and stars similar to it emit most of their radiation. It is probably not a coincidence that the human eye is sensitive to the wavelengths that the sun emits most strongly. Visible light and near-infrared light is typically absorbed and emitted by electrons in molecules and atoms that move from one energy level to another. The light we see with our eyes is really a very small portion of the electromagnetic spectrum. A rainbow shows the optical (visible) part of the electromagnetic spectrum; infrared would be located just beyond the red side of the rainbow with ultraviolet appearing just beyond the violet end.

Electromagnetic radiation with a wavelength between approximately 400 nm and 700 nm is detected by the human eye and perceived as visible light. Other wavelengths, especially near infrared (longer than 700 nm) and ultraviolet (shorter than 400 nm) are also sometimes referred to as light, especially when the visibility to humans is not relevant. If radiation having a frequency in the visible region of the electromagnetic spectrum reflects off an object and then strikes our eyes, this results in our visual perception of the scene. Our brain's visual system processes the multitude of reflected frequencies into different shades and hues, and through this not entirely understood psychophysical phenomenon. At most wavelengths, however, the information carried by electro-magnetic radiation is not directly detected by human senses. Natural sources produce electromagnetic radiation across the spectrum, and technology can also manipulate a broad range of wavelengths.

4.4.6. Ultraviolet light

Next in frequency comes ultraviolet (UV). This is radiation whose wavelength is shorter than the violet end of the visible spectrum. Being very energetic, UV can break chemical bonds making molecules unusually reactive or ionizing them, in general changing their mutual behavior. Sunburn, for example, is caused by the disruptive effects of UV radiation on skin cells, which can even cause skin cancer, if the radiation damages the complex DNA molecules in the cells. The Sun emits a large amount of

UV radiation, which could quickly turn Earth into a barren desert; however, most of it is absorbed by the atmosphere's ozone layer before reaching the surface.

4.4.7. X-rays

After UV come X-rays. Hard X-rays have shorter wavelengths than soft X-rays. X-rays are used for seeing through some things and not others, as well as for high-energy physics and astronomy. X-rays will pass through most substances, and this makes them useful in medicine and industry. X-rays are given off by stars, and strongly by some types of nebulae. An X-ray machine works by firing a beam of electrons at a target. If the electrons are fired with enough energy, X-rays will be produced.

4.4.8. Gamma rays

After hard X-rays come gamma rays, which were discovered by Paul Ulrich Villard in 1900. These are the most energetic photons having no defined lower limit to their wavelength. They are useful to astronomers in the study of high energy objects or regions. The wavelength of gamma rays can be measured with high accuracy by means of Compton scattering.

4.5. THE DOPPLER EFFECT

The *Doppler Effect* refers to the apparent shift in the wavelength and frequency of a wave when there is relative motion between the source or emitter of the wave and an observer. A common example of this effect is the sound made by a passing ambulance. As the ambulance approaches you the sound waves are apparently shifted to a shorter wavelength hence higher frequency. You perceive this increased frequency as a higher pitched sound. Once the ambulance passes you it is now receding from you so the sound waves are shifted to a longer wavelength and lower frequency and hear this as a lower pitched sound.

When discussing electromagnetic waves, the term *Doppler shift* to describe this effect. If we obtain a spectrum from an object at rest to us then there is no Doppler shift in the spectrum. The spectral lines for the Balmer series, for instance, would appear at the same wavelengths as those from a hydrogen discharge tube in the laboratory or observatory. No relative motion therefore no apparent shift in lines. If an object such as a star was moving away from us then the spectral lines would appear to shift towards the longer wavelength or redder part of the spectrum. Such a shift is termed a *red shift*. The amount of the shift would depend on the relative velocity between source and observer, the greater the recession velocity, the greater the shift in the lines.

In cases where an object is moving towards us, its spectral lines will appear to shift to the shorter wavelength, bluer part of the spectrum this is called a blue shift. Again, the amount of blue shift will depend upon the relative velocity between source and observer. Fig. 4.7 showing comparison of position of spectral lines for a source at rest (top), red shift where the source is moving away from the observer (middle) and blue shift where the source is moving towards the observer (bottom). Note that the motion between source and observer is relative.

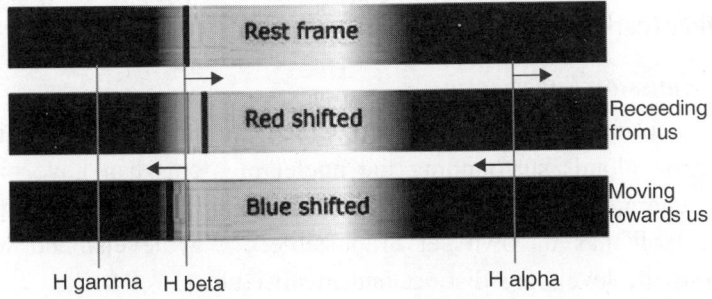

Fig. 4.7: Comparison of position of spectral lines

If a star moves relative to an observer on Earth with a radial velocity v and emits a certain wavelength λ then the Earth based observer will see a Doppler shifted wavelength of λ'. These two are related by the following formula:

$$(\lambda' - \lambda)/\lambda = v/c \qquad \textit{... eq. 4.4}$$

Where c = speed of light = 3.0×10^8 m.s^{-1} and v is non-relativistic in speed. By convention v is positive for receding objects and negative for those approaching.

4.6. SPECTRA TYPES

Typically Spectra can be classified into two types (1) Continuous spectra - For a continuous spectrum, the light is composed of a wide, continuous range of colors (energies). (2) Discrete spectra - With discrete spectra, one sees only bright or dark lines at very distinct and sharply defined colors (energies). Discrete spectra with bright lines are called emission spectra those with dark lines are termed absorption spectra.

4.6.1. Continuous spectra

Continuous spectra arise from dense gases or solid objects which radiate their heat away through the production of light. Such objects emit light over a broad range of wavelengths, thus the apparent spectrum seems smooth and continuous. Stars emit light in a predominantly continuous spectrum. Other examples of such objects are incandescent light bulbs, electric cooking stove burners, flames and cooling fire embers.

4.6.2. Discrete spectra

Discrete spectra are the observable result of the physics of atoms. There are two types of discrete spectra, emission (bright line spectra) and absorption (dark line spectra).

4.6.2.1. Emission line spectra

Unlike a continuous spectrum source, which can have any energy it wants, the electron clouds surrounding the nuclei of atoms can have only very specific energies dictated by quantum mechanics. Each element on the periodic table has its own set of possible energy levels, and with few exceptions the levels are distinct and identifiable.

Atoms will also tend to settle to the lowest energy level. This means that an excited atom in a higher energy level must dump some energy. The way an atom dumps that energy is by emitting a wave of light with that exact energy. In Fig. 4.8, a hydrogen atom drops from the 2nd energy level to the 1st giving off a wave of light with an energy equal to the difference of energy between levels 2 and 1. This energy corresponds to a specific color or wavelength of light and thus we see a bright line at that exact wavelength. An emission spectrum is born tiny changes of energy in an atom generate photons with small energies and long wavelengths, such as

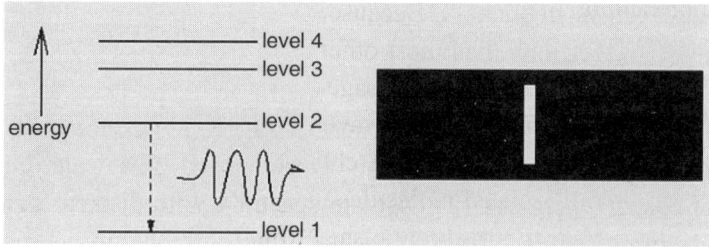

Fig. 4.8: An excited Hydrogen atom relaxes from level 2 to level 1 yielding a photon, this result in a bright emission line

radio waves. Similarly, large changes of energy in an atom will mean that high energy; short wavelength photons (UV, X-ray and gamma-rays) are emitted.

4.6.2.2. Absorption line spectra

The atom could absorb that especially energetic photon and would become excited, jumping from the ground state to a higher energy level. If a star with a continuous spectrum is shining upon an atom, the wavelengths corresponding to possible energy transitions within that atom will be absorbed and therefore an observer will not see them. In this way, a dark line absorption spectrum is born (Fig. 4.9).

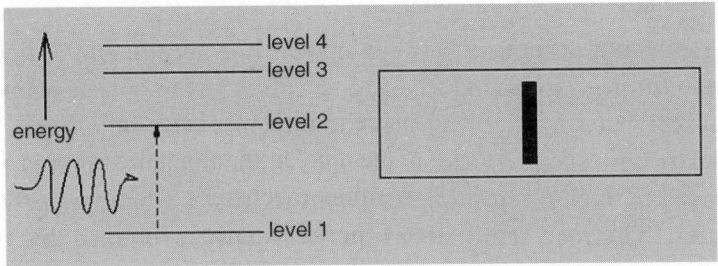

Fig. 4.9: A hydrogen atom in the ground state is excited by a photon of exactly the right energy needed to send it to level 2, absorbing the photon in the process this results in a dark absorption line.

4.6.5. Bathochromic shift

Shift of a spectral band to lower frequencies (longer wavelengths) owing to the influence of substitution or a change in environment. It is informally referred to as a red shift and is opposite to hypsochromic shift (blue shift). Bathochromic shift is a change of spectral band position in the absorption, reflectance, transmittance or emission spectrum of a molecule to a longer wavelength (lower frequency). Because the red color in the visible spectrum has a higher wavelength than most other colors, this effect is also commonly called a red shift, although this usage is considered informal and has no relation to Doppler shift or other wavelength independent meanings of red shift. This can occur because of a change in environmental conditions, a series of structurally related molecules in a substitution series can also show a bathochromic shift. Bathochromic shift is a phenomenon seen in molecular spectra not atomic spectra; it is thus more common to speak of the movement of the peaks in the spectrum rather than lines.

$$\Delta\lambda = \lambda_{observed}^{state2} - \lambda_{observed}^{state1} \text{ and } \lambda_{observed}^{state2} > \lambda_{observed}^{state1} \quad ... \text{ } eq. \text{ } 4.5$$

Where λ is the wavelength of the spectral peak of interest and Bathochromic shift is typically demonstrated using a spectrophotometer, colorimeter or spectro-radiometer.

4.6.6. Pressure broadening

The broadening of spectral lines, particularly in white dwarfs caused by the pressure of the stellar atmosphere which in turn is caused by the high surface gravity of the star. Emission or absorption at a discrete wavelength or frequency caused by a specific electron transition within an atom, molecule or ion. The dark lines in an absorption spectrum and the bright lines that make up an emission spectrum are caused by the transfer of an electron from one energy level to another.

The essential difference between optical line spectra and X-ray spectra is that the former correspond to energy changes in the outer electrons in an atom and the latter to energy changes in the inner electron orbitals. Gamma rays usually correspond to energy changes in the nucleus. Infrared radiation is produced by high-*n* transitions of atoms or by the vibration or rotation of molecules. Thermal radio emission is usually produced by higher-*n* transitions.

4.7. SPECTROSCOPY

4.7.1. Introduction

Spectroscopy is a technique that uses the interaction of energy with a sample to perform an analysis. The data that is obtained from spectroscopy is called a spectrum. A spectrum is a plot of the intensity of energy detected versus the wavelength or mass or momentum or frequency of the energy. Spectrum can be used to obtain information about atomic and molecular energy levels, molecular geometries, chemical bonds, interactions of molecules, and related processes. Often, spectra are used to identify the components of a sample (qualitative analysis). Spectra may also be used to measure the amount of material in a sample (quantitative analysis). There are several instruments that are used to perform a spectroscopic analysis. Spectroscopy requires an energy source commonly a laser, but this could be an ion source or radiation source and a device for measuring the change in the energy source after it has interacted with the sample.

Spectroscopy is used to refer to the broad area of science dealing with the absorption, emission or scattering of electromagnetic radiation by molecules, ions, atoms or nuclei. Spectroscopic techniques are useful in determining both the identity of unknown substances and their concentration in solution.

Different regions of the electromagnetic spectrum such as infrared, visible, ultraviolet or X-ray radiation can be used to interact with matter. The electrons and nuclei of atoms and molecules may exist in only certain specific energy levels, they area quantized. They can absorb only photons having certain energies or wavelengths. The energies of light absorbed by a molecule can be related to motions (energy modes) of the molecule (Table 4.3).

Table 4.3: Examples for energy modes

S.No	Molecules mode	Radiation wavelength	Motions
1.	Molecular motion	Electromagnetic radiation	Energy
2.	Rotation	Microwave and Infrared	Low
3.	Vibration	Infrared	Moderate
4.	Electron transitions	Visible and Ultraviolet	High

The instruments that are used to measure the interaction of various regions of electromagnetic radiation with matter differ a great deal in design and operation they all contain the same basic components. The source provides the electromagnetic radiation that will be absorbed by the sample, it is often some sort of light bulb or lamp. The monochromator selects one particular energy or wavelength or color of light from the source. A prism, diffraction grating or a colored filter together with slits and mirrors can serve as a monochromator. The detector measures the amount of light that passes through the sample. A phototube or photo cell or photomultiplier is often used as a detector. Light falling on the surface of the detector causes current to flow in a surrounding electrical circuit. The amount of current in the circuit is proportional to the amount of light striking the detector.

All the parts of the instrument work together as follows: Light from the source passes through the monochromator producing a beam with a single energy or a narrow band of energies. The intensity of this beam Io is measured by the detector. The sample is then placed in the beam between the monochromator and the detector. If some of the light is absorbed by the sample, the intensity of the beam reaching the detector I will be less than Io. The detector compares the two intensities and reports the result as either percent transmittance (%T) or absorbance (A). These terms are defined to be:

$$\%T = I/\ Io \times 100 \qquad \qquad ...\ eq.\ 4.6$$

Where T is the fraction of Io that gets through the samples is called Transmittance

$$A = -\log T = -\log(I/Io)$$
$$= 2 - \log(\%T) \qquad \qquad ...\ eq.\ 4.7$$

If the monochromator is a prism or a diffraction grating, all of the energies or wavelengths are available and may be varied. Molecules do not absorb all wavelengths equally well. Consider a colored object human sight is the brain's interpretation of photons of electromagnetic radiation in the Visual range (light) entering the eye. If all the energies (wavelengths, colors) are mixed they are perceived as white light. If no photons at all enter the eye we see black.

Table 4.4: Visible spectrum and Complementary colors

S.No.	Wavelength (nm) (Absorbed)	Color (Observed)	Complementary
1.	400-435	Violet	Yellow–green
2.	435-480	Blue	Yellow
3.	480-490	Green-blue	Orange
4.	490-500	Blue-green	Red
5.	500-560	Green	Purple
6.	560-580	Yellow-green	Violet
7.	580-595	Yellow	Blue
8.	595-610	Orange	Green-blue
9.	610-780	Red	Blue-green

A color is perceived if only energy of one photon (monochromatic light) enters the eye or if photons of a complementary color are missing from the usual white light mix. A white object then appears to be white because it does not absorb any of the light that strikes it. A black object looks black because it absorbs the entire incident light. A rose looks red if it absorbs all the light except the red or if it absorbs the light of the color complementary to red that is blue-green. Table 4.4 shows a brief list of colors absorbed to give observed colors.

If a solution is green-blue, it will absorb orange light and the wavelength of maximum absorbance, λmax, will fall between 595nm and 610nm (Table 4.4). Amounts of light absorbed, even in the case of concentrated solutions is very small compared to the amount of light available from the source. Most spectrometers have variable monochromators such as a prism or diffraction grating, making it is easy to select the complementary color from the source by stepping through the available wavelengths and plotting the absorbance or percent transmittance versus the wavelength manually or let the instrument scan through the range automatically. From the plot (spectra) we can determine the exact wavelength of the complementary color, which is the wavelength of the maximum absorbance (λmax).

Spectroscopy is a type of chemical analysis done by shining light on a sample to determine what is inside. Chemists commonly measure the absorbance, how much light is absorbed by the sample or the transmittance, how much light passes through the sample. In a chemical analysis, many different kinds (wavelengths or energies) of light (a spectrum) are shone through a sample. Some of the light is absorbed, by knowing what wavelengths of light are absorbed or eaten by the sample, we know what is inside. If we are looking for a specific molecule or characteristic and shine the wrong wavelengths of light through a sample, no matter how much light we put through, we will never learn anything about the sample. Because different molecules and characteristics of molecules absorb at different energies of light, there is a need for different forms of spectroscopy. Each different spectroscopy uses a different part of the electromagnetic spectrum to investigate specific characteristics of a sample. For example, infrared spectroscopy (IR) is used to investigate how the molecules in a sample vibrate, while ultraviolet spectroscopy (UV) is used to investigate how certain chemical bonds in a molecule are arranged. Remember, IR spectroscopy can not be used to investigate the information that UV provides and vice-versa.

4.7.2. Spectroscopy Types

There are as many different types of spectroscopy as there are different type of energy sources. The following are the example of Spectrocopy type:

1. **Atomic absorption spectroscopy:** Energy absorbed by the sample is used to assess its characteristics. Sometimes absorbed energy causes light to be released from the sample, which may be measured by a technique such as fluorescence spectroscopy.

2. **Electron paramagnetic spectroscopy:** This is a microwave

technique based on splitting electronic energy fields in a magnetic field. It is used to determine structures of samples containing unpaired electrons.

3. **Electron spectroscopy:** There are several types of electron spectroscopy; all associated with measuring changes in electronic energy levels.

4. **Fourier transforms spectroscopy:** This is a family of spectroscopic techniques in which the sample is irradiated by all relevant wavelengths simultaneously for a short period of time. The absorption spectrum is obtained by applying a mathematical analysis to the resulting energy pattern.

5. **Gamma – ray spectroscopy:** Gamma radiation is the energy source in this type of spectroscopy, which includes activation analysis and Mossbauer spectroscopy.

6. **Infrared spectroscopy:** The infrared absorption spectrum of a substance is sometimes called its molecular fingerprint. Although frequently used to identify materials, infrared spectroscopy also may be used to quantify the number of absorbing molecules.

7. **Laser spectroscopy:** Absorption spectroscopy, fluorescence spectroscopy, Raman spectroscopy, and surface-enhanced Raman spectroscopy commonly use laser light as an energy source. Laser spectroscopes provide information about the interaction of coherent light with matter. Laser spectroscopy generally has high resolution and sensitivity.

8. **Mass spectroscopy:** A mass spectrometer source produces ions. Information about a sample may be obtained by analyzing the dispersion of ions when they interact with the sample, generally using the mass-to-charge ratio.

9. **Raman spectroscopy:** Raman scattering of light by molecules may be used to provide information on a sample's chemical composition and molecular structure.

10. **X-ray spectroscopy:** This technique involves excitation of inner electrons of atoms, which may be seen as X-ray absorption. An X-ray fluorescence emission spectrum may be produced when an electron falls from a higher energy state into the vacancy created by the absorbed energy.

4.8. UV–VISIBLE SPECTROSCOPY

Ultraviolet – Visible spectroscopy ($\lambda 200 - 800$ nm) studies the changes in electronic energy levels within the molecule arising due to transfer of electrons from π- or non-bonding orbitals. It commonly provides the knowledge about π-electron systems, conjugated un-saturations, aromatic compounds and conjugated non-bonding electron systems. This absorption spectroscopy uses electromagnetic radiations between 190 nm to 800 nm and is divided into the ultraviolet (190-400 nm) and visible (400-800 nm) regions. Since the absorption of ultraviolet or visible radiation by a molecule leads transition among electronic energy levels of the molecule, it is also often called as electronic spectroscopy.

4.8.1. Nature of Electronic Transitions

The total energy of a molecule is the sum of its electronic, its vibrational energy and rotational energy. Energy absorbed in the UV region produces changes in the electronic energy of the molecule. As a molecule absorbs energy, an electron is promoted from an occupied molecular orbital usually a non-bonding n or bonding π orbital to an unoccupied molecular orbital an antibonding π^* and σ^* orbital of greater potential energy. For most molecules, the lowest-energy occupied molecular orbital's are σ orbital's, which correspond to σ bonds. The π orbital's lie at relatively higher energy levels than σ orbital's and the non-bonding orbital's that hold unshared pairs of electrons lie even at higher energies.

The saturated aliphatic hydrocarbons (alkanes) exhibit only $\sigma \rightarrow \sigma^*$ transitions but depending on the functional groups the organic molecules may undergo several possible transitions which can be placed in the increasing order of their energies.

$$n \rightarrow \pi^* < n \rightarrow \sigma^* < \pi \rightarrow \pi^* < \sigma \rightarrow \pi^* < \sigma \rightarrow \sigma^*$$

Since all these transitions require fixed amount of energy (quantized), an ultraviolet or visible spectrum of a compound would consist of one or more well defined peaks, each corresponding to the transfer of an electron from one electronic level to another. If the differences between electronic energy levels of two electronic states are well defined i.e. if the nuclei of the two atoms of a diatomic molecule are held in fixed position, the peaks accordingly should be sharp. However, vibrations and rotations of nuclei occur constantly and as a result each electronic state in a molecule is associated with a large number of vibrational and rotational states.

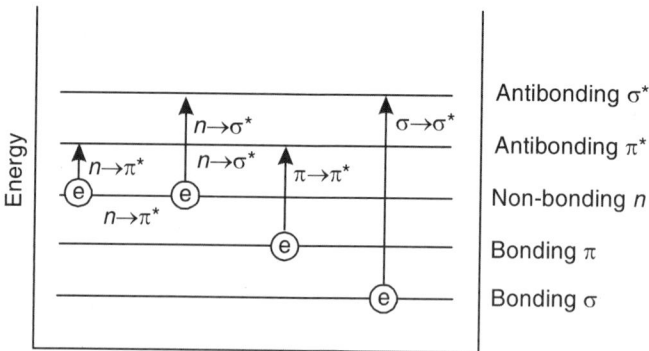

Fig. 4.10: Possible electronic transitions of p, s and *n* electrons

At room temperature, the molecules in the ground state will be in the zero vibrational level (GU_0). The transition of an electron from one energy level to another is thus accompanied by simultaneous change in vibrational and rotational states, and causes transitions between various vibrational and rotational levels of lower and higher energy electronic states. Therefore many radiations of closely placed frequencies are absorbed and a broad absorption band is obtained. When a molecule absorbs ultraviolet or visible light of a defined energy, an assumption is made that only one electron is excited form bonding orbital or non-bonding orbital to an anti-bonding orbital and time i.e. of the order of 10^{-15} seconds. In accordance with Franck-Condon principle, during electronic excitation the atoms of the molecule do not move.

The most probable transition would appear to involve the promotion of one electron from the highest occupied molecular orbital to the lowest unoccupied molecular orbital, but in many cases several transitions can be observed, giving several absorption bands in the spectrum.

Alkanes can only undergo $\sigma \rightarrow \sigma^*$ transitions. These are high energy transitions and involve very short wavelength ultraviolet light (<150 nm). These transitions usually fall out-side and generally available measurable range of UV-visible spectrophotometers (200-1000 nm). The $\sigma \rightarrow \sigma^*$ transitions of methane and ethane are at 122 and 135 nm, respectively. In alkenes amongst the available $\sigma \rightarrow \sigma^*$ and $\pi \rightarrow \pi^*$ transitions, the $\pi \rightarrow \pi^*$ transitions are of lowest energy and absorb radiations between 170-190 nm. In saturated aliphatic ketones the lowest energy transition involves the transfer of one electron of the nonbonding electrons of oxygen to the relatively low-lying π^* anti-bonding orbital. This $n \rightarrow \pi^*$ transition is of lowest energy (~280 nm) but is of low intensity as it is symmetry forbidden. Two other available transitions are n→π* and π→π*. The most intense band for these compounds is always due to π→π* transition.

In conjugated dienes the π→π* orbital's of the two alkene groups combine to form new orbital's – two bonding orbital's named as π_1 and π_2 and two antibonding orbital's named as π_3* and π_4*. It is apparent that a new π→π* transition of low energy is available as a result of conjugation. Conjugated dienes as a result absorb at relatively longer wavelength than do isolated alkenes.

4.8.2. Principle of Absorption Spectroscopy

Many compounds absorb ultraviolet (UV) or visible light. The Fig. 4.11 below shows a beam of monochromatic radiation of radiant power P_0, directed at a sample solution. Absorption takes place and the beam of radiation leaving the sample has radiant power P.

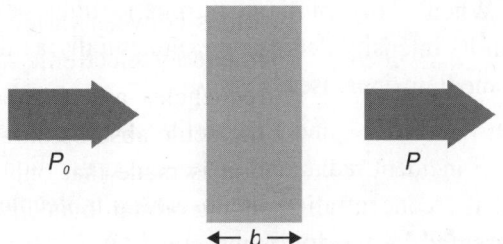

Fig. 4.11: A beam of monochromatic radiation.

The amount of radiation absorbed may be measured in a number of ways:

$$\text{Transmittance (T)} = P/P_0 \qquad \qquad \textit{... eq. 4.8}$$

$$\%\ \text{Transmittance} \ \%T = 100\ T \qquad \qquad \textit{... eq. 4.9}$$

$$\text{Absorbance} \qquad A = \log_{10} P_0\ /\ P \qquad \qquad \textit{... eq. 4.10}$$

$$A = \log_{10} 1\ /\ T \qquad \qquad \textit{... eq. 4.11}$$

$$A = \log_{10} 100\ /\ \%T \qquad \qquad \textit{... eq. 4.12}$$

$$A = 2 - \log_{10} \%T \qquad \qquad \textit{... eq. 4.13}$$

The 4.13 equation, $A = 2 - \log_{10} \%T$ is worth remembering because it allows to easily calculate absorbance from percentage transmittance data. The relationship between absorbance and transmittance is illustrated in fig. 4.12. So, if all the light passes through a solution without any absorption, then absorbance is zero and percent transmittance is 100%. If all the light is absorbed, then percent transmittance is zero, and absorption is infinite.

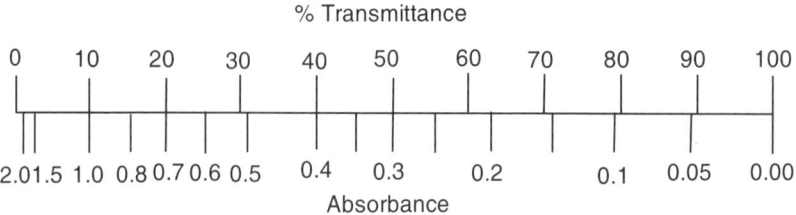

Fig. 4.12: The relationship between absorbance and transmittance.

4.8.2.1. Beer-lambert absorption law

Lambert's law: When a ray of monochromatic light passes through an absorbing medium its intensity decreases exponentially as the length of the absorbing medium increases.

Beer's law: When a ray of monochromatic light passes through an absorbing medium its intensity decreases exponentially as the concentration of the absorbing medium increases.

Combine both Beer's Law and Lambert's Law to get the Beer-Lambert Law: the fraction of incident radiation absorbed is a function of the thickness (path length) and the concentration of absorbing molecules in its path. It can also be represented by the following equation:

$$A = abc \qquad \qquad ... eq.\ 4.14$$

When the concentration (c) is expressed in moles/litre the expression becomes:

$$A = \varepsilon bc \qquad \qquad ... eq.\ 4.15$$

Where the molar absorptivity and c is is the concentration in moles/litre.

There are important assumptions to this law; Beer-Lambert's law indicates a direct proportionality between A and c only if (1) Incident radiation is monochromatic (2) Each molecule in solution acts as an independent absorbing species in solution (3) Absorption takes place in a solution of uniform cross-section.

The relationship between the amounts of light a solution transmits or absorbs and its concentration is given by Beer's Law:

$$\log \frac{I_0}{I} = \varepsilon c l \qquad \qquad ... eq.-4.16$$

Where I_0 is the intensity of the light incident on the solution, I is the intensity of the light transmitted by the solution, C is the concentration (moles/litre) of the solution, l is the path length (cms), i.e. the length of the path of solution through which the light passes, Σ is the decadic molar extinction coefficient, a characteristic of the components of the solution and

$$\log \frac{I_0}{I}$$

is the absorbance of the solution.

If b is 1 cm and c is 1 mol/liter then the absorbance is equal to k, the molar extinction coefficient, which is characteristic for a compound. The molar extinction coefficient k is thus the extinction given by 1 mol/liter in a light path of 1 cm and is usually written E, it has the dimension of liter mol^{-1} cm^{-1}.

4.8.2.1.1. Limitations

Sometimes, a non-linear plot is obtained of extinction against concentration and this is probably due to one or other of the following conditions not being fulfilled.

1. Light must be of narrow wavelength range and preferable monochromatic.

2. The wavelength of light used should be at the absorption maximum of the solution, this also gives the greatest sensitivity.

3. There must be no ionization, association, dissociation or solvation of the solute with concentration or time.

4. The solution is too concentrated, giving an intense colour. The law only holds up to a threshold maximum concentration for a given substance.

4.8.2.1.2. Deviations

The fundamental concept of Beer-Lambert Law is that every photon of light striking the detector must have an equal chance of absorption. Thus, every photon must have the same absorption coefficient alpha, must pass through the same absorption path length (L) and the same absorber concentration (c), anything that upsets these conditions will lead to an apparent deviation from the Beer's law. Most common apparent deviations are due to (1) Chemical reasons arising when the absorbing compound, dissociates, associates or reacts with a solvent to produce a product having a different absorption spectrum. (2) The presence of stray radiation. (3) The polychromatic radiation.

4.8.2.1.2.1 Stray light

The source of instrumental non-ideality is stray light, which is any light striking the detector whose wavelength is outside the spectral band pass of the monochromator or which has not passed through the sample. Since in most cases the wavelength setting of the monochromator is the peak absorption wavelength of the analyte, it follows that any light outside this range is less absorbed. The most serious effect is caused by stray light that is not absorbed by the analyte at all; this is called unabsorbed stray light. This effect also leads to a concave down curvature of the analytical curve, but the effect is relatively minor at low absorbances and increases quickly at high absorbances. Unabsorbed stray light results in a flat plateau in the analytical curve at an absorbance of $-\log(fsl)$, where *fsl* is the fractional stray light.

4.8.2.1.2.2. Polychromatic radiation

Strict adherence to Beer's law is observed only with truly monochromatic radiation. Monochromators are used to isolate portions of the output from continuum light sources hence a truly monochromatic radiation never exists and can only be approximated i.e. by using a very narrow exit slit on the monochromator. Let's assume that the incident radiation consists of two wavelengths λ' and λ'' with powers P_0' and P_0''. Considering that $A = -\log(P/P_0)$, then the power of the radiation to come out from (P) the cell for each wavelength would be:

$$P' = P_0'10^{-\varepsilon'bC}$$

$$P'' = P_0''10^{-\varepsilon''bC} \qquad \qquad \textit{... eq. 4.17}$$

Where ε' and ε'' are the molar absorptivities for each wavelength. Therefore the measured absorbance A_m will be:

$$A_m = -\log\left(\frac{P'+P''}{P_0'+P_0''}\right)$$

$$= \log\left(\frac{P_0'+P_0''}{P_0'10^{-\varepsilon'bC}+P_0''10^{-\varepsilon''bC}}\right) \qquad \textit{... eq. 4.18}$$

The equation 4.18 indicates a nonlinear relation between A_m and C. The proportionality between A_m and C is restored only if $\varepsilon' = \varepsilon''$. The same situation occurs when a radiation consists of many wavelengths (Fig. 4.13).

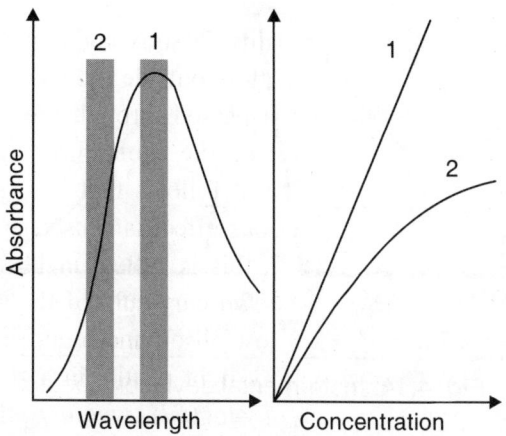

Fig. 4.13: Polychromatic radiation absorption peaks

When a polychromatic radiation beam is used, its preferable position in terms of central wavelength is on the top of a relatively wide absorption peak (Fig. 4.13, position 1). In this case the proportionality between absorbance (A) and concentration (C) is maintained, because the molar absorptivities are practically the same for all wavelengths. On the other hand, marked deviations are expected when the band is positioned in spectral regions such as the sides of absorption peaks (Fig. 4.13, position 2) where a wide range of molar absorptivity values is expected. It has been found experimentally that for absorbance measurements at the maximum of narrow peaks deviations from Beer's law are insignificant, if the effective bandwidth of the incident beam is less than 1/10 of the width of the absorption peak at half height.

4.8.2.1.2.3. Instrumental deviations

The Fig. 4.14 on the left of the window shows the absorption spectrum of the analyte in red over a wavelength range from 200 – 400 nm. The blue line is the transmitted intensity it shows the spectrum of light emerging from the exit slit of the monochromator and passing through the absorbing sample. A monochromator never really passes a single color or wavelength of light it actually passes a small range of wavelengths, this range of wavelengths is called the spectral band pass.

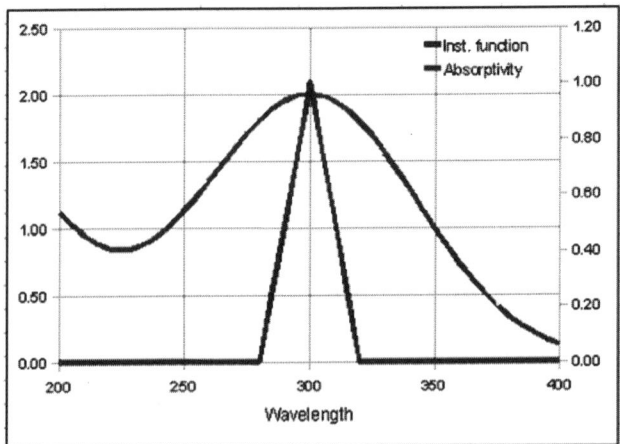

Fig. 4.14: Instrumental deviation spectra

In laboratory instruments, the spectral band pass is controlled by the slit width, which is adjustable by the experimenter on many instruments. In this simulation vary the slit width of the simulated instrument from 10 nm to 100 nm by using the slit width control above the graph (Fig. 4.14), but it cannot be set below 10 nm. The peak of the slit function falls at the wavelength setting of the monochromator and control the wavelength setting by using the wavelength setting control this is equivalent to turning the wavelength knob on the spectrometer.

Every monochromator passes a small amount of white light as a result of scattering off optical surfaces within the monochromator (mirrors, lenses, windows and the diffraction grating). Stray light is a very small fraction of the light intensity within the spectral band pass, but it is important because it can lead to a significant source of deviation from Beer's Law.

In most cases the monochromator is tuned to the wavelength of maximum absorption of the analyte in order to achieve the greatest sensitivity of analysis means that stray light is less absorbed than the light within the spectral band pass. The worst offender is stray light that is not at all absorbed by the analyte unabsorbed stray light usually expressed as a percentage of the light intensity within the spectral band pass. In the simulation, this is set by the unabsorbed stray light control. Typical monochromators have stray light rating in the 0.01 – 1% ranges depending on the wavelength setting and the type of light source used. The stray light is always worse at wavelengths where the light source is least intense and where the detector is least sensitive.

4.8.2.1.2.4. Other potential sources

There are other potential sources of deviation that are not included in this simulation either because they are usually not so serious under the conditions of typical laboratory applications of absorption spectrophotometry or because they can be avoided by proper experiment design. The following are the other potential sources for deviation of Beer's law.

1. Unequal light path lengths across the light beam. In most laboratory applications, the samples are measured in square cuvettes to insure a constant path length for all photons. When round test tube sample cells are used, the light beam passing through the sample is restricted to a central region of the sample tube in order to minimize this effect.

2. Unequal absorber concentration across the light beam. Solution samples are carefully mixed before measurement to insure homogeneity.

3. Changes in refractive index at high analyte concentration most analytical applications operate at lower concentrations.

4. Shifts in chemical equilibrium as a function of concentration. Solutions may need to be buffered to prevent this or the measurement can be made at a multi-component analysis may be performed if the spectra of all the species in equilibrium can be determined.

5. Fluorescence of the sample, in which some of the absorbed light is re-emitted and strikes the detector. Most analytes are not fluorescent, but if so, this error can be reduced by using a spectrophotometer that places the sample between the light source and the monochromator, such as a photodiode-array spectrometer.

6. Light scattering by the sample matrix, especially in turbid samples this is a common source of variable background absorption, which can be reduced by using a spectrophotometer that places the sample cuvette right up against the face of the detector so that it captures and detects a large fraction of the scattered light.

7. If the light intensity is extremely high (focused laser), it's possible to observe non-linear optical effects, which are a fundamental failure of the Beer-Lambert Law for example, as the absorber approaches optically saturation equal populations in the ground and excited states, in which case it no longer absorbs light.

4.8.2.2. Principle

The principle of spectrophotometer may be illustrated in the Fig. 4.15. Source of radiation may be hydrogen discharge tube in the ultraviolet region and an incandescent tungsten lamp in the visible region. Light from the radiation source is allowed to pass, by means of lens, through a narrow slit and then by means of a mirror to an optical grating which divides light into narrow spectral region corresponding to different wavelengths. The light of a desired wavelength emerging from the grating is allowed to pass through the reference solution under examination and from there to a photomultiplier detector. The intensity of light which can be measured with the help of the photomultiplier tube is recorded. Now sample solution allowed the same light to pass through it and then to photomultiplier tube. The intensity of light is again measured by the deflection in the detector.

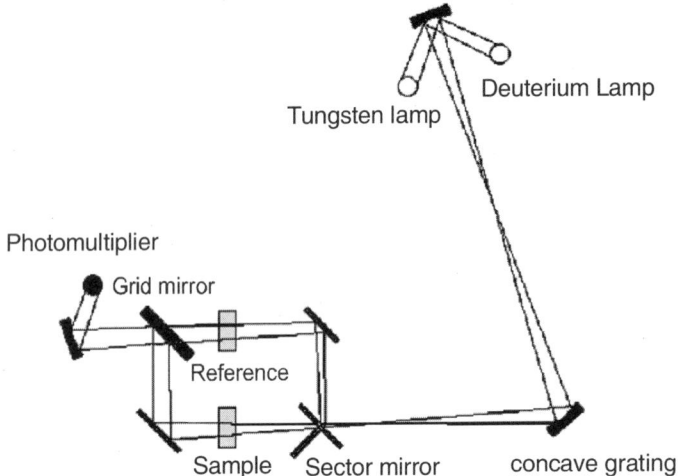

Fig. 4.15: Schematic UV-Visible spectrophotometer.

4.8.3. Instrumentation

Instruments for measuring the absorption of UV or Visible radiation are made up of the following components (Fig. 4.15):

1. Sources (UV and visible)

2. Wavelength selector (monochromator)

3. Sample containers

4. Detector

5. Signal processor and readout

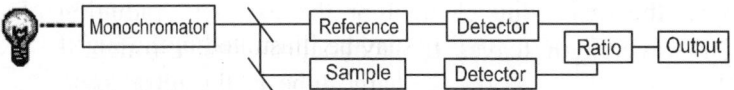

Fig. 4.15: Components of UV-Visible Spectrophotometer

4.8.3.1. Sources of UV and Visible radiation

It is important that the power of the radiation source does not change abruptly over its wavelength range. The electrical excitation of Deuterium or Hydrogen at low pressure produces a continuous UV spectrum. The mechanism for this involves formation of an excited molecular species, which breaks up to give two atomic species and an ultraviolet photon. This can be shown as:

$$D_2 + \text{ electrical energy} \rightarrow D_2^* \rightarrow D' + D'' + h\nu \qquad \textit{... eq. 4.19}$$

Both Deuterium and Hydrogen lamps emit radiation in the range 160 – 375 nm. Quartz windows must be used in these lamps because glass absorbs radiation of wavelengths less than 350 nm.

The Tungsten filament lamp is commonly employed as a source of visible light. This type of lamp is used in the wavelength range of 350 – 2500 nm. The energy emitted by a Tungsten filament lamp is proportional to the fourth power of the operating voltage this means that for the energy output to be stable, the voltage to the lamp must be very stable indeed. Electronic voltage regulators or constant voltage transformers are used to ensure this stability.

Tungsten or Halogen lamps contain a small amount of iodine in a quartz envelope which also contains the Tungsten filament. The iodine reacts with gaseous tungsten formed by sublimation, producing the volatile compound WI_2. When molecules of WI_2 hit the filament they decompose, re-depositing tungsten back on the filament. The lifetime of a Tungsten/Halogen lamp is approximately double that of an ordinary Tungsten filament lamp. Tungsten/ Halogen lamps are very efficient and their output extends well into the ultra-violet. They are used in many modern spectrophotometers.

4.8.3.2. Wavelength selector or Monochromator

All the monochromators contain component parts (a) an entrance slit (b) a collimating lens (c) a dispersing device (usually a prism or a grating) (d) a focusing lens (e) an exit slit. Polychromatic radiation (radiation of more than one wavelength) enters the monochromator through the entrance slit. The beam is collimated and then strikes the dispersing element at an angle. The beam is split into its component wavelengths by the grating or prism.

By moving the dispersing element or the exit slit, radiation of only a particular wavelength leaves the monochromator through the exit slit (Fig. 4.16).

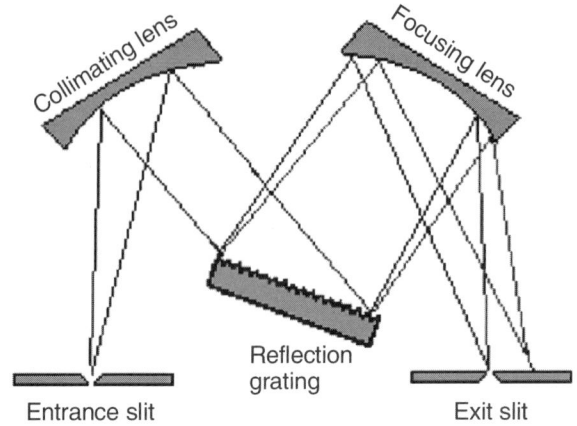

Fig. 4.16: Czerney-Turner grating monochromator.

4.8.3.3. Cuvettes

The containers for sample and reference solution must be transparent to the radiation which will pass through them. Quartz or fused silica cuvettes are required for spectroscopy in the UV region. These cells are also transparent in the visible region. Silicate glasses can be used for the manufacture of cuvettes for use between 350 and 2000 nm.

4.8.3.4. Detectors

The photomultiplier tube is a commonly used detector in UV-Visible spectrophotometer. It consists of a photo-emissive cathode which emits electrons when struck by photons of radiation, several dynodes which emit several electrons for each electron striking them and an anode.

Photomultipliers are very sensitive to UV and Visible radiation and they have fast response times. Intense light damages photomultipliers, they are limited to measuring low power radiation. PMTs are similar to phototubes. They consist of a photocathode and a series of dynodes in an evacuated glass enclosure (Fig. 4.17 and 4.18). A photon of radiation entering the tube strikes the cathode causing the emission of several electrons. Instead of collecting these few electrons at an anode like in the phototubes, the electrons are accelerated towards a series of additional electrodes called dynodes (Fig. 4.18). These electrons are accelerated towards the first dynode which is 90V more positive than the cathode. The electrons strike the first dynode causing the emission of several electrons for each incident electron. These electrons are then accelerated towards the second dynode to produce more electrons which are accelerated towards dynode three and so on (Fig. 4.17). Eventually, the electrons are collected at the anode. By this time, each original photon has produced $10^6 - 10^7$ electrons for each photon hitting the first cathode depending on the number of dynodes and the

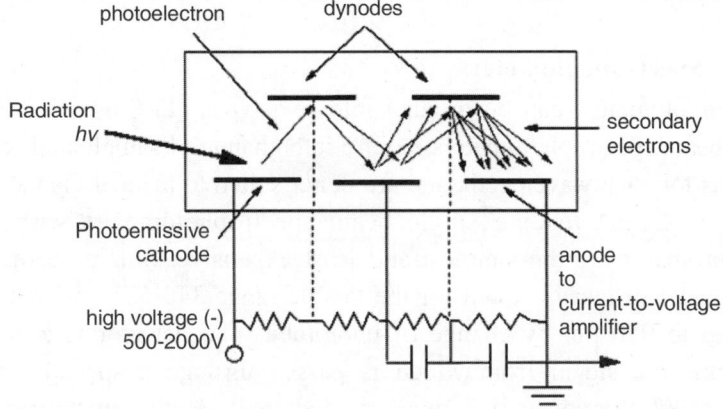

Fig. 4.17: A Photomultiplier tube

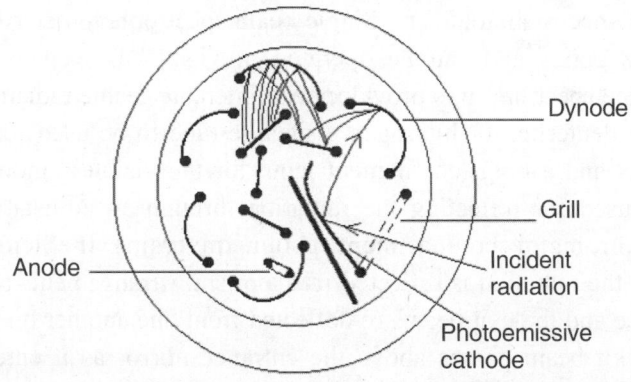

Fig. 4.18: Cross section of a Photomultiplier tube.

accelerating voltage. This amplified signal is finally collected at the anode where it can be measured (Fig. 4.17).

The linear photodiode array is an example of a multi-channel photon detector. These detectors are capable of measuring all elements of a beam of dispersed radiation simultaneously. Linear photodiode arrays comprise many small silicon photodiodes formed on a single silicon chip. There can be sensor elements on a chip, the most common being 1024 photodiodes. For each diode, there is also a storage capacitor and a switch. The individual diode capacitor circuits can be sequentially scanned. In use, the photodiode array is positioned at the focal plane of the monochromator after the dispersing element such that the spectrum falls on the diode array. They are useful for recording UV-Visible absorption spectra of samples that are rapidly passing through a sample flow cell, such as in an HPLC detector.

Type of Spectrophotometers

Spectrophotometers can be divided into two types (1) **Single beam type:** Single beam spectrophotometers require interchange of sample and reference solutions for each wavelength, and are better suited to manual than automatic operation. Direct reading single beam spectrophotometers with grating monochromator are the simplest and least expensive. Single beam type of spectrometer primarily suited for the visible range 340-625 nm, can also be extended to 950 nm by change of phototube. The detector is a gas filled phototube, the signal from which is passed through a special amplifier, compensated to reduce non-linearity and drift. Both transmittance and absorbance are indicated on the large panel meter, wavelength selection, zero adjustment and 100% adjustment are the three important controls in this type of spectrophotometers. Single beam spectrophotometer is equipped with quartz optics and can be operated in visible as well as ultraviolet regions of the spectrum. It is provided with interchangeable radiation sources, including a deuterium or hydrogen discharge tube to be used for the lower wavelengths and a tungsten filament lamp for the visible region. A pair of mirrors is used for reflecting the radiation through an adjustable slit into the monochromator compartment. After traversing the length of the instrument, the radiation is reflected into a prism. Arrangements are so made that entrance and the exit beams re deflected from one another on the vertical axis. The exit beam passes above the entrance mirror as it enters the cell compartment. Cell compartment can hold as many as 4 rectangular 1 cm cells. Compartments holding cylindrical cells and cells up to 10 cm length are also available. (2) **Double beam type:** Double beam instruments using high quality double monochromator are complex as well as expensive. A double spectrophotometer for the ultraviolet and visible regions normally uses some sort of beam splitter between the exit slit of the monochromator and two cuvettes. After passing thorough the cuvettes, the beams may fall on the same detector or on two matched detectors. In all modern instruments, the radiation is chopped at some point by a rotating shutter. The chopper and beam splitter are often combined in a single component. The optical system includes two monochromator, the first with a silica prism and the second with a plane reflecting grating. Radiation from the source either a tungsten lamp or hydrogen tube enters the monochromator through slit S1 and is dispersed by the littrow mounted quartz prism. After passing through slit S2, it is further dispersed by a reflecting grating and then leaves the monochromator via slit S3. The double monochromator produced very narrow band widths and has freedom from stray radiation. The beam is then alternately passed through the reference cell and the sample cell by means of a semicircular mirror arrangement. A photomultiplier tube used as a detector, receives alternate pulses or radiation, first from the reference beam and then from the sample beam (Fig. 4.19).

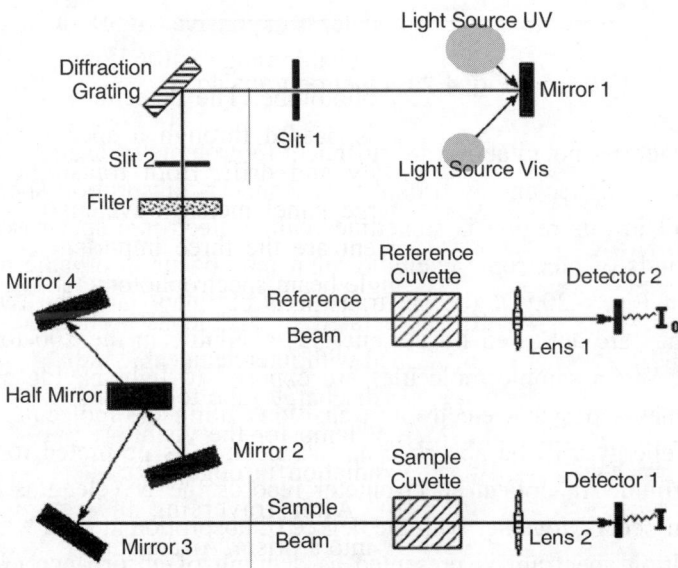

Fig. 4.19: The path way of double beam of UV-Visible spectrophotometer

4.8.4. UV-Visible absorption spectra

To understand why some compounds are colored and others are not, and to determine the relationship of conjugation to color, we must make accurate measurements of light absorption at different wavelengths in and near the visible part of the spectrum. Commercial optical spectrometers enable such experiments to be conducted with ease, and usually survey both the near ultraviolet and visible portions of the spectrum.

The visible region of the spectrum comprises photon energies of 36 – 72 kcal/mole and the near ultraviolet region, out to 200 nm, extends this energy range to 143 kcal/mole. Ultraviolet radiation having wavelengths less than 200 nm is difficult to handle, and is seldom used as a routine tool for structural analysis.

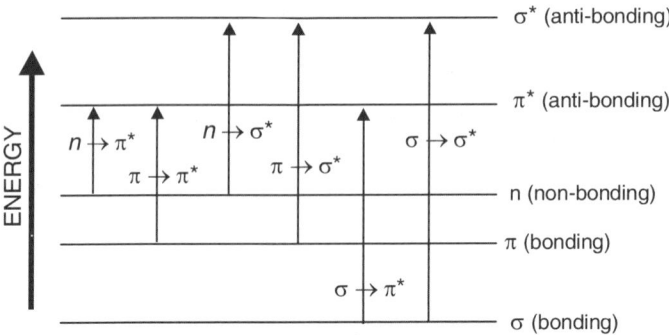

Fig. 4.20: Electron transitions

The energies noted above are sufficient to promote or excite a molecular electron to a higher energy orbital. Consequently, absorption spectroscopy carried out in this region is sometimes called electronic spectroscopy. The various kinds of electronic excitation that may occur in organic molecules are shown Fig. 4.20. Of the six transitions outlined, only the two lowest energy ones are achieved by the energies available in the 200 to 800 nm spectrum. When sample molecules are exposed to light having an energy that matches a possible electronic transition within the molecule, some of the light energy will be absorbed as the electron is promoted to a higher energy orbital. An optical spectrometer records the wavelengths at which absorption occurs, together with the degree of absorption at each wavelength. The resulting spectrum is presented as a graph of absorbance (A) versus wavelength, as in the isoprene spectrum Fig. 4.21. Since isoprene is colorless, it does not absorb in the visible part of the spectrum and this region is not displayed on the graph. Absorbance usually ranges from 0 (no absorption) to 2 (99% absorption), and is precisely defined in context with spectrometer operation.

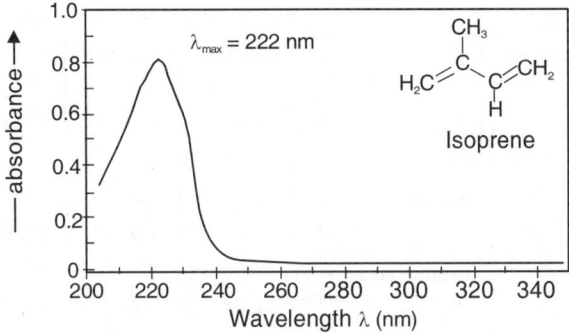

Fig. 4.21: Graph of absorbance (A) vs. wavelength (λ); as in the isoprene spectrum

The absorbance of a sample will be proportional to the number of absorbing molecules in the spectrometer light beam for example their molar concentration in the sample tube, it is necessary to correct the absorbance value for this and other operational factors if the spectra of different compounds are to be compared in a meaningful way. The corrected absorption value is called molar absorptivity and is particularly useful when comparing the spectra of different compounds and determining the relative strength of light absorbing functions (chromophores). Molar absorptivity (ε) is defined as:

Molar Absorptivity, $\varepsilon = A/c\, l$ *... eq. 4.20*

Where A = absorbance, c = sample concentration in moles/liter and l = length of light path through the sample in cm.

If the isoprene spectrum was obtained from a dilute hexane solution in a 1 cm sample cuvette, a simple calculation using the above formula indicates a molar absorptivity of 20,000 at the maximum absorption wavelength. From the Table 4.5 it should be clear that the only molecular moieties likely to absorb light in the 200 to 800 nm region are pi-electron functions and hetero atoms having non-bonding valence-shell electron pairs. Such light absorbing groups are referred to as chromophores. The oxygen non-bonding electrons in alcohols and ethers do not give rise to absorption above 160 nm. Consequently, pure alcohol and ether solvents may be used for spectroscopic studies.

Table 4.5: Chromophore examples

S.No.	Chromophore	Example	Excitation	λ_{max},nm	ε	Solvent
1.	C=C	Ethene	$\pi \to \pi^*$	171	15,000	hexane
2.	C=C	1-Hexyne	$\pi \to \pi^*$	180	10,000	hexane
3.	C=C	Ethanol	$n \to \pi^*$	290	15	hexane
			$\pi \to \pi^*$	180	10,000	hexane
4.	N=O	Nitromethane	$n \to p^*$	275	17	ethanol
			$\pi \to \pi^*$	200	5,000	ethanol
5.	C–X X=Br	Methyl bromide	$n \to \sigma^*$	205	200	hexane
6.	X=I	Methyl Iodide	$n \to \sigma^*$	255	300	hexane

The presence of chromophores in a molecule is best documented by UV-Visible spectroscopy, but the failure of most instruments to provide absorption data for wavelengths below 200 nm makes the detection of isolated chromophores problematic. Fortunately, conjugation generally moves the absorption maxima to longer wavelengths, as in the case of isoprene, so conjugation becomes the major structural feature identified by this technique. Molar absorptivities may be very large for strongly absorbing chromophores (>10,000) and very small if absorption is weak (10 to 100). The magnitude of ε reflects both the size of the chromophore and the probability that light of a given wavelength will be absorbed when it strikes the chromophore.

4.8.5. Importance of conjugation

A comparison of the absorption spectrum of 1-pentene, λ_{max} = 178 nm with that of isoprene clearly demonstrates the importance of chromophore conjugation. The spectrum Fig. 4.22 illustrates that conjugation of double and triple bonds also shifts the absorption maximum to longer wavelengths. From the polyene spectra displayed in the Fig. 4.23, it is clear that each additional double bond in the conjugated pi-electron system shifts the absorption maximum about 30 nm in the same direction. Also, the molar absorptivity (ε) roughly doubles with each new conjugated double bond. Extending conjugation generally results in bathochromic and hyperchromic shifts in absorption.

The appearance of several absorption peaks or shoulders for a given chromophore is common for highly conjugated systems, and is often solvent dependent. This fine structure reflects not only the different conformations but also electronic transitions between the different vibrational energy levels possible for each electronic state. Vibrational fine structure of this kind is most pronounced in vapor phase spectra, and is increasingly broadened and obscured in solution as the solvent is changed from hexane to methanol.

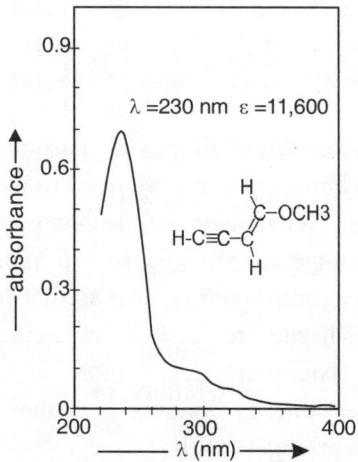

Fig. 4.22: Conjugation of double and triple bonds shifts the absorption maximum to longer wavelengths

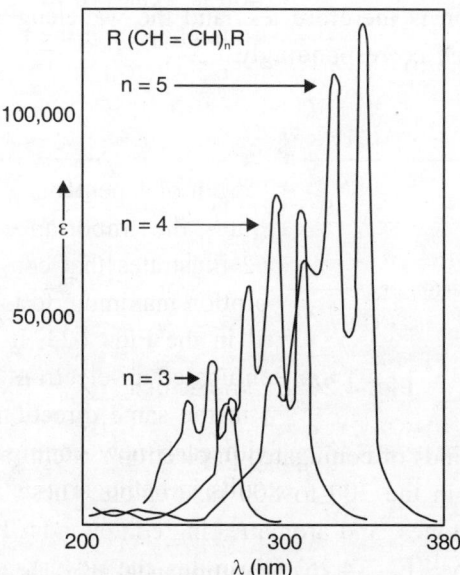

Fig. 4.23: The polyene spectra

Table 4.6: Terminology for absorption shifts

S.No.	Nature of shift	Descriptive term
1.	To Longer Wavelength	Bathochromic
2.	To Shorter Wavelength	Hypsochromic
3.	To Greater Absorbance	Hyperchromic
4.	To Lower Absorbance	Hypochromic

To understand why conjugation should cause bathochromic shifts in the absorption maxima of chromophores, we need to look at the relative energy levels of the pi-orbital. When two double bonds are conjugated, the four p-atomic orbital's combine to generate four pi-molecular orbitals i.e. two are bonding and two are anti-bonding. In a similar manner, the three double bonds of a conjugated triene create six pi-molecular orbitals, half bonding and half antibonding. The energetically most favorable $\pi \to \pi^*$ excitation occurs from the highest energy bonding pi-orbital (HOMO) to the lowest energy antibonding pi-orbital (LUMO). The following Fig. 4.24 illustrates this excitation for an isolated double bond only two pi-orbital for a conjugated diene and triene. In each case the HOMO is colored blue and the LUMO is colored magenta. Increased conjugation brings the HOMO and LUMO orbital's closer together. The energy (ÄE) required to effect the electron promotion is therefore less and the wavelength that provides this energy is increased correspondingly.

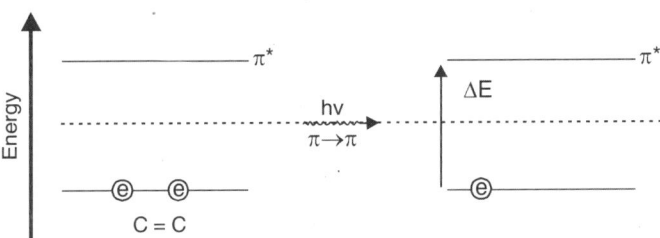

Fig. 4.24: Examples of $\pi \to \pi^*$ Excitation.

Many other kinds of conjugated pi-electron systems act as chromophores and absorb light in the 200 to 800 nm region. These include unsaturated aldehydes and ketones, and aromatic ring compounds. The spectrum of the unsaturated ketone (Fig. 4.25) illustrates the advantage of a logarithmic display of molar absorptivity. The $\pi \to \pi^*$ absorption located at 242 nm is very strong, with a $\varepsilon = 18,000$. The weak $n \to \pi^*$ absorption near 300 nm has a $\varepsilon = 100$.

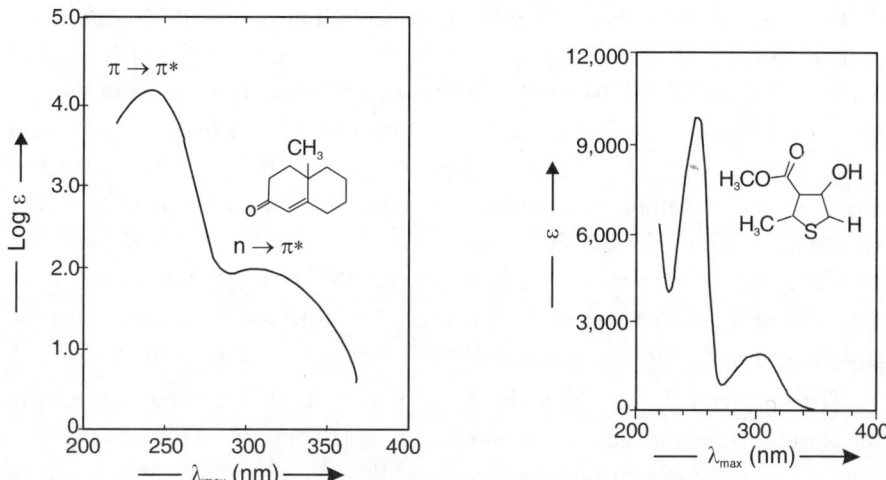

Fig. 4.25: UV-Visible Benzene spectra.

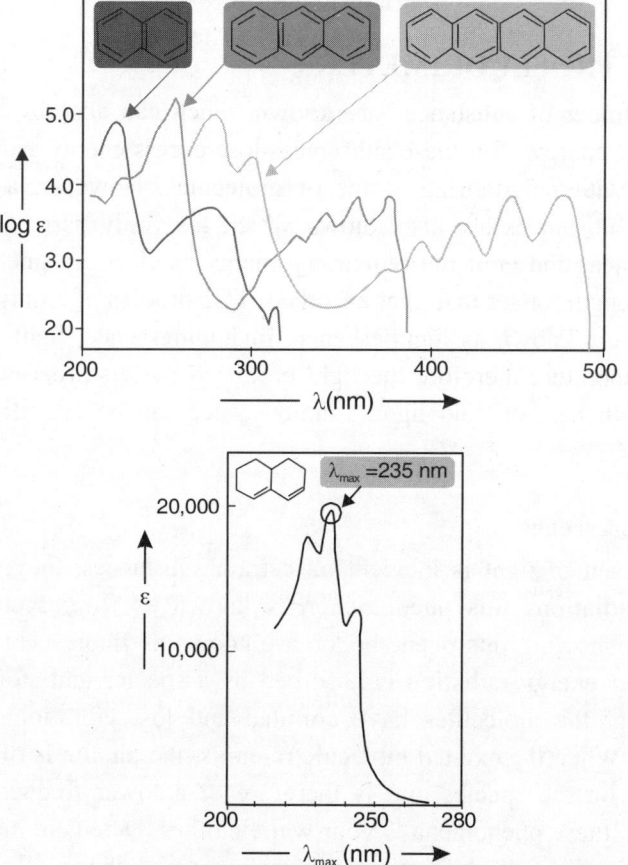

Fig. 4.26: The spectrum of bicyclic diene

Benzene exhibits very strong light absorption near 180 nm ($\varepsilon > 65,000$) weaker absorption at 200 nm ($\varepsilon = 8,000$) and a group of much weaker bands at 254 nm ($\varepsilon = 240$). Only the last group of absorptions is completely displayed because of the 200 nm cut off characteristic of most spectrophotometers. The added conjugation in naphthalene, anthracene and tetracene causes bathochromic shifts of these absorption bands (Fig. 4.26). All the absorptions do not shift by the same amount, so for anthracene and tetracene the weak absorption is obscured by stronger bands that have experienced a greater red shift. As might be expected from their spectra, naphthalene and anthracene are colorless, but tetracene is orange.

The spectrum of the bicyclic diene Fig. 4.26 shows some vibrational fine structure, but in general is similar in appearance to that of isoprene, shown above. Closer inspection discloses that the absorption maximum of the more highly substituted diene has moved to a longer wavelength by about 15 nm, this substituent effect is general for dienes and trienes.

4.9. SPECTROFLUORIMETER

A large number of substances are known which can absorb ultraviolet or visible light energy. But these substances lose excess energy as heat through collisions with neighboring atoms or molecules. However, a number of important substances are also known which lose only part of this excess energy as heat and emit the remaining energy as electromagnetic radiation of a wavelength longer than that absorbed. This process of emitting radiation is collectively known as luminescence. In luminescence, light is produced at low temperature therefore the light produced by this process is regarded as light with heat or cold light. Luminescence can be classified into two types:

4.9.1. Fluorescence

When a beam of light is incident on certain substances, they emit visible light or radiations this phenomenon is known as fluorescence and the substances showing this phenomenon are known as fluorescent substances. Here higher energy radiation is absorbed by a species and stored for long enough that the molecules have collided and lost vibrational energy in collisions. When the excited molecule re-emits, the photon is of less energy depending on the species and is therefore of a lower frequency. A good example of these phenomena is your white clothes. Detergent manufacturers put fluoresces in with their mixture to make your clothes really whiter than white as they absorb UV radiation in sunlight and emit it in the visible region causing the net light reflected from your clothes to be higher than expected. The same fluoresces are to blame for your white clothes glowing blue under UV lights as the re-emitted wavelength falls in the blue portion of the spectrum. Fluorescence is normally emitting from a singlet state whereas phosphorescence occurs from triplet states.

4.9.2. Phosphorescence

When light radiation is incident on certain substances, they emit light continuously even after the incident light is cut off this type of delayed fluorescence is called Phosphorescence and the substances are called Phosphorescent substances. Here light is absorbed by a molecule which becomes excited to a higher singlet state. The light can either be re-emitted straight away or very-rarely, the electron can switch to a triplet state, this is technically forbidden but can happen often enough through a process known as intersystem crossing. Transitions between different multiplicities are forbidden by quantum selection rules and as the ground state of the molecule will often be a single state, the molecule is not allowed to emit a photon that will get it directly to the ground state. The molecule must wait a long time (in quantum chemistry) until it can make the forbidden transition. Fluorescence reemit excess radiation within 10^{-6} to 10^{-4} sec. and materials exhibiting phosphorescence reemit excess radiation within 10^{-4} to 20 seconds or longer. Thus, the life time of phosphorescence is much longer than fluorescence.

4.9.3. Singlet and Triplet states

Singlet and Triplet define the number of unpaired electrons in the absence of a magnetic filed. If there are n numbers of unpaired electrons, it means the (n + 1) fold equal energy states will be associated with the electron spin, regardless of the molecular orbital occupied. Thus, if no unaired electrons are present (n = 0), there is only n+1 or 0+1 or 1 spin state. Such a state is called a singlet state. Similarly, systems having 1,2,3,4... unpaired electrons refer to doublet, triplet, quartet, quintet, etc., states respectively. Most of the molecules in their ground state do not have unpaired electrons (singlet state). When such a molecule absorbs ultraviolet or visible radiation of the proper frequency, one or more of the paired electrons get raised to an excited singlet state. In this excited state, the spin of the electron does not undergo any change and the net spin is still zero. One more possibility is that one set of electron spins may have undergone un-pairing, resulting in two unpaired electron which make an excited triplet state.

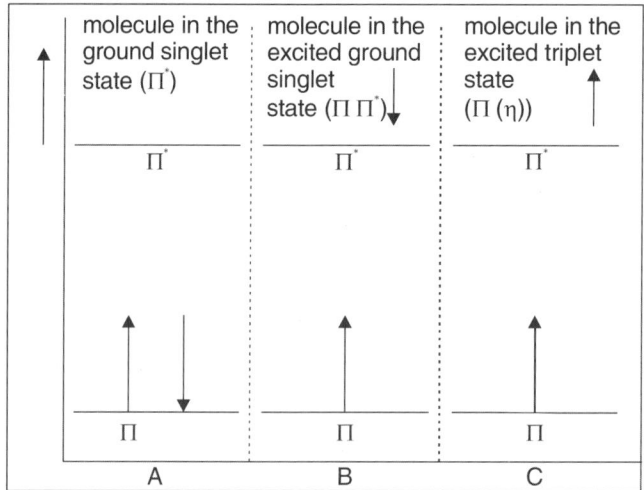

Fig. 4.27: A : Represents a molecule in the ground singlet state.
B : Exhibits the molecule in the excited singlet state.
C : Represents a molecule in an excited triplet state.

4.9.4. Excited state processes in molecules

When molecules are irradiated with light of the appropriate frequency, it will be absorbed in about 10^{-15} second. In the process of absorption, the molecules may move from the ground to the first excited singlet electronic state. Although at room temperature the molecules may be present in their ground vibration level, after absorption the excitational molecules can end up in any one of the vibrational levels in the first excited electronic state. From the excited singlet state, one of the following three phenomena will probably occur, depending on the molecule involved and the conditions.

1.   The first possibility is that the excited singlet state is relatively unstable. In such a situation, the excited molecules will return to the ground state by collision deactivation with emitting any radiation.

2.   The second possibility is that the molecule in the excited singlet state may emit an ultraviolet or visible light photon. This process is known as fluorescence.

   Compare with the absorption and fluorescence spectrum of the same compound, they do not superimpose on each other as expected but they are mirror images of each other with the fluorescence spectrum shifter to longer wavelengths, most of the excess vibrational energy will be given to the surroundings and then the excited molecules will decay in their ground vibrational levels.

3. The third possibility is that the molecule with relatively stable excited singlet state may undergo transition to a meta-stable triplet state and some time thereafter returns to the ground state, usually and the process of crossing from a singlet (no unpaired electron) to a triplet state (two unpaired electrons) is termed as intersystem crossing. The decay from the triplet to the ground state singlet is forbidden by spin symmetry and is therefore slow. Thus, the life time of phosphorescence is much longer than fluorescence.

4.9.5. Principle

Fluorescence is an emission phenomenon. The energy transmission from a higher to lower state within the molecule concerned being measured by the detection of this emitted radiation rather than the absorption. The transmission from higher to lower states to occur by absorption of electromagnetic radiation, must have taken place. The wavelengths of radiation must be at lower values (higher energy) than the emitted wave length. The difference between their two wavelengths is known as stokes shift. The emitted radiation appears as bond spectra because there are many closely related values for the wavelengths dependent upon the final vibrational and rotational energy levels attained. These bond spectra are usually independent of the wavelength of the exciting radiation and have a mirror image relationship with the absorption peak with the greatest wavelength.

The emission has long decay time and usually persists when the exciting energy is no longer applied. The light emission usually occurs at longer wavelengths than does fluorescence. The fluorescence spectra give information about events that occur in less than 10^{-8} sec.

$$\text{The ratio} \times Q = \frac{\text{Quanta fluorescenced}}{\text{Quanta absorbed}} \qquad \textit{... eq. 4.21}$$

Q is the quantum efficiency and usually independent of the exciting wavelength. At low concentrations, the intensity of fluorescence (I_f) is related to the intensity of the incident radiation (I_0)

$$I_f = 2.3\ I_0\ \varepsilon\lambda\quad cdQ \quad \text{i.e } I_f \alpha c \qquad \textit{... eq. 4.22}$$

Where c is the concentration of the fluorescing solution (molar), d is the light path in fluorescing solution, $\varepsilon\lambda$ is the molar extinction coefficient for the absorbing material at wavelength $\lambda(dm^3.mol^{-1}.cm^{-1})$

The technique of spectrofluorometry is most accurate at very low concentrations, where as absorption spectrophotometry is least accurate at these concentrations. For example 100pg of catecholamines or NaOH may be measured fluoremetrically where as absorption spectrophotometry requires. This is due to increased sensitivity, which is easily adjustable over a large range by amplification of the detector signal. The technique allows great spectral selectivity because owing to stokes shift. Two monochromators may be used, one for the exciting wavelength and the other for the emitted fluorescence.

Susceptibility pH, temperature, solvent polarity and the inability to predict whether a particular compound will fluorescence are disadvantages but the major one is the phenomenon of quenching. This occurs because energy that has been emitted as fluorescence is lost to other molecules by interaction method. This explaining the increased sensitivity and accuracy in low concentrations because there are molecules the effects of solvent may not be neglected. Many materials such as detergents, stopcock, grease, filter paper and some tissues may cause interference by the release of fluorescing agents.

4.9.6. Instrumentation

The direct relationship between fluorescent intensity and concentration allows relatively simple electronics and optics to be used. The essential components of an instrument for measuring fluorescence are source, monochromator, detector, amplifier and recorder. There are two monochromators and that sample emission is monitored by the detector at an angle of 90° to the exciting beam. Accessories such as lenses for transmitting exciting and emitted radiation efficiently through the system and there may be other features such as scanning monochromator motor drives which are also emitted.

4.9.6.1. Source

The best general purpose source is Xenon-Arc lamp. The electrical discharge in the gas is initiated by high voltage of source of the Xenon atoms, after which the current maintenance itself about 75A and 20vdc (150w). The pressure is high enough to broaden the Xenon emission lines to a continuous, which is useful for excitation in the spectrofluoremetry region down to slightly about 200nm. High pressure DC Xenon–Arc lamps are used in nearly all commercial stable or radiation that extends from 300 – 1300 nm. Several long emission lines present between 800 – 11000 nm. Sometimes an alternative source used is a high pressure Mercury Vapor Lamp; this emits with super imposed mercury emission lines.

4.9.6.2. Monochromators

For atomic scans of both excitation and emission wavelength, with regarding of the detector signal, monochromators may used in this spectroscopy. The monochromator between the sample and the detector is then set at a wavelength in the blue region of the spectrum i.e. 450 nm. Now use a motor drive on the excitation monochromator to record the detector signal as a function of exciting wavelength. So, obtain a graph showing maximum signal at 200 nm. Now set the excitation monochromator at 260 nm and scan the fluorescent emission as a function of wavelength. Find several emission bonds with the largest at 428 nm and re-determine the wavelength that gives the maximum signal, then excite wavelength distribution on the emission side. Now end up with excitation and emission wavelength and at these settings, ready to measure detector response as a function of concentration with a series of standard solution and unknown samples (Fig. 4.28).

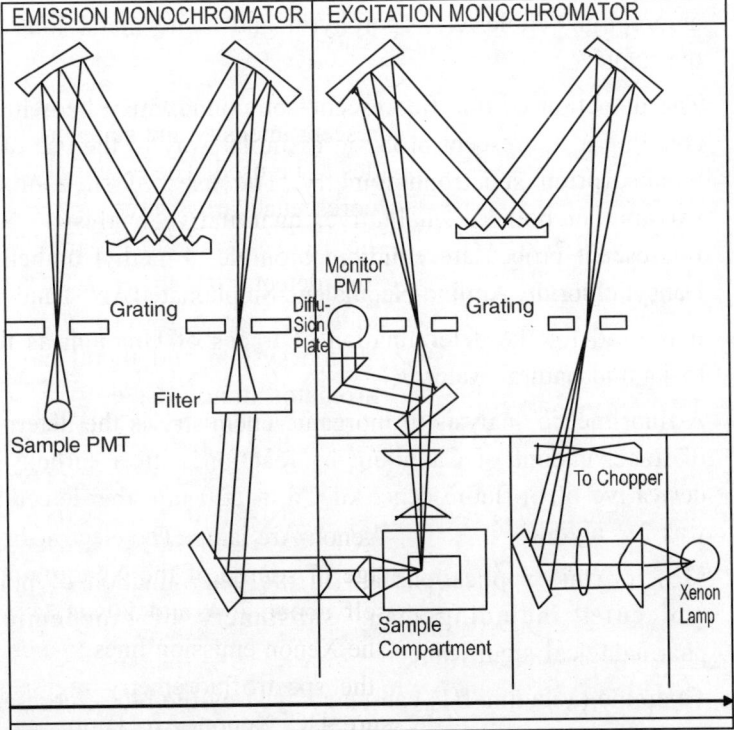

Fig. 4.28: Design of PERKIN ELMER model MPF-44b spectrofluorimeter

4.9.6.3. Detector

The detector is usually a photo multiplies tube (PMT). Detectors have been devices which rapidly yield a three dimensional data set. The output is a fluorescence intensity surface upon a base of excitation and emission wavelengths which is particularly advantages with mixtures of fluorescent compounds.

4.9.7. Applications

1. Fluorimetry is noted for its great sensitivity, of $10^{-2} - 10^4$ times greater than comparable absorption techniques. It is also more selective since favor compounds, fluorescence efficiently than the absorbed radiation.

2. The major use of fluorimetry in biochemistry is in quantitative determination of materials present in concentrations too low for absorption spectrophotometer. For examples, assays of vitamin-B in food stuffs, NADH, hormones, drugs, pesticides, carcinogens, chlorophyll, cholesterol, porphyrin and some metal ions indicate the range.

3. The detection of non-fluorescent compounds may be achieved by coupling a fluorescent probe in a similar way to the use of groups in absorption spectrophotometry. The uses of such probes are valuable in both in qualitative, quantitative analysis. Some of fluorescent probes are ethidium bromide, 4-methyl umbelliferone, Dansyl chloride, Anilino-Napthalene-Sulphonate (ANS) and Arsenic.

4. It is used for the determination of traces of Uranium as found in rocks and natural water.

5. A fluorimetric analysis in inorganic chemistry is the determination of trace amount of Cadmium by reaction with a carboxy methyl derivative of bi-fluorescence of Cd at 520 nm, the detection limit was 2.2 ng/ml.

6. Most organic applications are in the determination of polycyclic molecules including many substances of biochemical and pharmalagical significance.

7. Thiamine (Vitamin–B1) can be assayed by the blue fluorescence of its oxidation product thiochrome.

8. Riboflavin can also be determined fluroimetrically, the fluorescence power is dependent to a large degree on the exact conditions and on the nature and amount of impurities present. The procedure also

takes advantage of the fact that riboflavin can be oxidized to a non-fluorescent compounds that can easily be reduced to generate riboflavin quantitatively.

9. Enzyme assays and kinetic analysis: The enzymes are group specific hydrolysis and then kinetics may be studied by fluorescent measurement. The anion of 4-methyl umbelliferone which fluorescent at 450 nm, extinction is usually at 350-450 nm wavelengths and virtually all the fluorescence measured between 450-500 nm is due to the anion product. Spectrofluorimetry can be applied widely in metabolic studies where NAD forms are involved as cofactors, this arises because NADH and NADPH are fluorescent.

10. Protein structure: The presence of tryptophan and FAD as cofactor allows fluorescence to be measured in proteins. The binding and releasing of cofactors inhibitors and substances at sides close to the flow cause changes in associated fluorescence spectra suitable flow or probe such as Dansyl chloride, Anilino naphthalene sulphonate (ANS) and derivatives of Rhodamine.

12. Membrane structure: The fluorescent properties of a molecule are affected by its mobility and environment, particularly the polarity of the matter. These effects in the presence of a fluorescent probe may used by measuring changes in fluorescence. Various probes are hydrophobic regions (ANS, MNS), (N-methyl 2 anilino 6-Naphthalene Sulphonate) and to orient themselves across liquid interfaces may use to study membrane structure and information about the properties of such interfaces. Phospholipids containing 12-(9-anthoanoyl)-stearic acid and 2-(9-antheoanoyl)-palmitic acid into membranes yields information about regions 0.5 nm and 1.5 nm, respectively from the phosphate head group of the lipid bilayer. Changes in the mitochondrial membranes during energy transduction have also been monitored using an ANS probe.

4.10. ATOMIC ABSORPTION SPECTROSCOPY

4.10.1. Introduction

All atoms and their components have energy, the energy level at which an atom exists is referred to as its state. Under normal conditions, atoms exist in their most stable states and refer to that most stable level as the ground state. Although, cannot measure the precise energy state for an atom and can usually measure changes to its energy relative to its ground state. Certain processes can change the energy state for an atom. For example, adding thermal energy (heat) can cause an atom to increase to a higher energy state, refer to energy states which are higher than the ground state as excited states. In theory, there are infinite excited states; however there are decreasing numbers of atoms from a population that reach higher excited states.

The laws of quantum mechanics tell that atoms do not increase their energy levels gradually. An atom goes directly from one state to another without going through intermediates refer to these quantum leaps as transitions. The transition from the ground state (E_o) to the first excited state (E_1) requires some form of energy input. This energy is absorbed by the atom, that energy absorption is equal to ΔE. When this energy absorption takes place in the presence of ultraviolet light, some of that light will be absorbed, this UV absorption occurs at a specific wavelength. Each element in the periodic table will have a specific E that will absorb a specific wavelength of UV light. The relationship between the energy transition and the wavelength (l) can be described by:

$$\Delta E = h/\lambda \qquad\qquad\qquad ... eq.\ 4.23$$

Where h is Planck's constant and λ is the wavelength.

Atomic absorption uses this relationship to determine the presence of a specific element based on absorption in a specific wavelength. For example, calcium absorbs light with a wavelength of 422.7nm and Iron absorbs light at 248.3nm. Atomic absorption is a very common technique for determination of microelements (Cu, Mn, Zn and Fe) and macro elements (Ca, Mg and K) in agriculture sector from various species of plant leaves and different fields of soils.

4.10.2. Principle

The principle used in atomic absorption spectroscopy was first discovered in 1802 by Wollaston however, the Australian physicist Alan Walsh in 1955 reported the first application of atomic absorption spectra to chemical analysis. The technique is based on the fact that ground state metals absorb light at specific wavelengths. Metal ions in a solution are converted to atomic state by means of a flame light. A solution containing a metallic compounds are aspirated into a flame for example, acetylene burning in air, a vapour which contains atoms of the metal, may be formed (Fig. 4.29). These gaseous metal atoms may be raised to an energy level, which is sufficiently high to permit the emission of radiation characteristic of the metal. Gaseous metal atoms will normally remain in an unexcited state or in the ground state. These ground state atoms are capable of absorbing radiant energy of their own specific resonance wavelength, which in general is the wavelength of the radiation that the atoms would emit if excited from the ground state. Hence if light of the resonance wavelength is passed through a flame containing the atoms, then part of the light will be absorbed and the extent of absorption will be proportional to the number of ground state atoms present in the flame, this is the underlying principle of Atomic Absorption Spectroscopy (AAS).

The characteristic wavelengths are element specific and accurate to 0.01-0.1nm. To provide element specific wavelengths, a light beam from a lamp whose cathode is made of the element being determined is passed through the flame. A device such as photon multiplier can detect the amount of reduction of the light intensity due to absorption by the analyte and this can be directly related to the amount of the element in the sample (Fig. 4.29).

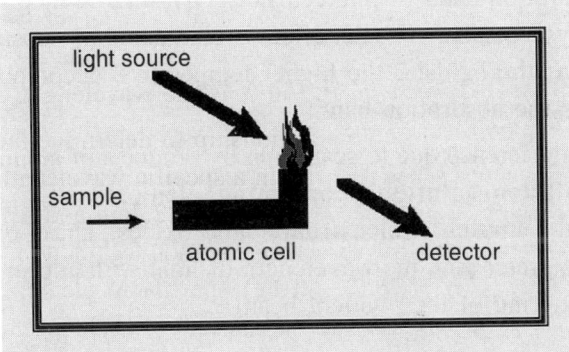

Fig. 4.29: Principle of an Atomic Absorption Spectrometer.

4.10.3. Interferences

The concentration of the analyte element is considered to be proportional to the ground state atom population, in the flame any factor that affects the ground state population of the analyte element can be referred as interference. Factors that may affect the ability of the instrument to read this parameter can also be classified as interference. The following are the most common interferences:

4.10.3.1. Spectral interferences

Spectral interferences occur when particulate matter from the atomization process scatters the incident radiation from the source or when the absorption by an interfering species is close to the analyte wavelength that overlap of absorption peaks occurs. Interference due to overlapping lines is rare because the emission lines of hollow cathode sources are very narrow. Such as interference occur if the separations between two lines are on the order of 0.01nm, for example Vanadium line at 308.211nm interferes in an analysis based upon the Aluminum absorption line at 308.215nm, the interference avoided by selecting a different aluminum line i.e. 309.27nm.

Spectral interferences also result from the presence of molecular combustion products that exhibit broadband absorption or particulate products that scatter radiation. Problem is encountered when the source of absorption or scattering originates in the sample matrix. In this type of interference, the power of the transmitted beam P is attenuated by the matrix components but the incident beam power P_0 is not, a positive error in absorbance. For example, Potential matrix interference due to absorption occurs in the determination of Barium in alkaline earth mixture. The wavelength of the Barium line used for atomic absorption analysis appears in the center of the broad absorption band for molecular CaOH, interference by Calcium in a Barium analysis results. The net effect is eliminated by substituting nitrous oxide for air as the oxidant, the higher temperature decompose the CaOH and eliminates the absorption band.

Spectral interference due to scattering by products of atomization occurs when concentrated solutions containing elements such as Titanium, Zirconium and Tungsten, which from stable oxides. These oxide particles with diameters grater than the wavelength of light with appear to be formed and cause scattering of the incident beam.

4.10.3.2 Chemical interferences

Chemical interferences can be minimized by a suitable choice of operation conditions. Chemical interference results from various chemical processes occurring during atomization that alter the absorption characteristic of the analytes and these can be minimized by suitable choice of condition. The chemical interferences are classified into two categories (1) Anion interference: The most common type of chemical interference is by anions that form compounds of low volatility with the analyte and decrease the rate at which it is atomized. For example the decrease in Calcium absorbance observed with increasing concentrations of sulfate or phosphate ions of which form anion volatile compounds with Calcium ion. (2) Cation interference: The example of this type of interference is the formation of a heat stable

Aluminum or Magnesium compound that caused two results in the determination of Aluminum. Above two interferences which are due to the formation of compounds of low volatility can be avoided by using high temperature flames or by the use of releasing agents, which are cations that react with the interfering anions or by use of protective agents which prevent interference by forming stable but volatile species with the analytic.

4.10.4. Broadening of a spectra line

Broadening of a spectral line can occur due to a number of factors. The most common line width broadening effects are: (a) Doppler effect (see 4.5) (b) Lorentz effect: Lorentz effect occurs as a result of the concentration of foreign atoms present in the environment of the emitting or absorbing atoms. The magnitude of the broadening varies with the pressure of the foreign gases and their physical properties. (c) Quenching effect: In a low pressure spectral source, quenching collision can occur in flames as the result of the presence of foreign gas molecules with vibrational levels very close to the excited state of the resonance line. (d) Self absorption or Self reversal effect: The atoms of the same kind as that emitting radiation will absorb maximum radiation at the centre of the line than at the wings, resulting in the change of shape of the line as well as its intensity. This effect becomes serious if the vapour, which is absorbing radiation, is considerably cooler than that which is emitting radiation.

4.10.5 Instrumentation

The Atomic Absorption Spectroscopy requires the following components. (1) Hollow cathode lamp (2) Atom cell (3) Monochromator (4) Detector (5) Amplifier (6) Signal display (7) Data station (Fig. 4.30 and 4.31).

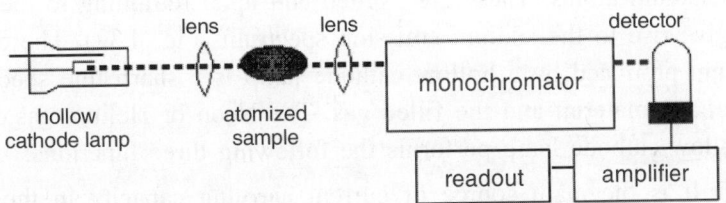

Fig. 4.30: Schematic diagram of an Atomic Absorption Spectroscopy

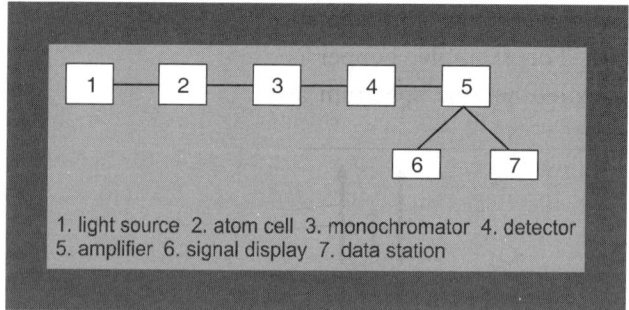

Fig. 4.31: Schematic of basic instrumental parts of
Atomic Absorption Spectrometer

4.10.5.1. Radiation source

The radiation source for atomic absorption spectrophotometer should emit stable and intense radiation of the element to be determine usually a resonance line of the element. The resonance spectral lines should be narrow as compared with the width of the absorption lines to be measured. These lines should not be interfered from other spectral lines which are not resolved by spectrophotometer. The problem of using such narrow spectral lines has been solved by adopting a hollow cathode lamp as the radiation source.

Hollow cathode consists of a hollow cup. In the cup the element which is to be determines in this case sodium and the anode is a tungsten wire. The two electrodes are housed in a tube containing an inert gas. The lamp window is constructed of quartz, silica or glass. The exact material depends upon the wavelength which is to be transmitted (Fig. 4.32). When a potential is applied between the two electrodes, a current in the milliampere range arise, the inert gas is charged at the anode and charged gas is attracted at high velocity to the cathode. The impact with the cathode vaporized some of the sodium atoms. These are excited and upon retraining to the ground state give rise to the sodium emission spectrum (Fig. 4.32). The emission spectrum produced by a hollow cathode lamp is a sharp line spectrum of the cathode material and the filled gas. The Neon or Helium gas filled in the hollow cathode lamp performs the following three functions:

    i. It is the main source of current carrying capacity in the hollow cathode.

    ii. It dislodges atoms from the surface of the cathode.

    iii. It is primarily responsible for excitation of the ground state metal atoms.

The pressure maintained in the hollow cathode lamp is 1 to 5 torr.

Each hollow cathode lamp emits the spectrum of that metal which is used in the cathode. For example, copper cathode emits the copper spectrum; zinc cathode emits the zinc spectrum and so on.

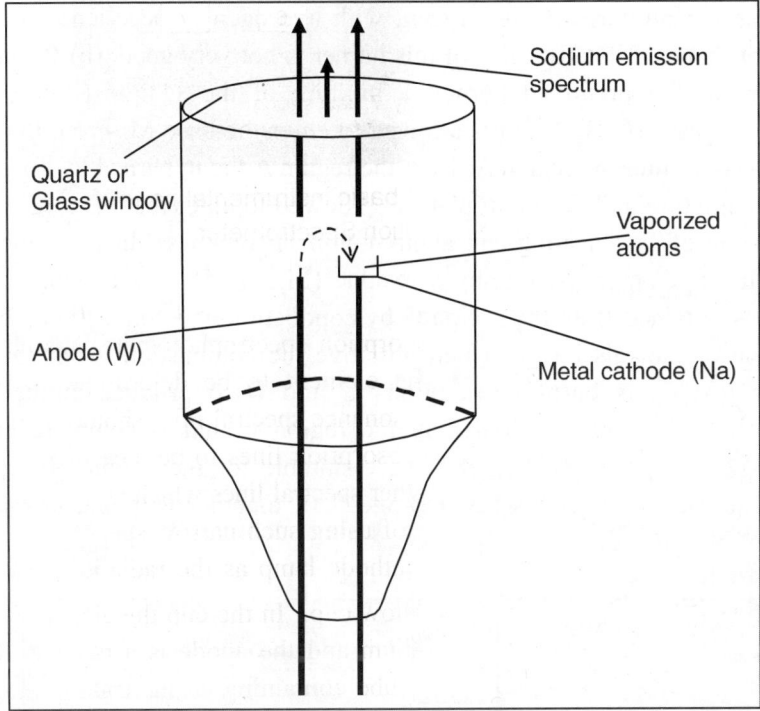

Fig. 4.32: Schematic diagram of a hollow cathode source lamp for atomic absorption of sodium.

4.10.5.2. Atomizers

In order to achieve absorption of atoms, it becomes necessary to reduce the sample to the atomic state. This is done by using either Flame atomizers or Non-Flame atomizers.

4.10.5.2.1. Flame atomizers

The common atomization devices consist of a nebulizer and a burner. The nebulizer produces a flame spray or aerosol from the liquid sample, which is then fed into the flame. The flame is used for converting the liquid sample into the gaseous state and also for conversion of the molecular entities into an atomic vapour. Commonly two types of burners are used they are

(a) Total consumption burner: In the total consumption burner, the sample solution, fuel and oxidizing gases are passed through separate passages to meet at the opening of the base of the flame (Fig. 4.33). As the sample containing metallic element to be estimated by atomic spectroscopy is a liquid, the flame breaks up the liquid sample into droplets which are then evaporated or burnt, leaving the residue which is reduced to atoms. Total consumption burners do use oxygen, with hydrogen or acetylene and give very hot flames. The efficiency of this burner is not very good. (b) **Premixed burner:** In the premixed burner, a mixture of the sample (liquid) and premixed gases ($C_2H_2 + O_2$) is allowed to enter the base M. From the base M, the gases enter the region A; from the region A the unburned hydrocarbon gaseous mixture and liquid droplets are allowed to enter the region B which is a region of free heating and about 1 mm in thickness. In the region B, the liquid is evaporated leaving a residue (Fig. 4.34). The heating in this region is obtained from the region C by conduction and convection, and by diffusion of radicals into it which initiative the combustion. After this the sample residue is burnt into regions C and D to produce atoms. The production of atoms is initiated in the region C and is complete in the region D. The premixed burner is very suitable for the atomic absorption studies of metals of groups I A, I B and II B, together with Ga, In, Ti, Pb, Te, Mn, Ni and Pd.

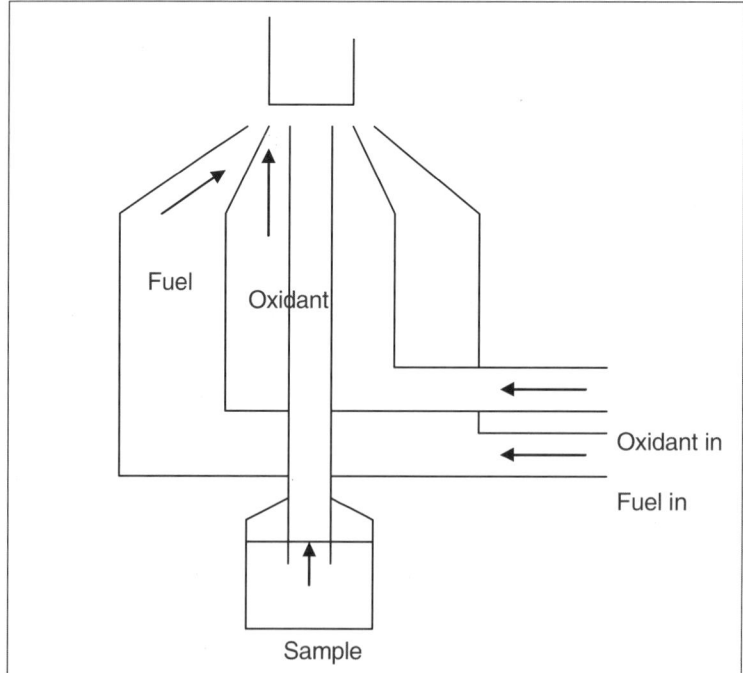

Fig. 4.33: A total consumption burner.

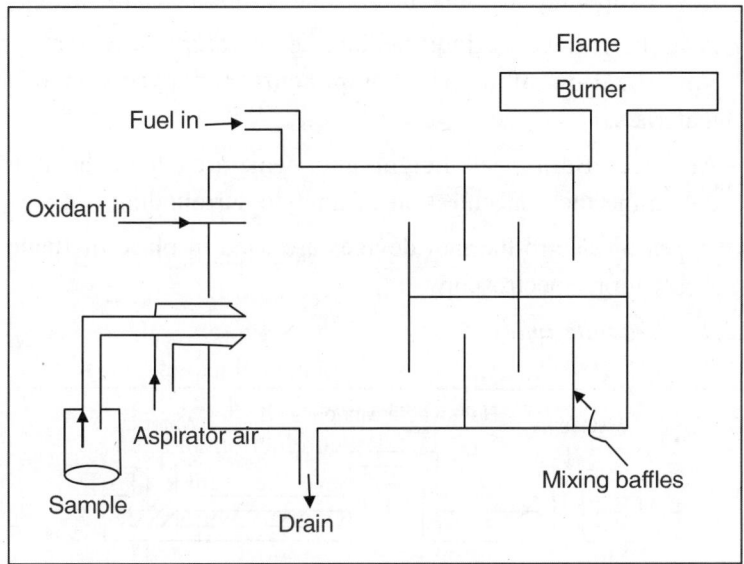

Fig. 4.34: A Premixed type

4.10.5.2.2. Non-flame atomizers

In Atomic Absorption Spectroscopy electro-thermal devices used as non-flame atomizers for reduce the sample to the atomic state. Features of electro-thermal devices in Atomic Absorption Spectroscopy are (i) High sensitivity (ii) The ability to handle small sample volumes of liquids (iii) The ability to analyze solid samples directly without pretreatment (iv) Low noise from the furnace. Electro thermal atomizers are much more difficult to use than the flame atomizers, but using computer to control of these difficulties. This atomization consists of three components. They are:

a. Work head is replaces the burner–nebulizer assembly in the AAS spectrometer.

b. The Power unit is supplies the operating current at the proper voltage to the work head.

c. The controls for the inert gas supply are provides for metering and control of the flow inert gas around the exterior and through the work head during the analysis.

This technique requires fast recording of the absorbance signal, since the peaks appear and disappears rapidly. After insertion or injection of the sample into the electro-thermal atomizer, a heating sequence is initiated to take the sample through the following three steps.

1. Dry: Evaporate any solvent or volatile matrix components.

2. Ash: Performed at intermediate temperature necessary process. Volatilization of matrix components and pyrolysis of matrix materials.

3. Atomize: Both peak height and peak area have been used to determine the concentration of analyte surrounding.

Two types of electro-thermal devices are used in place of flame in the Atomic Absorption Spectroscopy.

4.10.5.2.1.1 Graphite analyzer

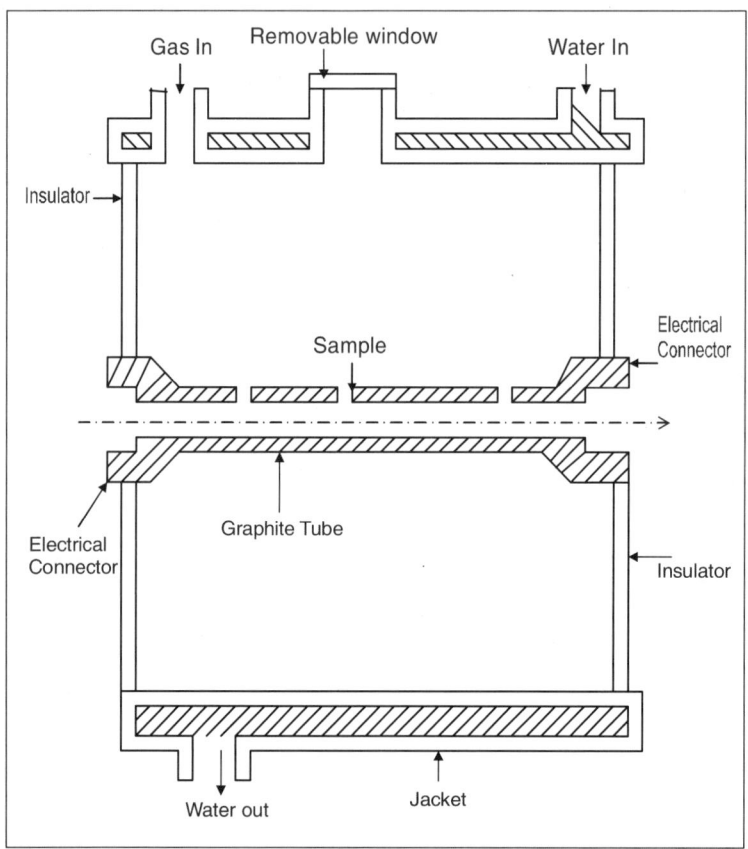

Fig. 4.35: Cross section of a Heated Graphite Atomizer

Graphite analyzer length is 28 mm and diameter is 8 mm. Radiation passed through the centre of the cylinder (Hollow cathode tube). The interior of the cylinder coated with pyrolytic graphite. Electrodes at the end of the cylinder are connected with low voltage and high current power supply 3.6

KW to the cylinder walls. Liquid samples are injected with a micro-syringe through the small opening in the top of the removable window. In this furnace inert gas is Argon (Fig. 4.35).

4.10.5.2.1.2 Carbon rod atomizers

A small quantity of sample, usually of the order of 2-30µl is loaded onto the carbon atomizer. It is then warmed gently to remove the solvent. The temperature is then increased under controlled conditions to ash the sample and removes most of the organic material present. Finally the sample is heated rapidly to very high temperatures to cause atomization. The free atoms are vaporized from the carbon atomizer into the optical light path where their absorption measured.

4.10.5.3. Background corrector

When utilizing a source lamp, the measured energy is that from the unabsorbed transmitted light plus that contributed form the flame background. Consider the substitution of a deuterium lamp for the hollow cathode lamp. The emission wavelengths from the deuterium lamp are not the same as those required for sample excitation, no absorption can occur and the measured intensity is the due solely to the flame background. When the intensity of the signal from the deuterium source lamp is subtracted from the intensity of the element cathode lamp, the intensity due to the sample alone is obtained. Utilization of this principle has led to the development of background corrector. The ray from a deuterium lamp is pulsed alternately with those from the element cathode lamp. Continuous corrected absorbance values are obtained on the readout (Fig. 4.36).

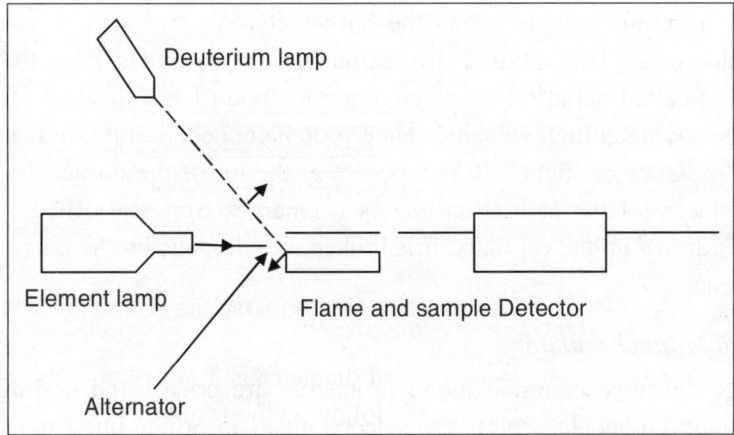

Fig. 4.36: Diagram of a back ground corrector

4.10.5.4. Flame

The greatest source of uncertainly in Atomic Absorption Spectroscopy is the behavior of the flame. Flame temperature depends upon the ratio of fuel and oxidant. Properties of flame are (i) **Base**-In this part, the sample enters in the form of minute droplets (ii) **Inner cone**-In this zone, some of the sample enters as solid vaporization and decomposition to the atomic sates. Absorption process also occurs in this part of zone. (iii) **Reaction zone**-The atoms are converted into oxides and then pass out to outer mantle.

Different flames can be using with different mixtures of gases, depending on the desired temperature and burning velocity. Some elements can only be converted to atoms at high temperatures. Even at high temperatures, if excess oxygen is present, some metals form oxides that do not re-dissociate into atoms. To inhibit their formation, conditions of the flame may be modified to achieve a reducing, non-oxidizing flame. Table 4.7 shows the characteristics of various flames.

Table 4.7: Characteristics of various flames.

S.No.	Fuel	Oxidant	Temperature (°C)
1.	Natural Gas	Air	1700 -1900
2.	Natural Gas	Oxygen	2750
3.	Hydrogen	Air	2000-2050
4.	Hydrogen	Oxygen	2550-2700
5.	Acetylene	Air	2100-2400

4.10.5.5. Nebulisation

Before the liquid sample enters the burner, it is first of all converted into small droplets. This method, formation of small droplets from the liquid sample is called nebulisation. A common method of nebulisation is by use of a gas moving at high velocity, called pneumatic nebulisation. In the burner, a back pressure of about 250 torr occurs at the tip of the burner due to the high velocity of the aspiration gas as it emerges from the office. As the liquid is drawn up the capillary, it is broken into droplets by the high velocity gas stream.

4.10.5.6 Monochromator

In AAS, the most common monochromators are prisms and gratings. The function of a monochromator is to select a given absorbing line from spectral lines emitted from the hollow cathode. When the cathode in the hollow cathode lamp is made up of transition metals, the emission spectrum from the hollow cathode is so complicated that high dispersion is essential. For such cases large dispersion and high resolving monochromators are advantageous for resolving spectra.

4.10.5.7.Detector

For AAS the photomultiplier tube is most suitable detector. It has good stability if used with a stable power supply.

4.10.5.8. Amplifier

The electric current from the photomultiplier detector is fed to the amplifier which amplifies the electric current may times.

4.10.5.9. Read out device

In most of the AAS, chart recorders are used as read out devices. A chart recorder is a potentiometer using a servomotor to move the recording pen. The displacement is directly proportional to the input voltage.

4.10.6 Applications

Atomic Absorption Spectroscopy finds valid applications in every branch for chemical analysis. The technique is already established procedure in analytical chemistry, mineralogy, biochemistry, water suppliers and soil analysis.

1. Qualitative analysis: In AAS, a different hollow cathode lamp is to be used for each element to be tested. It means that an element which is used in the construction of cathode of the hollow cathode lamp can be detected only that particular element. As qualitative analysis involves the checking of one element at a time, it means that the process is very laborious.

2. Quantitative analysis: The technique of quantitative analysis is based on the determination of the amount of radiation absorbed by the sample. If the value of radiation absorbed, the number of absorbing atoms in the light path can be determined.

3. The atomic absorption spectroscopy is becoming a very important tool for the determination of trace metals in biological materials.

4. Copper, Zinc and Nickel are the most common toxic elements these elements can be determine by using Atomic Absorption Spectroscopy.

5. Determination of Calcium, Magnesium, Sodium and Potassium in blood serum is of vital importance in diagnosing many pathological

conditions, diabetes and primary aldostriosnism. AAS is a most suitable method for determination of all these elements in blood serum.

6. In petrol the two anti-knocking additives are tetraethyl and tetra methyl lead. In order to apply AAS to the analysis of lead in petrol, the analytical chemist must know which of these or what mixture of the two has been included in the sample to be analyzed.

7. Atomic Absorption Spectroscopy is also plays a vital role in soil, water and leaf analysis for quantitative estimation of macro- and micro-elements.

4.11. ATOMIC EMISSION SPECTROSCOPY

4.11.1. Principle

Atomic Emission Spectroscopy (AES) uses quantitative measurement of the optical emission from excited atoms to determine analyte concentration. Analyte atoms in solution are aspirated into the excitation region where they are desolvated, vaporized and atomized by a flame, discharge or plasma. The high temperature atomization sources provide sufficient energy to promote the atoms into high energy levels and the atoms decay back to lower levels by emitting light. Since the transitions are between distinct atomic energy levels, the emission lines in the spectra are narrow. The spectra of multi-elemental samples can be very congested and spectral separation of nearby atomic transitions requires a high resolution spectrometer. Since all atoms in a sample are excited simultaneously, they can be detected simultaneously and is the major advantage of AES compared to Atomic Absorption Spectroscopy (AAS).

4.11.2. Instrumentation

Instrumentation of AES may be divided into three major components:

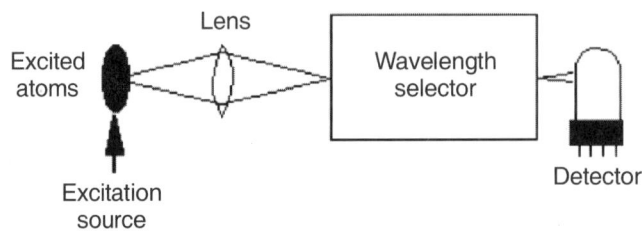

Fig. 4.37: Instrumentation of Atomic Emission Spectroscopy.

1. The sampling device and source
2. The spectrometer
3. The detector and readout device

4.11.2.1. Sampling devices and sources

Wide variety of sources can be selected in accordance with the requirements of the desired analysis in Atomic Emission Spectroscopy. Factors that influence the selection of an excitation source are the concentration of the elements being determined, the vapor pressures or volatilities of these elements, the excitation potentials of the atomic lines used in the analysis and the physical condition of the sample.

In AES, the sample must be converted to free atoms, usually in a high temperature excitation source. Liquid samples are nebulized and carried into the excitation source by a flowing gas. Solid samples can be introduced into the source by slurry or by laser ablation of the solid sample in a gas stream. Solids can also be directly vaporized and excited by a spark between electrodes or by a laser pulse. The excitation source must desolvate, atomize and excite the analyte atoms. Different types of excitation sources are used in AES.

1. Direct-current plasma (DCP)
2. Flame excitation source
3. Inductively-coupled plasma (ICP)
4. Laser-induced breakdown (LIBS)
5. Laser-induced plasma
6. Microwave-induced plasma (MIP)
7. Spark or Arc emmission source

4.11.2.1.1. Direct current plasma excitation source

A direct-current plasma (DCP) is created by an electrical discharge between two electrodes. A plasma support gas is necessary, commonly used plasma support gas is Ar. Samples can be deposited on one of the electrodes, or if conducting can make up one electrode. Insulating the solid samples are placed near the discharge so that, the ionized gas atoms sputter the sample into the gas phase where the analyte atoms are excited. This sputtering process is often referred to as glow discharge excitation (Fig. 4.38).

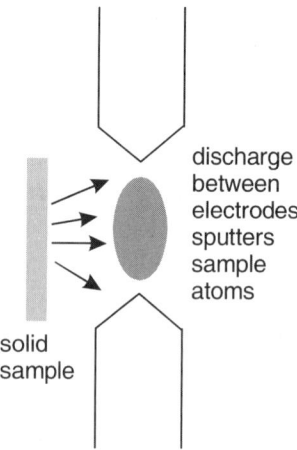

Fig. 4.38: Direct current plasma source

4.11.2.1.2. Flame excitation source

A flame provides high temperature source for desolvating and vaporizing a sample to obtain free atoms for spectroscopic analysis. In atomic absorption spectroscopy ground state atoms are desired. For atomic emission spectroscopy the flame must excite the atoms to higher energy levels. The Table 4.8 lists the temperatures that can be achieved in some commonly used flames.

Table 4.8: Temperature of some common flames.

S.No.	Fuel	Oxidant	Temperature (K)
1.	H_2	Air	2000-2100
2.	C_2H_2	Air	2100-2400
3.	H_2	O_2	2600-2700
4.	C_2H_2	N_2O	2600-2800

The Fig. 4.39 shows a total consumption burner in which the sample solution is directly aspirated into the flame and this flame design is common for atomic emission spectroscopy. All desolvation, atomization and excitation occur in the flame. Other flame designs nebulize the sample and premix it with the fuel and oxidant before it reaches the burner. Atomic absorption instruments almost always use a nebulizer and also use a slot burner to increase the path length for the sample absorption. Different types of flames are produced and used in Atomic Emission Spectroscopy for determination of various fields of elements (1) Cooler flames produced by Air-Propane are used for alkali metals (Na, K). (2) The acetylene flame suitable for alkaline earth metals (Ca, Mg). (3) A lean is used for elements that form stable compounds but not the refractory oxides. Example: Cu, Fe, Mn etc. (4) A rich flame is used for compounds which form refractory oxides with radicals present in the flame. Example: Ca, Mg.

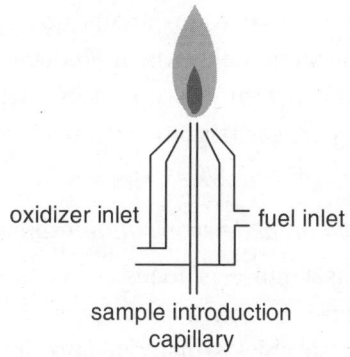

oxidizer inlet   fuel inlet

sample introduction
capillary

Fig. 4.39: Flame excitation source

4.11.2.1.3 Inductively coupled plasma (ICP) excitation source

Inductively coupled plasma (ICP) is a very high temperature (7000-8000K) excitation source that efficiently desolvates, vaporizes, excites and ionizes atoms. Molecular interferences are greatly reduced with this excitation source

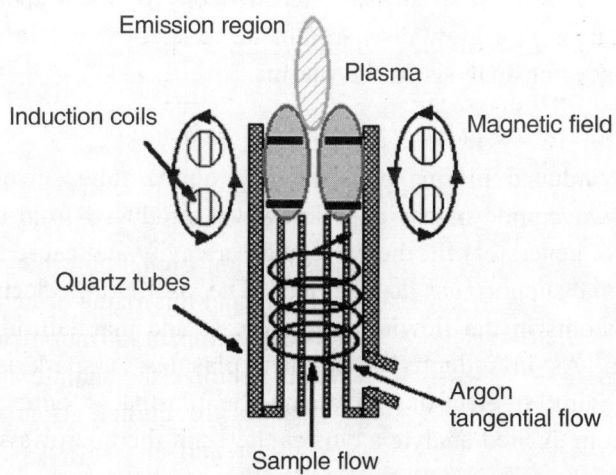

Emission region

Plasma

Induction coils

Magnetic field

Quartz tubes

Argon
tangential flow

Sample flow

Fig. 4.40: Inductively coupled plasma (ICP) excitation source

but are not eliminated completely. ICP sources are used to excite atoms for atomic emission spectroscopy and to ionize atoms for mass spectrometry. The sample is nebulized and entrained in the flow of plasma support gas, which is typically Ar. The plasma torch consists of concentric quartz tubes. The inner tube contains the sample aerosol and Ar support gas and the outer tube contains flowing gas to keep the tubes cool (Fig. 4.40). A radiofrequency (RF) generator (typically 1-5 kW) produces an oscillating current in an induction coil that wraps around the tubes. The induction coil creates an oscillating magnetic field, which produces an oscillating magnetic field The magnetic field in turn sets up an oscillating current in the ions and electrons of the support gas (argon). The ions and electrons are colliding with other atoms in the support gas.

4.11.2.1.4. Laser induced breakdown excitation source

When a high energy laser pulse is focused into a gas or liquid or onto a solid surface, it can cause dielectric breakdown and create hot plasma. For solids the laser pulse also ablates material into the gas phase. The energy of the laser created plasma can atomize, excite and ionize analyte species, which can then be detected and quantified by atomic emission spectroscopy or mass spectrometry.

4.11.2.1.5. Laser induced plasma excitation source

A high power CO_2 laser that is focused into a support gas, such as Ar can maintain hot plasma. The energy of the plasma can atomize, excite and ionize analyte species present in the support gas, which can then be detected and quantified by atomic emission spectroscopy or mass spectrometry. It can also be used in a glow discharge mode to sputter analyte atoms off of a solid surface for analysis in the plasma.

4.11.2.1.6. Microwave induced plasma excitation source

Microwave induced plasma consists of a quartz tube surrounded by a microwave waveguide or cavity. Microwaves produced from a magnetron (a microwave generator) fill the waveguide or cavity and cause the electrons in the plasma support gas to oscillate. The oscillating electrons collide with other atoms in the flowing gas to create and maintain high temperature plasma. As in inductively coupled plasmas, a spark is needed to create some initial electrons to create the plasma. Atomic emission is measured from excited analyte atoms as they exit the microwave waveguide or cavity.

4.11.2.1.7. Spark and arc emission sources

Spark and arc excitation sources use a current pulse (spark) or a continuous electrical discharge (arc) between two electrodes to vaporize and excite analyte atoms (fig.-4.41). The electrodes are either metal or graphite. If the sample to be analyzed is a metal, it can be used as one electrode. Non-conducting samples are ground with graphite powder and placed into a cup-shaped lower electrode. Arc and spark sources can be used to excite atoms for atomic-emission spectroscopy or to ionize atoms for mass spectrometry. Arc and spark excitation sources have been replaced in many applications with plasma or laser sources, but are still widely used in the metals industry.

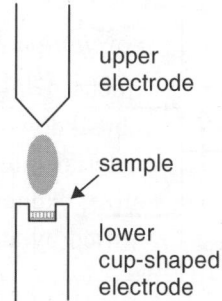

upper electrode

sample

lower cup-shaped electrode

Fig. 4.41: Spark and Arc emission sources

4.11.2.2. Introduction of liquid samples into plasma sources

Liquid sample introduction is important in plasma methods. The analyses can be only as good as the sample introduction. Optimum conditions for sample introduction into plasma sources differ markedly from those of flame atomic absorption sources. The transport efficiency for ICP is usually lower than that for AAS. The time required for washout of the analyte from ICP spray chambers is longer that required for AES due to the substantially lower gas and liquid flow rates (1 l/min and 1ml/min) compared with those of AAS systems (18 l/min and 6-8 ml/min). There are three practical considerations when introducing liquid samples into plasma sources (1) The maximum drop size for introducing the sample into the atomizer. (2) The rate of solvent flow must fall within a specific range of values. (3) The analytical precision of an ICP system is strongly dependent on careful control of these parameters over the long term.

Any significant change in either the drop size or the solvent loading that reached the atomizer reduces both the accuracy and precision of the ICP system. Pneumatic nebulation is the method used in most ICP and DCP determinations. Systems for DCP sample introduction generally tolerate both suspended particles and high concentrations of dissolved solids much better than ICP systems. The excitation characteristics of DCP sources are affected by high concentrations of easily ionized elements such as Na and Ca. Generally cross-flow nebulizer is used for introduction of samples into the AES, clogging is avoided because the liquid sample stream can interact at right angles with the argon gas to produce the aerosol (Fig. 4.42).

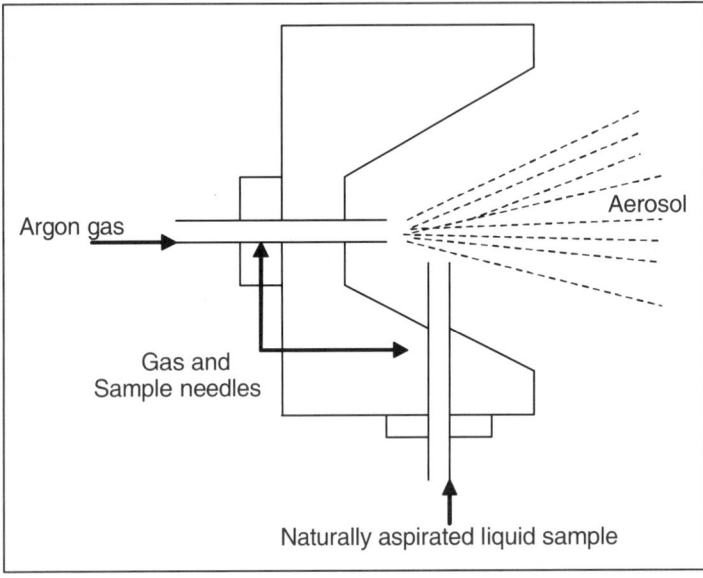

Fig. 4.42: Cross-Flow nebulizer

4.11.2.3. Hollow cathode discharge lamp

Hollow cathode discharge lamps are widely used as radiation sources for both atomic absorption and atomic fluorescence spectrometry, their use in atomic emission spectroscopy has been limited despite some attractive characteristics. The major limitation of this emission source is the requirement of low pressure operation.

The hollow cathode lamp consists of two coaxial cylinders (Fig. 4.43). The inner graphite cylinder is the cathode and contains the sample material. The discharge material is helium. The radiation is emitted from the negative glow, which is confined to the cathode cavity. At low temperatures of the discharge and the low operating pressures produce spectra that contain sharp, narrow lines compared with the lines obtained from a discharge source that operated at atmospheric pressure. These lamps are often used to determine elements with low boiling points in high melting point matrices.

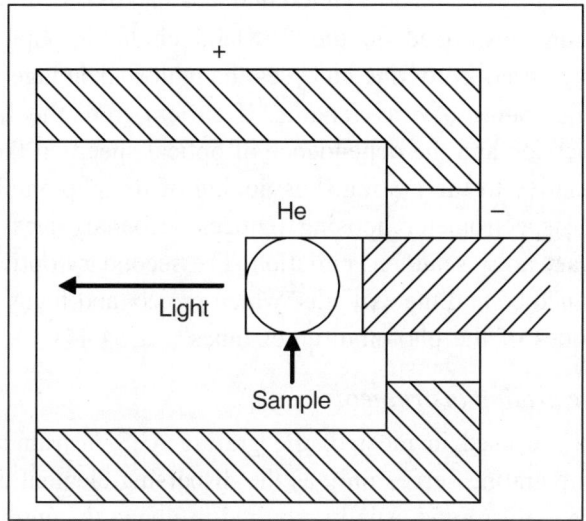

Fig. 4.43: Hollow cathode lamp the sample is placed in the cathode cavity.

4.11.2.4. Concave grating instruments

Concave holographic grating, known more specifically as the Paschen-Range mounting have the entrance slit, grating and focal plane lie on the circumference of the Roland circle. The Rowland circle has a radius of curvature half that of the grating. If the entrance slit and the grating are on

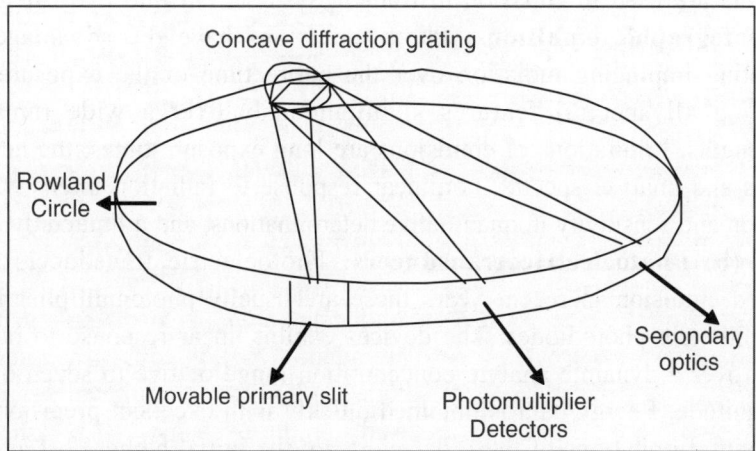

Fig. 4.44: Schematic diagram of holographic concave grating in the Rowland circle configuration.

the Rowland circle, the spectrum is focused on the circle. The grating is the only optical component between the entrance and exit slits. The Paschen-Range mounting, positioned on the Rowland circle, is popular for large concave gratings used in routine analyses. Its major advantage is the ability to cover a large range of wavelengths. If suffers from the limitations of nonlinear dispersion and the dependence of optical speed on the position of the exit slit relative to the grating. Positioning of the photomultiplier tubes external to the spectrometer housing reduces secondary array clutter and avoids secondary array scattered radiation. The secondary optics consists of mirrors positioned behind the exit slits, which project and focus the radiation onto the cathodes of the photomultiplier tubes (Fig. 4.44).

4.11.2.5. Plane grating instruments

Plane gratings are used in most of the grating AES instruments, in these instruments, the grating serves only as the dispersing element and therefore a pair of concave mirrors is usually required to image the entrance slit onto the focal plane. Different spectral regions are focused on the camera detector's film or different wavelengths on the exit slit by rotating the grating. The Ebert mounting is used in almost all large (3 m focal length) spectrometers that contain plane gratings. This mounting is nearly stigmatic and achromatic, so that light of all wavelengths is brought to focus on the detector without changing the detector to mirror distance. This makes it easy to change wavelengths by simply rotating the grating.

4.11.2.6. Detector

Detectors are used in emission instrument systems fall into two categories (a) **Photographic emulsions:** Photo emulsions have the advantages of integrating impinging radiation over the entire time of the exposure and recording all spectral features simultaneously over a wide range of wavelengths. Limitations of emulsions are long exposure times, the need to process and analyze spectra, nonlinear response to radiation intensity, low precision and sensitivity in quantitative determinations, and a limited dynamic range. (b) **Photoelectric transducers:** Photoelectric transducers have replaced emulsion in recent years these are usually photomultiplier tubes and solid state photodiodes. The devices exhibit linear response to radiant energy over a dynamic analyte concentration range of five to seven orders of magnitude. Energy data is obtained quickly with excellent precision and sensitivity. Limitations of these detectors are the initial higher cost and the ability to measure only a single spectral line at any given instant.

4.11.3. Applications

1. Atomics Emission Spectroscopy has used for analysis of ferrous and nonferrous alloys.

2. Determination of trace metal impurities in alloys, metals, reagents and solvents.

3. Analysis of metals in geological, environmental and biological materials and Water analysis.

4. AES is used in the most determination of Na, K, Li and Ca.

4.12. CIRCULAR DICHROISM (CD)

4.12.1. Dichroism

Dichroism has two related but distinct meanings in optics. A dichroic material is either one which causes visible light to be split up into distinct beams of different wavelengths (colors) or one in which light rays having different polarizations are absorbed by different amounts. To understand CD, one has first to know what a circular polarized beam of light is, let us first consider a plane polarized light. As you may know, like any other electromagnetic radiation the light has a magnetic and an electric component. Neglect the magnetic one however keep in mind that what follows can be exactly applied to the magnetic component as well. Because we are considering a plane polarized light, we see an oscillation of the electric field vector intensity on a plane that has the same direction of the beam (red light). If we choose a point in the beam direction, we see that the intensity of the electric field vector (magenta) vary from a positive maximum (t_1) to a negative minimum (t_3) during time. By means of the usual vector decomposition rules, we can decompose the magenta vector in two green vectors of equal intensity. The change in intensity of the magenta electric field vector causes these two green vectors to rotate oppositely, each of them describing a circumpherence.

4.12.1.1. Understanding circular dichroism - the basics of polarization

A beam of polarized light is described by the set of each of the two green vectors along the original plane polarized light direction. As a consequence, we can also say that a plane polarized light can be seen as resulting from a right (clockwise) and left (anticlockwise) circularly polarized beams of light.

Linearly polarized light is light whose oscillations are confined to a single plane. All polarized light states can be described as a sum of two linearly polarized states at right angles to each other, usually referenced to the viewer as vertically and horizontally polarized light (Fig. 4.45 and 4.46).

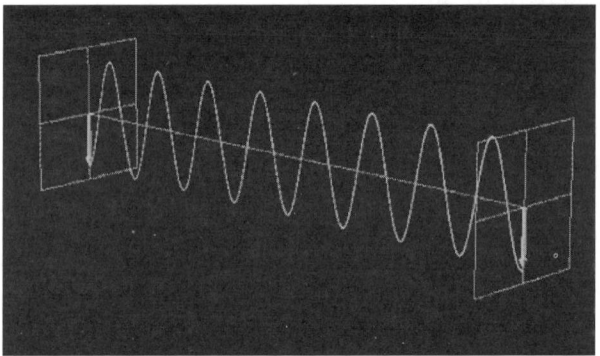

Fig. 4.45: Vertically polarized light

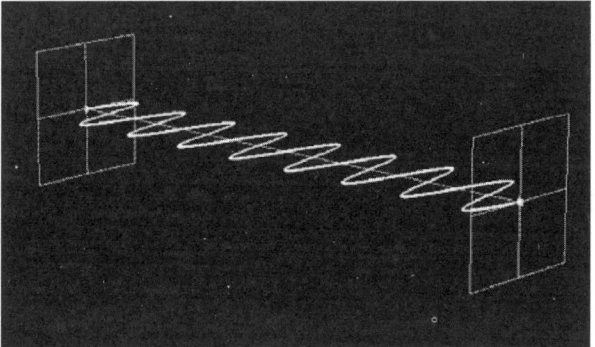

Fig. 4.46: Horizontally polarized light

If for instance we take horizontally and vertically polarized light waves of equal amplitude that are in phase with each other, the resultant light wave (blue) is linearly polarized at 45 degrees (Fig. 4.47).

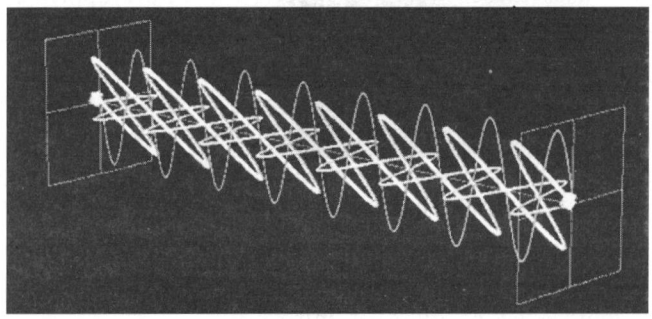

Fig. 4.47: Degree polarized light

If the two polarization states are out of phase, the resultant wave ceases to be linearly polarized. For example, if one of the polarized states is out of phase with the other by a quarter-wave, the resultant will be a helix and is known as circularly polarized light (CPL). The helices can be either right-handed (R-CPL) or left-handed (L-CPL) and are non-super imposable mirror images (Fig. 4.48 and 4.49).

The optical element that converts between linearly polarized light and circularly polarized light is termed a quarter-wave plate. A quarter-wave plate is birefringent, i.e. the refractive indices seen by horizontally and vertically polarized light are different. A suitably oriented plate will convert linearly polarized light into circularly polarized light by slowing one of the linear components of the beam with respect to the other so that they are one quarter-wave out of phase. This will produce a beam of either left- or right-CPL (Fig. 4.48 and 4.49).

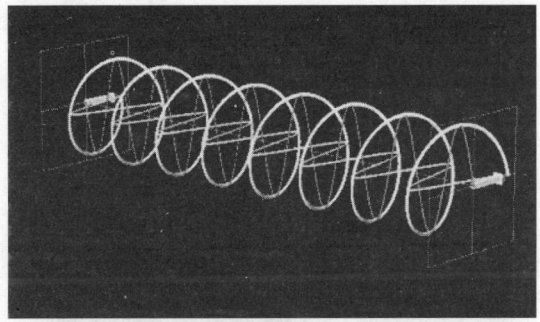

Fig. 4.48: Left Circularly polarized light

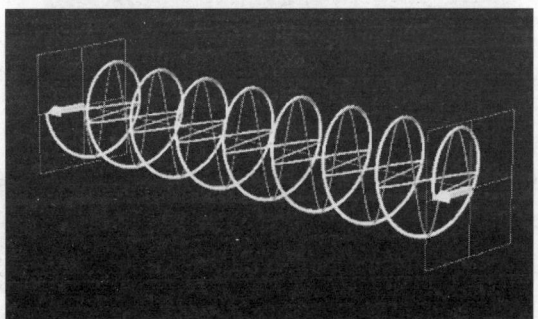

Fig. 4.49: Right circularly polarized light

4.12.1.2. Chiral molecules: the origin of optical activity

Chiral molecules exist as pairs of mirror image isomers. Mirror image isomers are not super-imposable and are known as enantiomers. The physical and chemical properties of a pair of enantiomers are identical with two exceptions: (a) the way that they interact with polarized light and (b) the way that they interact with other chiral molecules.

4.12.1.3Circular birefringence and optical rotation

Chiral molecules exhibit circular birefringence, which means that a solution of a chiral substance presents an anisotropic medium through which left circularly polarized (L-CPL) and right circularly polarized (R-CPL) propagate at different speeds. A linearly polarized wave can be thought of as the resultant of the superposition of two circularly polarized waves, one left-circularly polarized, the other right-circularly polarized. On traversing the circularly birefringent medium, the phase relationship between the circularly polarized wave's changes and the resultant linearly polarized wave rotates. This is the origin of the phenomenon known as optical rotation (fig.-4.50), which is measured using a polarimeter. Measuring optical rotation as a function of wavelength is termed optical rotatory dispersion (ORD) spectroscopy. Ex: 2-Butanol, 2-Chlorobutane.

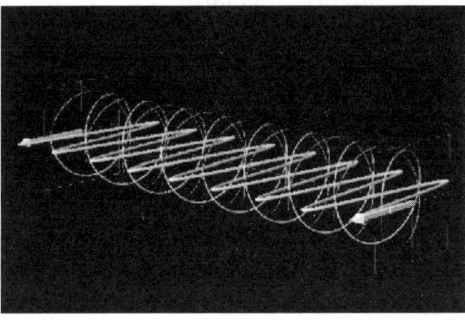

Fig. 4.50: Circular birefringence: the orange cuboid represents the sample

4.12.2. Circular Dichroism

Unlike optical rotation, circular dichroism only occurs at wavelengths of light that can be absorbed by a chiral molecule. At these wavelengths Left- and right-circularly polarized light will be absorbed to different extents. For instance, a chiral chromophore may absorb 90% of R-CPL and 88% of L-CPL. This effect is called circular dichroism and is the difference in absorption of L-CPL and R-CPL. Circular dichroism measured as a function of wavelength is termed circular dichroism (CD) spectroscopy and is the primary spectroscopic property measured by a circular dichroism spectrometer such as the Chirascan (Fig. 4.51). Optical rotation and circular dichroism stem from the same quantum mechanical phenomena and one can be derived mathematically from the other if all spectral information is provided.

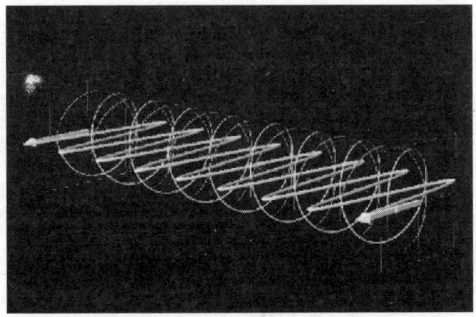

Fig. 4.51: Circular dichroism - the orange cuboids represents the sample

4.12.3 Principle or Operation

Circular dichroism (CD) is the difference in light absorbance between left- (L-CPL) and right-circularly polarized (R-CPL) light and circular dichroism spectrophotometers are highly specialized variations of the absorbance spectrophotometer. A circular dichroism spectrophotometer is also commonly termed a circular dichroism spectropolarimeter or a circular dichrograph.

Most modern circular dichroism instruments operate on the same principles there is a source of monochromatic linearly polarized light which can be turned into either left- or right-circularly polarized light by passing it through a quarter-wave plate whose unique axis is at 45 degrees to the linear polarization plane as described in the section about polarized light. Instead of a static quarter-wave plate, a circular dichroism spectrophotometer has a specialized optical element called a photo-elastic modulator (PEM). This is a piezoelectric element cemented to a block of fused silica. At rest, when the piezoelectric element is not oscillating, the silica block is not birefringent; when driven, the piezoelectric element oscillates at its resonance frequency (typically around 50 kHz), and induces stress in the silica in such a way that it becomes birefringent. The alternating stress turns the fused silica element into a dynamic quarter-wave plate, retarding first vertical with respect to horizontal components of the incident linearly polarized light by a quarter-wave and then vice versa, producing left- and then right-circularly polarized light at the drive frequency. The amplitude of the oscillation is tuned so that the retardation is appropriate for the wavelength of light passing through the silica block.

On the other side of the sample position there is a light detector. When there is no circularly dichroic sample in the light path, the light hitting the detector is constant. If there is a circularly dichroic sample in the light path, the recorded light intensity will be different for right- and left-CPL. Using a lock in amplifier tuned to the frequency of the PEM, it is possible to measure the difference in intensity between the two circular polarizations (vAC). The average total light intensity across many PEM oscillations (vDC) can be used to scale the size of the lock in amplifier signal to take into account variations in total light level. Both signals can be recorded and from them the circular dichroism signal can be calculated easily by dividing the vAC component by the vDC signal.

$$CD = \left(\frac{vAC}{vDC}\right) \cdot G$$ *eq.-4.24*

G is a calibration-scaling factor to provide either ellipticity or differential absorbance. The section about CD units and their inter conversion explains how ellipticity and differential absorbance are related.

4.12.4. Performance

The limit of detection of a circular dichroism (CD) spectrophotometer or any spectrophotometer is determined by its signal-to-noise (S/N) characteristics: the better S/N and the better its limit of detection. The signal-to-noise ratio is limited by photon shot noise, which is the statistical variation about an average in the number of photons per unit time detected by the light detector. The quantized nature of photons and their random arrival at the detector means that although the average number of photons detected per second may be say 5, the number in any particular one-second interval may be 0, 2, 7 or some other number. Thus, a measurement must be made over a sufficiently long period of time to determine the true average and the time taken to determine the true average will be inversely proportional to S/N. It is therefore important to design a circular dichroism spectrophotometer to maximize its S/N characteristics. A general relationship between the contributing factors to the signal-to-noise in an optical spectrometer can be written as:

$$\frac{Signal}{Noice} = \left(Q \cdot I \cdot t\right)^{1/2}$$ *... eq. 4.25*

Where

Q = Detector quantum efficiency

I = Light intensity

t = Time scale of the measurement.

From this, it is apparent there are three ways to improve the signal-to-noise of a circular dichroism spectrophotometer: increase the intensity of the incident linearly polarized monochromatic light, increase the quantum efficiency of the detector or spend more time collecting and averaging data points. The first two factors, light intensity and detector performance, are those that can be influenced by the design of a circular dichroism spectrophotometer and work together to lower the last factor, the time required to carry out a measurement. The higher the light throughput and better the detector efficiency, the less time it takes to collect quality data or, equally, the higher the quality of data that can be collected in time-limited experiments such as stopped-flow measurements.

Increasing the intensity of the incident light is the main avenue for increasing the performance of circular dichroism spectrophotometers and this finds its ultimate expression in the use of synchrotron light-sources for CD spectroscopy. Synchrotron facilities provide tremendous light intensity across a very wide spectral range of wavelengths but access to them is expensive and limited and their use is restricted to the more cutting-edge applications of circular dichroism. For the vast majority of CD experiments, a high-intensity bench-top source is the only practical option: Applied Photophysics Chirascan has been designed from the ground up to maximize the light throughput from its Xe arc-lamp source to the sample.

The Chirascan monochromator uses two synthetic, single-crystal quartz prisms instead of the diffraction gratings that most people are familiar with from normal absorbance spectrophotometers. Quartz prisms are more efficient than diffraction gratings for a very wide range of wavelengths, particularly in the UV.

4.12.5. Applications

1. Determining whether a protein is folded, and if so characterizing its secondary structure, tertiary structure and the structural family to which it belongs.

2. Comparing the structures of a protein obtained from different sources (Ex: species or expression systems) or comparing structures for different mutants of the same protein.

3. Studying the conformational stability of a protein under stress thermal stability, pH stability, and stability to denaturants and how this stability is altered by buffer composition or addition of stabilizers.

4. CD is excellent for finding solvent conditions that increase the melting temperature and/or the reversibility of thermal unfolding, conditions which generally enhance shelf life.

5. Determining whether protein-protein interactions alter the conformation of protein.

6. Determination of Protein Secondary Structure: Secondary structure can be determined by CD spectroscopy in the far-UV spectral region (190-250 nm). At these wavelengths the chromophore is the peptide bond and the signal arises when it is located in a regular, folded environment.

7. Protein Tertiary Structure: The CD spectrum of a protein in the near-UV spectral region (250-350 nm) can be sensitive to certain aspects of tertiary structure. At these wavelengths the chromophores are the aromatic amino acids and disulfide bonds, and the CD signals they produce are sensitive to the overall tertiary structure of the protein.

Signals in the region from 250-270 nm are attributable to phenylalanine residues, signals from 270-290 nm are attributable to tyrosine, and those from 280-300 nm are attributable to tryptophan. Disulfide bonds give rise to broad weak signals throughout the near-UV spectrum. If a protein retains secondary structure but no defined three dimensional structure *Eg.* an incorrectly folded or molten-globule structure, the signals in the near-UV region will be nearly zero. On the other hand, the presence of significant near-UV signals is a good indication that the protein is folded into a well defined structure. The near-UV CD spectrum can be sensitive to small changes in tertiary structure due to protein-protein interactions and/or changes in solvent conditions. The signal strength in the near-UV CD region is much weaker than that in the far-UV CD region. Near-UV CD spectra require about 1 ml of protein solution with an OD at 280 nm of 0.5 to 1.

4.13. SUMMARY

Radiation may be defined as energy traveling through space. Radiation sources are found in a wide range of occupational settings. If radiation is not properly controlled it can be potentially hazardous to the health of workers. Radiation can be classified into two types (1) Non-ionizing radiation - Lower frequency radiation, consisting of ultraviolet (UV), infrared (IR), microwave (MW), Radio Frequency (RF), and extremely low frequency (ELF). Ionizing radiation - All forms of ionizing radiation have sufficient energy to ionize atoms that may destabilize molecules within cells and lead to tissue damage. The higher frequencies of EM radiation, consisting of x-rays and gamma rays, are types of ionizing radiation.

The atoms of sub-atomic particles are protons, neutrons and electrons. The nucleus is at the centre of the atom and contains the protons and neutrons. Protons and neutrons are collectively known as nucleons. Atoms are electrically neutral and the positiveness of the protons is balanced by the negativeness of the electrons.

Electromagnetic waves are produced by the motion of electrically charged particles. These waves are also called electromagnetic radiation because they radiate from the electrically charged particles. As the name indicates, there are both electrical and magnetic components of the wave at right angles to each other. Electromagnetic radiation can be described in terms of a stream of photons, which are mass less particles each traveling in a wave like pattern and moving at the speed of light.

The electromagnetic (EM) spectrum is the range of all possible electromagnetic radiation frequencies. The electromagnetic spectrum of an object is the characteristic distribution of electromagnetic radiation from that particular object. The electromagnetic spectrum extends from below the frequencies used for modern radio at the long wavelength end through gamma radiation at the short wavelength end, covering wavelengths from thousands of kilometers down to a fraction the size of an atom.

The *Doppler Effect* refers to the apparent shift in the wavelength and frequency of a wave when there is relative motion between the source or emitter of the wave and an observer. Electromagnetic waves we use the term *Doppler shift* to describe this effect. If we obtain a spectrum from an object at rest to us then there is no Doppler shift in the spectrum. The spectral lines for the Balmer series, for instance, would appear at the same wavelengths as those from a hydrogen discharge tube in the laboratory or observatory.

Spectra types: Continuous spectra - For a continuous spectrum, the light is composed of a wide, continuous range of colors (energies).(2) Discrete spectra - With discrete spectra, one sees only bright or dark lines at very distinct and sharply defined colors (energies). Discrete spectra with bright lines are called emission spectra those with dark lines are termed absorption spectra.

Bathochromic shift is a change of spectral band position in the absorption, reflectance, transmittance, or emission spectrum of a molecule to a longer wavelength (lower frequency). Because the red color in the visible spectrum has a higher wavelength than most other colors, this effect is also commonly called a red shift, although this usage is considered informal and has no relation to Doppler shift or other wavelength independent meanings of red shift. This can occur because of a change in environmental conditions, a series of structurally related molecules in a substitution series can also show a bathochromic shift. Bathochromic shift is a phenomenon seen in molecular spectra not atomic spectra; it is thus more common to speak of the movement of the peaks in the spectrum rather than lines.

The broadening of spectral lines, particularly in white dwarfs caused by the pressure of the stellar atmosphere which in turn is caused by the high surface gravity of the star. Emission or absorption at a discrete wavelength or frequency caused by a specific electron transition within an atom, molecule or ion. The dark lines in an absorption spectrum and the bright lines that make up an emission spectrum are caused by the transfer of an electron from one energy level to another.

Spectroscopy is a technique that uses the interaction of energy with a sample to perform an analysis. The data that is obtained from spectroscopy is called a spectrum. A spectrum is a plot of the intensity of energy detected versus the wavelength or mass or momentum or frequency, etc. of the energy. Spectrum can be used to obtain information about atomic and molecular energy levels, molecular geometries, chemical bonds, interactions of molecules, and related processes. Spectroscopy is used to refer to the broad area of science dealing with the absorption, emission, or scattering of electromagnetic radiation by molecules, ions, atoms, or nuclei. Spectroscopic techniques are useful in determining both the identity of unknown substances and their concentration in solution.

Lambert's law: When a ray of monochromatic light passes through an absorbing medium its intensity decreases exponentially as the length of the absorbing medium increases.

Beer's law: When a ray of monochromatic light passes through an absorbing medium its intensity decreases exponentially as the concentration of the absorbing medium increases.

The fundamental concept of Beer-Lambert Law is that every photon of light striking the detector must have an equal chance of absorption. Thus, every photon must have the same absorption coefficient alpha, must pass through the same absorption path length (L) and the same absorber concentration (c), anything that upsets these conditions will lead to an apparent deviation from the Beer's law. Most common apparent deviations are due to (1) Chemical reasons arising when the absorbing compound, dissociates, associates or reacts with a solvent to produce a product having a different absorption spectrum. (2) The presence of stray radiation. (3) The polychromatic radiation.

All the monochromators contain component parts (a) an entrance slit (b) a collimating lens (c) a dispersing device (usually a prism or a grating) (d) a focusing lens (e) an exit slit . Polychromatic radiation (radiation of more than one wavelength) enters the monochromator through the entrance slit. The beam is collimated and then strikes the dispersing element at an angle. The beam is split into its component wavelengths by the grating or prism. By moving the dispersing element or the exit slit, radiation of only a particular wavelength leaves the monochromator through the exit slit.

When a beam of light is incident on certain substances, they emit visible light or radiations this phenomenon is known as fluorescence and the substances showing this phenomenon are known as fluorescent substances. Here higher energy radiation is absorbed by a species and stored for long enough that the molecules have collided and lost vibrational energy in collisions. When the excited molecule re-emits, the photon is of less energy depending on the species and is therefore of a lower frequency.

When light radiation is incident on certain substances, they emit light continuously even after the incident light is cut off this type of delayed fluorescence is called Phosphorescence and the substances are called Phosphorescent substances. Here light is absorbed by a molecule which becomes excited to a higher singlet state. The light can either be re-emitted straight away or very-rarely, the electron can switch to a triplet state, this is technically forbidden but can happen often enough through a process known as intersystem crossing. Transitions between different multiplicities are forbidden by quantum selection rules and as the ground state of the molecule will often be a single state (singlets), the molecule is not allowed to emit a photon that will get it directly to the ground state.

The greatest source of uncertainly in Atomic Absorption Spectroscopy is the behavior of the flame. Flame temperature depends upon the ratio of fuel and oxidant. Properties of flame are (i) **Base**-In this part, the sample enters in the form of minute droplets (ii) **Inner cone**-In this zone, some of the sample enters as solid vaporization and decomposition to the atomic sates. Absorption process also occurs in this part of zone. (iii) **Reaction zone**-The atoms are converted into oxides and then pass out to outer mantle.

Chiral molecules exist as pairs of mirror image isomers. Mirror image isomers are not super-imposable and are known as enantiomers. The physical and chemical properties of a pair of enantiomers are identical with two exceptions: (a) the way that they interact with polarized light and (b) the way that they interact with other chiral molecules.

Chiral molecules exhibit circular birefringence, which means that a solution of a chiral substance presents an anisotropic medium through which left circularly polarized (L-CPL) and right circularly polarized (R-CPL) propagate at different speeds. A linearly polarized wave can be thought of as the resultant of the superposition of two circularly polarized waves, one left-circularly polarized, the other right-circularly polarized.

Circular dichroism only occurs at wavelengths of light that can be absorbed by a chiral molecule. At these wavelengths Left-and right-circularly polarized light will be absorbed to different extents. For instance, a chiral chromophore may absorb 90% of R-CPL and 88% of L-CPL. This effect is called circular dichroism and is the difference in absorption of L-CPL and R-CPL. Circular dichroism measured as a function of wavelength is termed circular dichroism (CD) spectroscopy and is the primary spectroscopic property measured by a circular dichroism spectrometer such as the Chirascan.

Chapter 5
SPECTROSCOPY–II

5.1. INFRARED SPECTROSCOPY

5.1.1. Introduction

Infrared light lies between the visible and microwave portions of the electromagnetic spectrum. Infrared light has a range of wavelengths, just like visible light has wavelengths that range from red light to violet. Infra red covers the range of the electromagnetic spectrum between 0.78 and 1000 mm. The most useful I.R. region lies between $4000 - 670$ cm^{-1}. The frequency scale is given in units of reciprocal centimeters (cm^{-1}) rather than Hz, because the numbers are more manageable. The reciprocal centimeter is the number of wave cycles in one centimeter; whereas, frequency in cycles per second or Hz is equal to the number of wave cycles. Wavelength units are in micrometers, microns (µ), instead of nanometers for the same reason. Most infrared spectra are displayed on a linear frequency scale, but in some older texts a linear wavelength scale is used.

Infrared region divide into three sections (Fig. 5.1 and Table 5.1). (1) Near Infrared: Near infrared light is closest in wavelength to visible light of the electromagnetic spectrum and near infrared ones is the size of cells or are microscopic. Shorter, near infrared waves are not hot at all, in fact cannot even feel them. These shorter wavelengths are the ones used by TV's remote control. (2) Middle Infrared. (3) Far Infrared: far infrared is closer to the microwave region of the electromagnetic spectrum and far infrared wavelengths are about the size of a pin head. Far infrared waves are thermal and this type of infrared radiation every day in the form of heat. The heat that we feel from sunlight, a fire, a radiator or a warm sidewalk is infrared. Infrared light is even used to heat food sometimes special lamps that emit thermal infrared waves are often used in fast food restaurants.

Table 5.1: Sections of Infrared region

S. No.	Region	Wavelength range (mm)	Wave number range (cm⁻¹)
1.	Near	0.78 – 2.5	12800 – 4000
2.	Middle	2.5 – 50	4000 – 200
3.	Far	50 – 1000	200 – 10

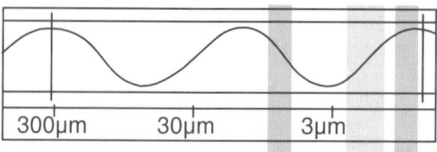

Fig. 5.1: Infrared Region of the Electromagnetic spectrum

Fig. 5.2: Thermal Infrared waves

Fig. 5.3: Near Infrared waves used by TV's remote control.

5.1.2. Principle

Infrared radiation does not have enough energy to induce electronic transitions as seen with Ultraviolet radiation. Absorption of Infrared is restricted to compounds with small energy differences in the possible vibrational and rotational states. For a molecule to absorb Infrared, the vibrations or rotations within a molecule must cause a net change in the dipole moment of the molecule. The alternating electrical field of the radiation interacts with fluctuations in the dipole moment of the molecule. If the frequency of the radiation matches the vibrational frequency of the molecule, then radiation will be absorbed, causing a change in the amplitude of molecular vibration.

5.1.2.1. Molecular rotations

Rotational transitions are of little use to the spectroscopic, rotational levels are quantized and absorption of IR by gases yields line spectra. However, in liquids or solids these lines broaden into a continuum due to molecular collisions and other interactions.

5.1.2.2. Molecular vibrations

The positions of atoms in a molecule are not fixed they are subject to a number of different vibrations. Vibrations fall into the two main categories (a) Stretching-Change in inter-atomic distance along bond axis (Fig. 5.4). (b) Bending-Change in angle between two bonds. These are again four types: (i) Rocking (ii) Scissoring (iii) Wagging (iv) Twisting (Fig. 5.5).

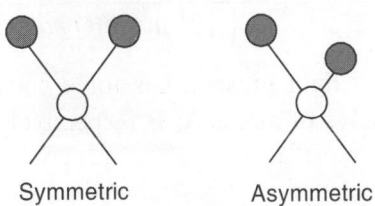

Symmetric   Asymmetric

Fig. 5.4: Stretching Vibrations

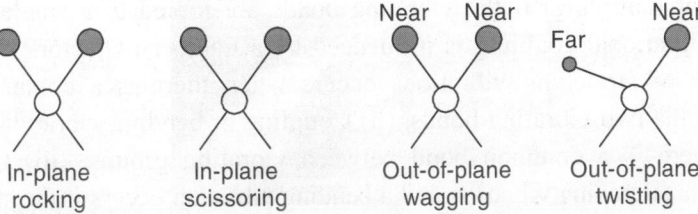

In-plane   In-plane   Out-of-plane   Out-of-plane
rocking   scissoring   wagging   twisting

Fig. 5.5: Bending Vibrations

The positions of atoms in molecules may be regarded as mean equilibrium positions and the bonds between atoms may be considered to springs, subject to stretching and bending. Each atom or group of atoms in a molecule swing about a point at which attraction of nuclei for electron balances, the repulsion of nuclei by nuclei and electrons carrying electric charges, the force required can be supplied by the swinging electric vector of an electromagnetic wave of frequency and phase which match those of a particular molecular vibration. Transfer of energy in this way is possible, if the vibration results in a change of molecules dipole movement, radiant energy is then absorbed and the intensity of radiation at this particular wavelength is decreased on passing through the compound. The intensity of absorption bonds depends upon the magnitude of the change in swinging dipole movement of the bonds during the transition and is directly proportional to the number of bonds in the molecular responsible for that particular absorption.

Hydrogen or carbon bonded to oxygen or nitrogen gas gives rise to strong infrared absorption, diatomic molecular (A-B) vibration that can occur in a periodic stretching, along the (A-B) bond. The masses of the two atoms and their connecting bond may be treated as two masses joined by a spring and Hooke's law may be applied. This leads to the expression for the frequency of vibration, $\bar{v}$ in wave number /cm.

$$\bar{v} = \frac{1}{2\prod c}\left(\frac{f}{mA.mB\,/\,mA+mB}\right)^{1/2} \quad \textbf{...\ eq 5.1}$$

Where C is Velocity of light (m/sec), f is force constant of the bonds (N/m), mA.mB is the masses of atoms A, B respectively and the value of *f* is *c*.

5.1.2.3. Vibration coupling

In addition to the vibrations mentioned above, interaction between vibrations can occur (coupling) if the vibrating bonds are joined to a single, central atom. Vibrational coupling is influenced by a number of factors (i) Strong coupling of stretching vibrations occurs when there is a common atom between the two vibrating bonds. (ii) Coupling of bending vibrations occurs when there is a common bond between vibrating groups. (iii) Coupling between a stretching vibration and a bending vibration occurs if the stretching bond is one side of an angle varied by bending vibration. (iv) Coupling is greatest when the coupled groups have approximately equal energies. (v) No coupling is seen between groups separated by two or more bonds.

5.1.2.4. Transitions in IR spectroscopy

In order for a mode to have an allowed IR transition, the mode must involve a change in the electric dipole moment of the molecule. Stated another way, if the mode is associated with a time varying dipole, the mode can interact with infrared electromagnetic energy to produce a vibrational energy transition. A permanent dipole moment is not required. Consider carbon dioxide, a molecule with no permanent electric dipole moment. There are 3 atoms ($N = 3$) and the molecule in its equilibrium geometry is linear. There are, 4 vibrational modes. The hypothetical non-oscillating molecule has no permanent dipole moment since the two opposed individual dipole vectors cancel. Since the dipole moment does not change during symmetric stretching, the symmetric stretch mode is inactive. Three modes do cause the dipole moment to change: two degenerate symmetric bends in plane and out-of-plane, and the asymmetric stretch. These modes, therefore, are infrared active.

5.1.3. Instrumentation

Infrared instrumentation is divided into two classes (a) Dispersive - The dispersive instruments use a prism or grating and are similar to ultraviolet resemblance dispersive spectrometer expect that in the infrared region different sources and detectors must be used. (b) Non-dispersive - Non-dispersive spectrometer may use interference, filters, laser sources or an interferometer in the very popular Fourier Transform Infrared Spectrometer (FTIR).

In the Fourier Transform Infrared Spectrometer (FTIR) the following components are present.

    i. Source

    ii. Interferometer

    iii. Detector

    iv. Back ground ferrogram

    v. Sample preparation

    vi. Laser

    vii. Optical bench

5.1.3.1. Source

The heated source emits infrared radiation. The radiation contains all the frequencies of interest but the intensity of frequencies is not all the same. Instead the intensities of the frequencies follow on energy distribution similar to the next one. The source in an FTIR is extremely stable. If allowed sufficient time heat up thoroughly it passes the same pattern of intensities over its defined frequency range, no matter when they are measured. In common with other types of absorption spectrometers, Infrared instruments require a source of radiant energy which provides a means for isolating narrow frequency band. The radiation source must emit IR radiation which must be intense enough for detection, steady and extend over the desired wavelengths. These radiations are continuous, only selected frequencies will be absorbed by the samples. The various sources of IR radiations are used in Fourier Transform Infrared Spectrometer (FTIR). (a) **Incandescent lamp:** In the Near Infrared instruments an ordinary incandescent lamp is generally used. (b) **Nernst glowers:** It consists of a hollow rod which is about 2 mm in diameter and 30 mm in length. The glower is composed of oxides such as Zirconia. Glower is generally heated to a temperature between 1000° to 1800° C. It provides maximum radiation at about 7100 /cm. The main disadvantage of Nernst glower is its mechanical failure. (c) **Globar source:** It is a rod of Silicon carbide which is about 50 mm in length and 4 mm in diameter. When it is heated to a temperature between 1300 and 1700° C, it strongly emits radiation in the IR region. It emits maximum radiation at 5200/cm and it is self starting.

5.1.3.2. Interferometer

The Infrared beam leaves the source and is deflected off a mirror the mirror directs the beam into the interferometer where the spectral encoding takes place. The path of Infrared beam as it travels through the interferometer. The Infrared radiation produced by the source is directed to a mirror which sends the beam into the interferometer. Inside the interferometer, the beam is immediately split into two paths by the beam splitter. Each path leads to a mirror, which sends the beam back to the beam splitter, where it is recombined. This configuration causes the beams to interfere with each other constantly as the beams are travels through the interferometer.

5.1.3.3. Detector

The infrared beam exits the interferometer and is deflected by a couple of mirrors before it reaches the detector, the detector produced an electrical signal is response to the encoded radiation striking it. The detector resorts the total intensity of infrared radiation reaching it across all frequency. This measurement is read many times per second to generate the interferogram.

In FTIR spectroscopy generally used detector is Photoconductivity cell. This is a non-thermal detector and have greater sensitivity. Photoconductivity cell has high sensitivity and good speed of response in Infrared detection, but it suffers from many practical disadvantages. When operated at room temperature, it has a very restricted range, usually limited to the near infrared. The range can be broadened by drastic cooling.

5.1.3.4. Interferogram

An interferogram is generated regarding the amount of radiation reaching the detector over time, we call this back ground interferogram because it shows the energy passing through the components of the optical bench. An interferogram can be generated regard less of whether a sample is present. As its name implies, the interferometer produces interference signals, which contain infrared spectral information generated after passing through a sample.

Most commonly used interferometer is a Michelson interferometer. It consists of three active components: (i) Moving mirror (ii) Fixed mirror (iii) Beam splitter. The two mirrors are perpendicular to each other (Fig. 5.6). The beam splitter is a semi reflecting device and is often made by depositing a thin film of Germanium onto a flat KBr substrate. Radiation from the broadband IR source is collimated and directed into the inter-

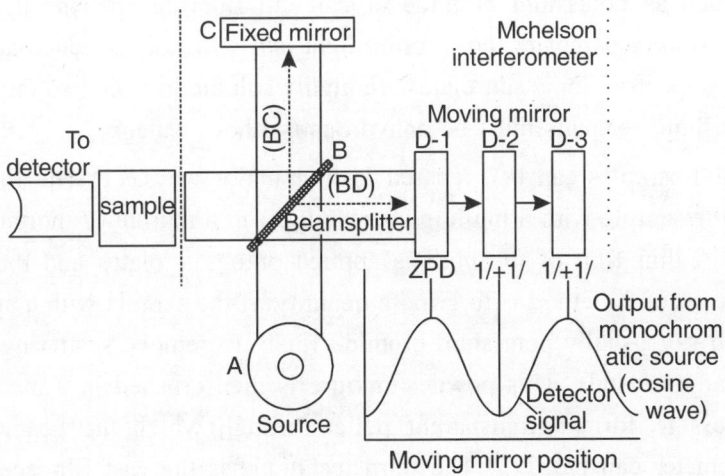

Fig. 5.6: Michelson interferometer of a typical FTIR spectroscopy

ferometer and impinges on the beam splitter. At the beam splitter half the IR beams is transmitted to the fixed mirror and the remaining half is reflected to the moving mirror. After the divided beams are reflected from the two mirrors, they are recombined at the beam splitter. Due to changes in the relative position of the moving mirror to the fixed mirror, an interference pattern is generated. The resulting beam then passes through the sample and is eventually focused on the detector (Fig. 5.6).

The interferogram contains information over the entire IR region to which the detector is responsive. A mathematical operation known as Fourier transformation converts the interferogram (a time domain spectrum displaying intensity versus time within the mirror scan) to the final IR spectrum, which is the familiar frequency domain spectrum showing intensity versus frequency. This also explains how the term Fourier transform infrared spectrometry is created.

5.1.3.5. Sample Preparation

Gaseous samples require little preparation beyond purification, but a sample cell with a long path length (5-10 cm) is normally needed as gases show relatively weak absorbances.

Liquid samples can be sandwiched between two plates of a high purity salt commonly sodium chloride or common salt, although a number of other salts such as potassium bromide or calcium fluoride are also used. The plates are transparent to the infrared light and will not introduce any lines onto the spectra. Some salt plates are highly soluble in water, so the sample and washing reagents must be anhydrous (without water).

Solid samples can be prepared in four major ways. (1) The first is to crush the sample with a mulling agent (nujol) in a marble or mortar with a pestle. A thin film of the mull is applied onto salt plates and measured. (2) The second method is to grind a quantity of the sample with a specially purified salt usually potassium bromide finely to remove scattering effects from large crystals. This powder mixture is then crushed in a mechanical die press to form a transparent pellet through which the beam of the spectrometer can pass. (3) The third technique is the cast film technique, which is used mainly for polymeric materials. The sample is first dissolved in a suitable non hygroscopic solvent. A drop of this solution is deposited on surface of KBr or NaCl cell. The solution is then evaporated to dryness and the film formed on the cell is analysed directly. Care is important to ensure that the film is not too thick otherwise light cannot pass through. This technique is suitable for qualitative analysis. (4) The final method is to use microtomy to cut a thin (20-100 μm) film from a solid sample. This is one of the most important ways of analyzing failed plastic products because the integrity of the solid is preserved.

It is important to note that spectra obtained from different sample preparation methods will look slightly different from each other due to differences in the samples physical states. There are many techniques for the measurement of IR spectra. The basic ones utilize the fact that the alkali salts are transparent in large portions of the IR part of the electromagnetic spectrum. Commonly used techniques for preparation of samples in FTIR spectroscopy are (a) **KBr pellet-** Good for powders, a few milligrams of the sample powder and an excess of KBr are finely ground and pressed under high pressure into a pellet. This is a useful and very general method for solids. (b) **Salt cells-** Good for organic liquids, the liquid is placed into a reservoir milled in alkali salt windows. (c) **Nujol mull-** The material of interest is suspended in oil, such as mineral oil and the resulting paste is spread thinly on a salt window to form a film. This is a good technique for oils and waxy solids that do not press well into pellets.

5.1.3.6. Laser

The laser produces a single frequency of red light that follows the same path as the Infrared radiation. The laser calibrates the instrument internally. A laser is an intense beam of light energy that is of one specific frequency. Lasers are created by passing a beam of light energy through molecules that have been artificially excited. The beam picks up energy from the charged molecules, which increased the beams intensity. The intensity of laser in optical bench, in low compared with lasers that are used for surgery. The frequency of the laser beam is known to great accuracy. As the laser beam passes thorough the interferometer and beam splitter, it undergoes the same interference that occurs with the IR beam.

5.1.3.7. *Optical bench*

The information contained in the background and sample interferograms is transmitted to the computer where further processing takes to produce the spectrum.

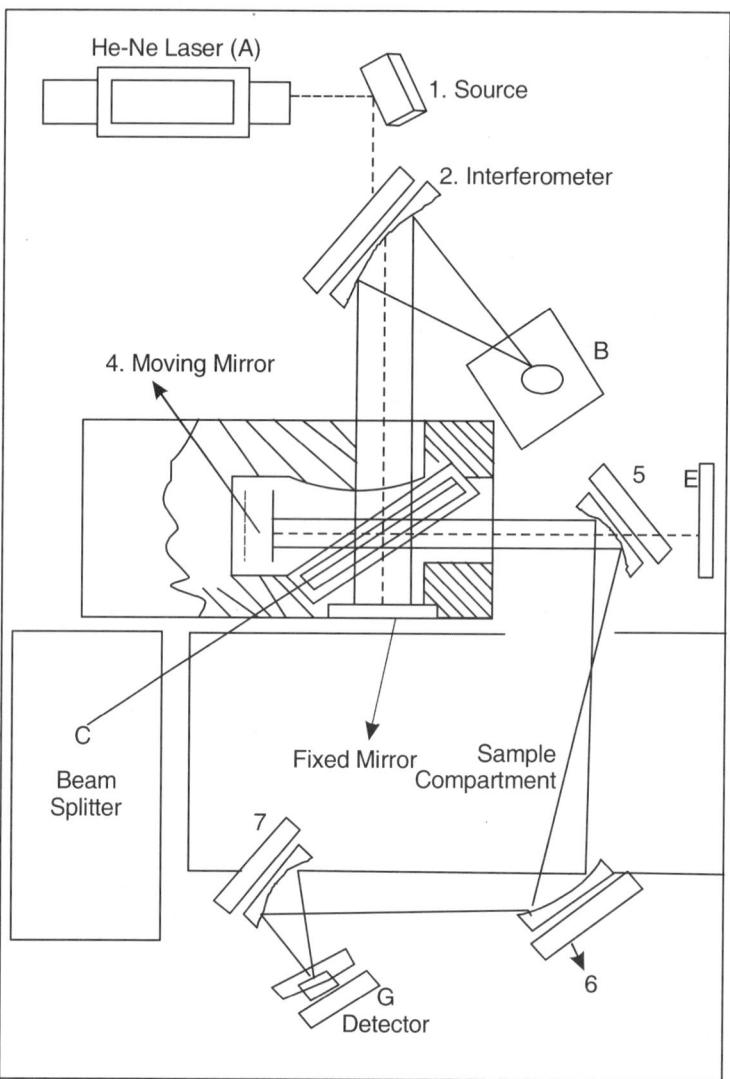

Fig. 5.7: FTIR optical path diagram

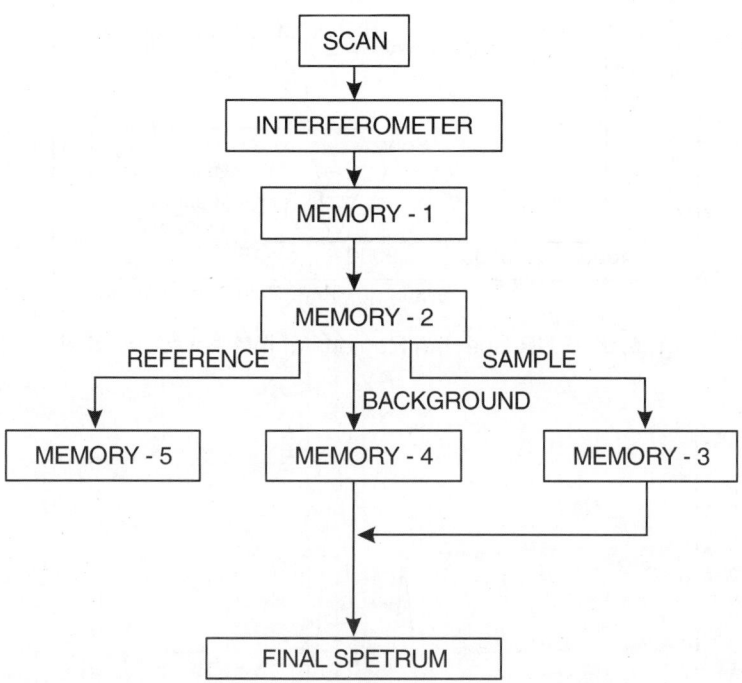

Fig. 5.8: Diagram of the FTIR instruments' function

5.1.4. Applications

1. FT-IR gas analyzer is used for detection of pollutants like CO, NOx, HC, NH_3, HCN, N_2O, detection limits less than 1 ppm and resolution is 0.09 cm^{-1}.

2. Approaches for lubricant QC total process checking: (i) Incoming ingredient testing -Base oil and add-pack lot consistency (ii) Blended product analysis - Additive levels (iii) Outgoing product verification-Assure product correct for shipment (iv) Used Oil Analysis - Engine Monitoring (v) Edible Oil Analysis - Product Consistency.

3. FTIR Spectroscopy is used for Integra oil analysis in used oils (Fig. 5.10).

4. FTIR spectroscopy is used to study of the melting process of polyethylene at ambient conditions (Fig. 5.11).

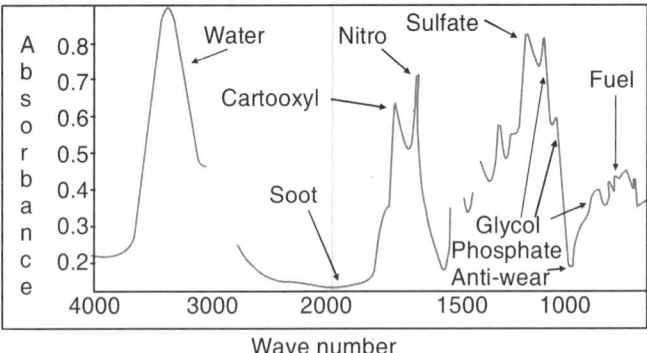

Fig. 5.10: FTIR Spectral Regions of Interest for used oil

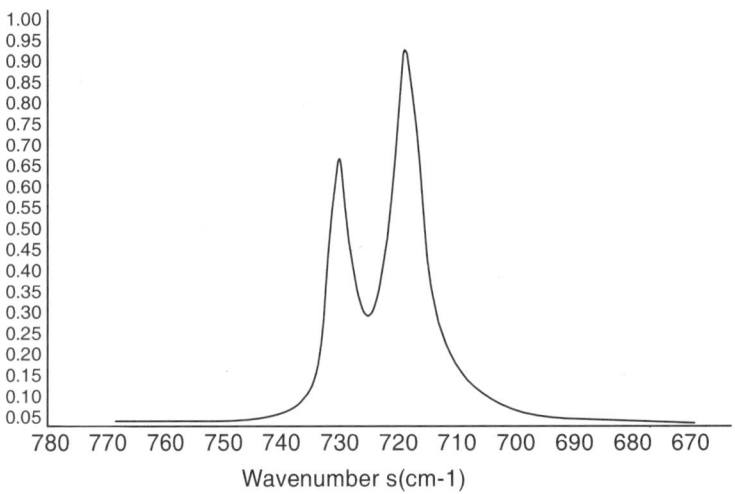

Fig. 5.11: IR Polyethylene at ambient conditions.

5. It is used for analysis of tire contaminant bloom qualitative Identifications of a contaminant creating a white bloom on the surface of finished tires.

6. Organic functional groups (atom groups bonded in particular ways) differ both in the strengths of the bond(s) and in the masses of the atoms involved. For instance, the O–H and C = O functional groups each contain atoms of different masses connected by bonds of different strengths. The O–H and C = O groups to absorb IR radiation at different positions in the spectrum. The presence of a strong, broad band between 3200 and 3400 cm^{-1} indicates the

presence of an O–H group in the molecule, while the presence of a strong band around 1700 cm^{-1} confirms the presence of a C=O group. For organic molecules, the infrared spectrum can be divided into three regions. Absorptions between 4000 and 1300 cm^{-1} are primarily due to specific functional groups and bond types. Those between 1300 and 909 cm^{-1}, the fingerprint region, are primarily due to more complex vibrational motions and those between 909 and 650 cm^{-1} are usually associated with the presence of benzene rings in the molecule.

7. Many inorganic compounds are in reality largely organic, and for these we look for the same functional group bands in the IR as we do for purely organic compounds. However, the infrared spectra of relatively simple, purely inorganic compounds are containing only a few atoms specifically, inorganic salts containing polyatomic ions are quite distinctive and can be used to rapidly identify the ions.

Inorganic salt, such as KNO_2, on the basis of the empirical formula, we might naively expect there to be a total of 3(4) − 6 = 6 normal modes of vibration associated with this material. However, this assumes that KNO_2 is covalent. In fact, KNO_2 consists of an ionic lattice of K^+ and NO_2^- ions arranged in an infinite and very regular array. The crystal consists of essentially isolated K^+ ions and NO_2^- ions. Thus we are able to consider the vibrational modes of the cation and anion independently of one another. In this case, since the potassium ions are monatomic, they have no vibrations (3(1) − 3 = 0), so we need only consider the nitrite anions.

5.2. MASS SPECTROSCOPY

5.2.1. History

In 1886, Eugen Goldstein observed rays in gas discharges under low pressure that travelled through the channels in a perforated cathode toward the anode, in the opposite direction to the negatively charged cathode rays. Goldstein called these positively charged anode rays Kanalstrahlen the standard translation of this term into English is canal rays.

Wilhelm Wien found that strong electric or magnetic fields deflected the canal rays in 1899, constructed a device with parallel electric and magnetic fields that separated the positive rays according to their charge-to-mass ratio (Q/m). Wien found that the charge-to-mass ratio depended on the nature of the gas in the discharge tube. English scientist J.J. Thomson later improved on the work of Wien by reducing the pressure to create a mass spectrograph.

Some of the modern techniques of mass spectrometry were devised by Arthur Jeffrey Dempster and F.W. Aston in 1918 and 1919 respectively. Francis William Aston won the 1922 Nobel Prize in Chemistry for his work in mass spectrometry. In 2002, the Nobel Prize in Chemistry was awarded to John Bennett Fenn for the development of electrospray ionization (ESI) and Koichi Tanaka for the development of soft laser desorption (SLD) in 1987. However earlier, matrix-assisted laser desorption/ionization (MALDI), was developed by Franz Hillenkamp and Michael Karas; this technique has been widely used for protein analysis.

5.2.2. Principle

Mass spectrometry is an analytical tool used for measuring the **molecular mass** of a sample. The Mass spectrometry (MS) principle consists of ionizing chemical compounds to generate charged molecules or molecule fragments and measurement of their mass-to-charge ratios. In a typical MS procedure, a sample is loaded onto the MS instrument, and its compounds are ionized by different methods by impacting them with an electron beam, resulting in the formation of charged particles (ions). The mass-to-charge ratio of the particles is then calculated from the motion of the ions as they transit through electromagnetic fields.

Mass spectrometry is based on slightly different principles to the other spectroscopic methods. The physics behind mass spectrometry is that a charged particle passing through a magnetic field is deflected along a circular path on a radius that is proportional to the mass to charge ratio, m/e. In an electron impact mass spectrometer, a high energy beam of electrons is used to displace an electron from the organic molecule to form a radical cation known as the molecular ion. If the molecular ion is too unstable then it can fragment to give other smaller ions (Fig. 5.12). The collection of ions is then focused into a beam and accelerated into the magnetic field and deflected along circular paths according to the masses of the ions. By adjusting the magnetic field, the ions can be focused on the detector and recorded. The most useful information should be able to obtain from a MS spectrum is the molecular weight of the sample. This will often be the heaviest ion observed from the sample provided this ion is stable enough to be observed.

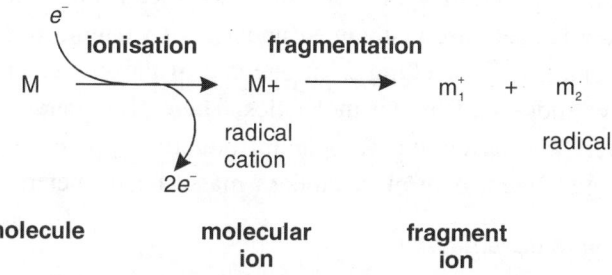

Fig. 5.12: Ionization process

5.2.3. Instrumentation

Mass spectrometers can be divided into three fundamental parts (Fig. 5.13):
(a) **Ionization source-**The sample has to be introduced into the ionization
source of the instrument. (b) **Analyzer-**Once inside the ionization source,
the sample molecules are ionized, these ions are extracted into the analyser
region of the mass spectrometer where they are separated according to their
mass(m)-to-charge (z) ratios (m/z). (c) **Detector-**The separated ions are
detected and this signal sent to a data system where the m/z ratios are
stored together with their relative abundance for presentation in the format
of a m/z spectrum.

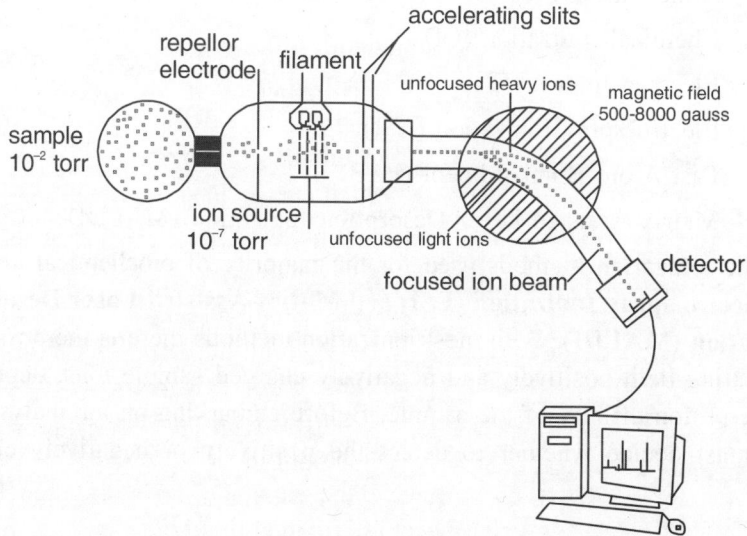

Fig. 5.13: Fundamental Parts of the Mass Spectrometry

The analyser and detector of the mass spectrometer, and often the ionization source too, are maintained under high vacuum to give the ions a reasonable chance of travelling from one end of the instrument to the other without any hindrance from air molecules. The entire operation of the mass spectrometer, and often the sample introduction process also, is under complete data system control in modern mass spectrometers.

5.2.3.1. Sample introduction

The method of sample introduction to the ionization source often depends on the ionization method being used, as well as the type and complexity of the sample. The sample can be inserted directly into the ionization source or can undergo some type of chromatography in route to the ionization source. This latter method of sample introduction usually involves the mass spectrometer being coupled directly to a high pressure liquid chromatography (HPLC), gas chromatography (GC) or capillary electrophoresis (CE) separation column hence the sample is separated into a series of components which then enter the mass spectrometer sequentially for individual analysis.

Many ionization methods are available and each has its own advantages and disadvantages. The ionization method should depend on the type of sample under investigation and the mass spectrometer available. *Ionization methods include the following:*

1.   Atmospheric Pressure Chemical Ionization (APCI)
2.   Chemical Ionization (CI)
3.   Electron Impact (EI)
4.   Electro spray Ionization (ESI)
5.   Fast Atom Bombardment (FAB)
6.   Matrix Assisted Laser Desorption Ionization (MALDI)

The ionization methods used for the majority of biochemical analyses are **Electro spray Ionization (ESI)** and **Matrix Assisted Laser Desorption Ionization (MALDI).** With most ionization methods there is the possibility of creating both positively and negatively charged sample ions, depending on the proton affinity of the sample. Before embarking on an analysis, the user must decide whether to detect the positively or negatively charged ions.

5.2.3.2.1. Atmospheric pressure chemical ionization (APCI)

Atmospheric pressure chemical ionization (APCI) is an analogous ionization method to chemical ionization (CI). APCI is not suitable for the analysis of thermally labile compounds. The general source set up (Fig. 5.14) shares a strong resemblance to electro spray ionization (ESI). In APCI, the analyte solution is introduced into a pneumatic nebulizer and desolvated in a heated quartz tube before interacting with the corona discharge creating ions.

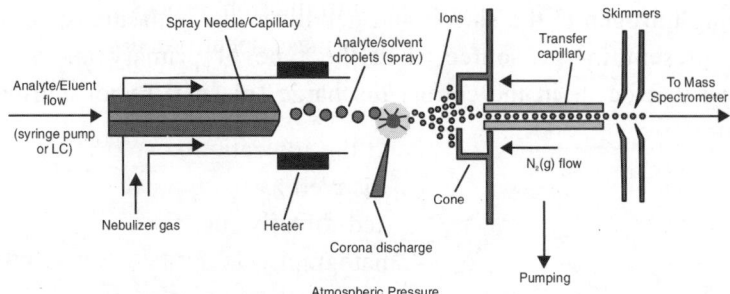

Fig. 5.14: A schematic of the components of an APCI source

The corona discharge replaces the electron filament in CI, the atmospheric pressure would quickly burn out any filaments, and produces primary $N_2^{\circ+}$ and $N_4^{\circ+}$ by electron ionization. These primary ions collide with the vaporized solvent molecules to form secondary reactant gas ions -

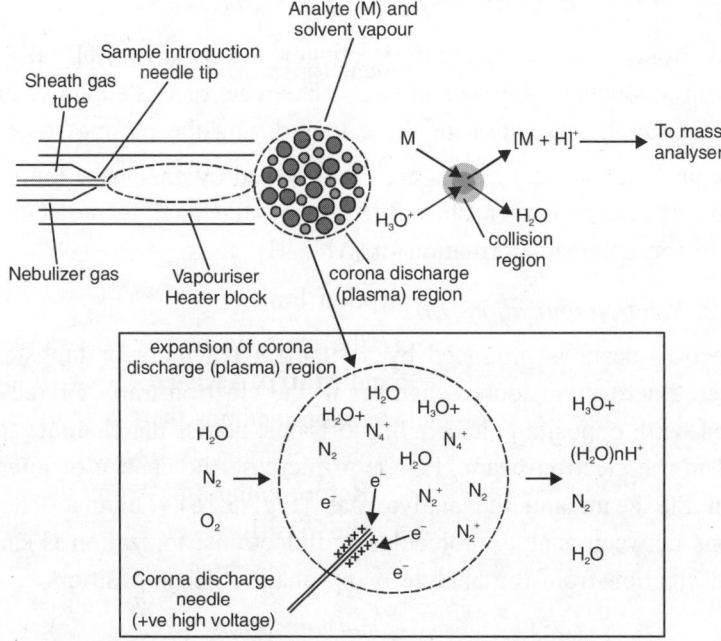

Fig. 5.15: A more detailed view of the mechanism of APCI

e.g. H_3O^+ and $(H_2O)_n H^+$ (Fig. 5.15). These reactant gas ions then undergo repeated collisions with the analyte resulting in the formation of analyte ions. Once the ions are formed, they enter the pumping and focusing stage.

Reactions in the plasma region:

Assuming nitrogen is the sheath and nebulizer gas with atmospheric water vapour present in the source, then the type of primary and secondary reactions that occur in the corona discharge (plasma) region during APCI are as follows:

$$N_2 + e \rightarrow N_2^+ + 2e$$

$$N_2^+ + 2N_2 \rightarrow N_4^+ + N_2$$

$$N_4^+ + H_2O \rightarrow H_2O^+ + 2N_2$$

$$N_4^+ + H_2O \rightarrow H_2O^+ + 2N_2$$

$$H_2O^+ + H_2O^+ \rightarrow H_3O^+ + OH$$

$$H_3O^+ + H_2O + N_2 \rightarrow H^+(H_2O)_2 + N_2$$

$$H^+(H_2O)_{n-1} + H_2O + N_2 \rightarrow H^+(H_2O)_n + N_2$$

The most abundant secondary cluster ion is $(H_2O)_2H^+$ along with significant amounts $(H_2O)_3H^+$ and H_3O^+. The reactions listed above are ways to account for the formation of these ions during the plasma stage.

The protonated analyte ions are then formed by gas-phase ion-molecule reactions of these charger cluster ions with the analyte molecules. This results in the abundant formation of $[M + H]^+$ ions.

5.2.3.2.2. Electron ionization (EI)

The electron beam is produced by a filament (rhenium or tungsten wire) and steered across the source chamber to the electron trap. A fixed magnet is placed, with opposite poles slightly off-axis, across the chamber to create a spiral in the electron beam. This is to increase the chance of interactions between the beam and the analyte gas (Fig. 5.16). There are no actual collisions between analyte molecules and electrons; ionization is caused by electron ejection from the analyte or by analyte decomposition.

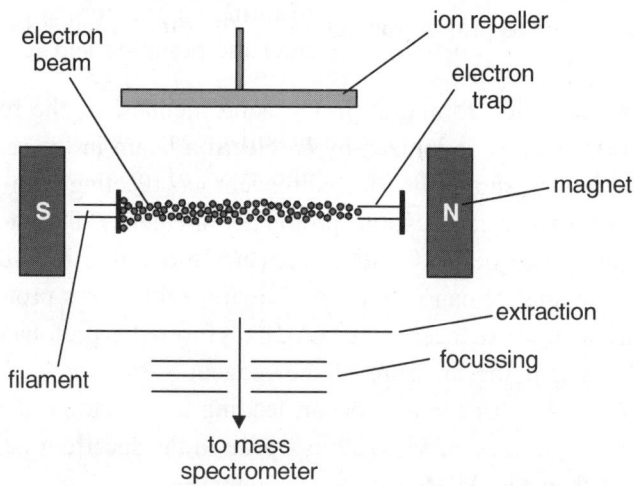

Fig. 5.16: Schematic side-view of an EI source

Fig. 5.17 shows some of the processes that can occur during the EI process. Consider the analyte molecule AB. The first two processes that might occur are the direct result of energy transfer from the electron beam to the analyte, causing primary fragmentation and the second main cause of fragment ions in the spectrum. The third process is electron ejection from the analyte to create the energized radical ion. This can then either lose energy through ion cooling and stabilize or lose energy through secondary fragmenting the third cause of fragment ions in the mass spectrum.

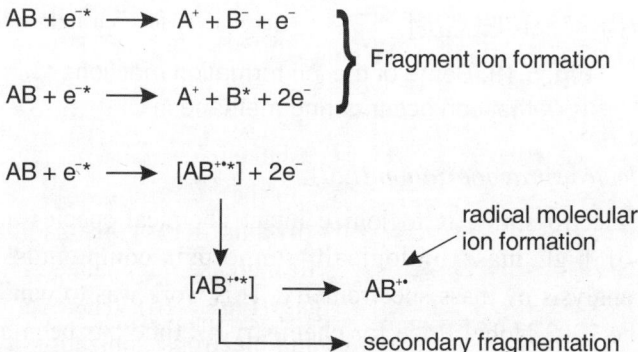

Fig. 5.17: Some of the ion formation reactions that can occur in EI

5.2.3.2.3. Chemical ionization (CI)

Chemical ionization is a lower energy alternative to EI for volatile analytes.

In CI, there is a reagent gas (ammonia or methane) in the ion chamber. In CI, ionization is due to proton transfer and is therefore a much lower energy process.

Fig. 5.18 shows ion formation in CI using methane as the reagent gas. In equation (a), methane is ionized by an electron beam in the same way as with EI. Equation (b) shows the ionized reagent gas reacting with un-ionized reagent gas to form the carbocation (protonated methane). This step requires the CI reagent gas to be at a critical pressure too low a pressure and no ionization of the analyte can take place. Equation (c) shows proton transfer from the carbocation to the analyte (AB) to form the protonated analyte molecule (ABH⁺). If the pressure of the reagent gas is too high, then the side reactions (d) and (e) can also occur, leading to formation of the analyte adduct ion, this is seen as an M_{AB}+29 m/z peak in the spectrum i.e. occurring 28 m/z higher than the ABH⁺.

(a) $CH_4 + e^{-*} \rightarrow CH_4^{+*} + 2e^-$ methane molecular ion formation

(b) $CH_4^{+*} + CH_4 \rightarrow CH_5^+ + CH_3^*$ carbocation formation

(c) $CH_5^+ + AB \rightarrow CH_4 + [AB + H]^+$ protonated analyte formation

(d) $CH_5^+ + AB \rightarrow CH_4 + [AB + H]^+$ alternative carbocation formation

(e) $CH_3^+ + M \rightarrow CH_4 + [AB - H]^+$ alternative analyte ion formation

(f) $CH_3^+ + M \rightarrow CH_4 + [AB - H]^+$ side reaction carbocation formation

(g) $C_2H_5^+ + AB \rightarrow [AB + C_2H_5]^+$ analyte adduct ion formation

Fig. 5.18: Some of the ion formation reactions
that can occur during methane in CI

5.2.3.2.4. Electro spray ionization (ESI)

The use of electro spray, is to ionize intact chemical species and for the ionization of high mass biologically important compounds and their subsequent analysis by mass spectrometry. This work was to win John Fenn a share of the 2002 Nobel Prize for chemistry 4th time has been awarded to mass spectrometry pioneers.

The analyte is introduced to the source in solution either from a syringe pump or as the eluent flow. Flow rates are typically of the order of 1 µl min⁻¹. The analyte solution flow passes through the electrospray needle that has a high potential difference applied to it. This forces the spraying of charged droplets from the needle with a surface charge of the same polarity to the charge on the needle. The droplets are repelled from the needle towards the source sampling cone on the counter electrode. As the droplets traverse the space between the needle tip and the cone and solvent evaporation occurs (Fig. 5.19).

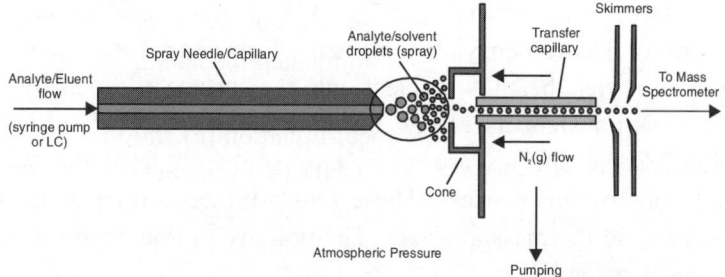

Fig. 5.19: A schematic of an *Electro spray ionization (*ESI) source

As the solvent evaporation occurs, the droplet shrinks until it reaches the point that the surface tension can no longer sustain the charge (Rayleigh limit) at which point a Coulombic explosion occurs and the droplet is ripped apart. This produces smaller droplets that can repeat the process as well as naked charged analyte molecules. These charged analyte molecules can be singly or multiply charged (Fig. 5.20). This is a very soft method of ionization as very little residual energy is retained by the analyte upon ionization. This is why ESI-MS is an important technique in biological studies where the analyst often requires that non-covalent molecule-protein

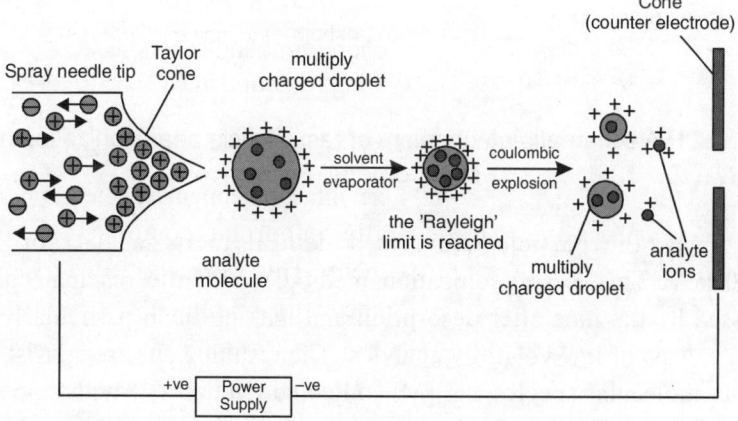

Fig. 5.20: A schematic of the mechanism of ion formation in ESI

or protein-protein interactions are representatively transferred into the gas phase.

5.2.3.2.5. Fast atom bombardment (FAB)

The technique of FAB is concept and design, in involve the bombardment of a solid spot of the analyte or matrix mixture on the end of a sample probe by a fast particle beam (Fig. 5.21). The matrix (a small organic species like glycerol or 3-nitro benzylalcohol) is used to keep a homogenous sample surface. The particle beam is incident onto the surface of the analyte/matrix spot, where it transfers its energy bringing about localized collisions and disruptions. Some species are ejected or sputtered from the surface as secondary ions by this process. These ions are then extracted and focused before passing to the mass analyzer. The polarity of ions produced depends on the source potentials.

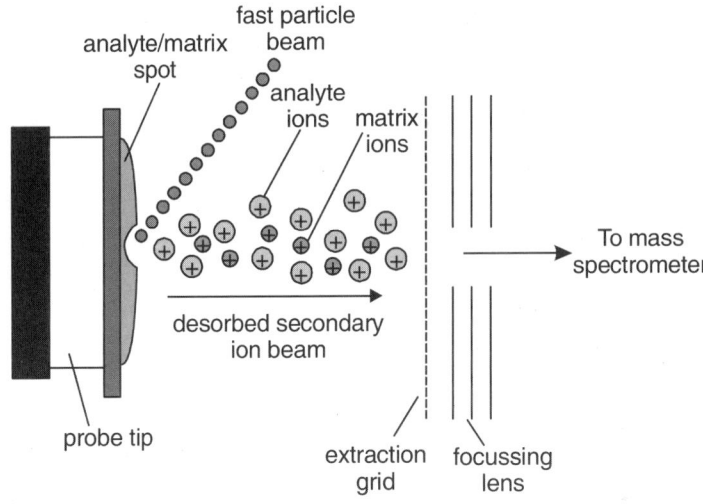

Fig. 5.21: A schematic mechanism of fast particle beam ionization mass spectrometry (FAB).

In FAB, the particle beam is a neutral inert gas (Ar or Xe) at 4 – 10 keV. This is soft ionization method, very little residual energy is possessed by the ions after desorption and making them particularly suited to the analysis of low volatility analytes. The resulting spectra consist largely of intact molecular species Eg. $[M + H]^+$ and $[M + Na]^+$) with some minor structural fragmentation. The low mass region of the spectra is, however, dominated by matrix and matrix/salt cluster ions.

5.2.3.2.6 *Matrix-assisted laser desorption/ionization (MALDI)*

A substantial burst of ions was produced with laser pulse. An unexpected side effect of the matrix was that it allowed for the laser incidence spot to be refreshed between each pulse, thus greatly enhancing shot-to-shot reproducibility. This was the foundation of matrix-assisted laser desorption/ ionization (MALDI). The application of MALDI is to a whole range of biological macromolecules.

The mechanism of MALDI is consisting of three basic steps: **(1) Formation of a Solid Solution:** It is essential for the matrix to be in access leading to the analyte molecules being completely isolated from each other. This ease the formation of the homogenous solid solution required to produce a stable desorption of the analyte. **(2) Matrix Excitation:** The laser beam is focused onto the surface of the matrix-analyte solid solution. The chromaphore of the matrix couples with the laser frequency causing rapid vibrational excitation, bringing about localized disintegration of the solid solution. The clusters ejected from the surface consist of analyte molecules surrounded by matrix and salt ions. The matrix molecules evaporate away from the clusters to leave the free analyte in the gas-phase. **(3) Analyte Ionization:** The photo-excited matrix molecules are stabilized through proton transfer to the analyte. Cation attachment to the analyte is also encouraged during this process in this way that the characteristic $[M + X]^+$ (X= H, Na, K etc.) analyte ions are formed. This ionization

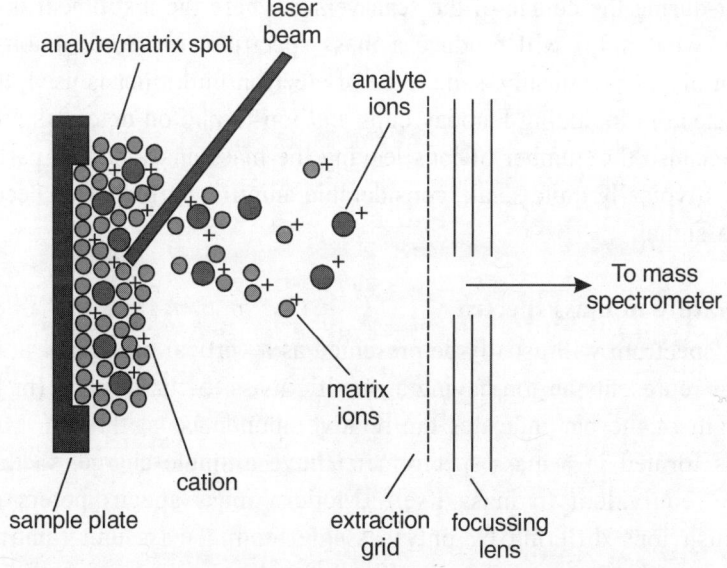

Fig. 5.22: A schematic diagram of the mechanism of MALDI

reaction takes place in the desorbed matrix-analyte cloud just above the surface. The ions are then extracted into the mass spectrometer for analysis.

5.2.3.3. Mass analyzer

Mass analyzers separate the ions according to their mass-to-charge ratio. The following two analyzers govern the dynamics of charged particles in electric and magnetic fields in vacuum. (1) Sector: A sector field mass analyzer uses an electric and/or magnetic field to affect the path and/or velocity of the charged particles in some way. Sector instruments bend the trajectories of the ions as they pass through the mass analyzer, according to their mass-to-charge ratios, deflecting the more charged and faster-moving, lighter ions more (Fig. 5.22). (2) Time of Flight: The time-of-flight (TOF) analyzer uses an electric field to accelerate the ions through the same potential and then measures the time they take to reach the detector. If the particles all have the same charge, the kinetic energies will be identical and their velocities will depend only on their masses. Lighter ions will reach the detector first.

5.2.3.4. Detector

The final element of the mass spectrometer is the detector. The detector records either the charge induced or the current produced when an ion passes by or hits a surface. In a scanning instrument, the signal produced in the detector during the course of the scan versus where the instrument is in the scan (at what m/Q) will produce a mass spectrum, a record of ions as a function of m/Q. Typically, some type of electron multiplier is used, though other detectors including Faraday cups and ion-to-photon detectors are also used. Because the number of ions leaving the mass analyzer at a particular instant is typically quite small, considerable amplification is often necessary to get a signal.

5.2.4. Nature of mass spectra

A mass spectrum will usually be presented as a vertical bar graph, in which each bar represents an ion having a specific mass-to-charge ratio (m/z) and the length of the bar indicates the relative abundance of the ion. Most of the ions formed in a mass spectrometer have a single charge, so the m/z value is equivalent to mass itself. Modern mass spectrometers easily distinguish ions differing by only a single atomic mass unit (amu), thus provide completely accurate values for the molecular mass of a compound. The highest mass ion in a spectrum is normally considered to be the molecular ion and lower mass ions are fragments from the molecular ion, assuming the sample is a single pure compound. The following Fig. 5.23 displays the mass spectra of three simple gaseous compounds, carbon dioxide, propane and cyclopropane. The molecules of these compounds are similar in size, CO and C_3H_8 both have a nominal mass of 44 amu and C_3H_6 has a mass of 42 amu. The molecular ion is the strongest ion in the spectra of CO_2 and C_3H_6 and it is moderately strong in propane. The unit mass resolution is readily apparent in these spectra and the separation of ions having m/z = 39, 40, 41 and 42 in the cyclopropane spectrum. Even though these compounds are very similar in size, it is a simple matter to identify them from their individual mass spectra.

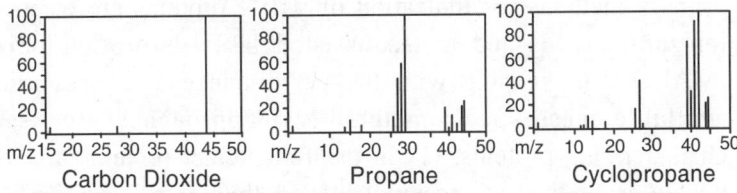

Fig. 5.23: Carbon Dioxide, Propane and Cyclopropane Mass spectrua

Since a molecule of carbon dioxide is composed of only three atoms, its mass spectrum is very simple. The molecular ion is also the base peak and the only fragment ions are CO (m/z = 28) and O (m/z = 16). The molecular ion of propane also has m/z = 44, but it is not the most abundant ion in the spectrum. Cleavage of a carbon-carbon bond is gives methyl and ethyl fragments, one of which is a carbocation and the other a radical. Both distributions are observed, but the larger ethyl cation (m/z = 29) is the most abundant, possibly because its size affords greater charge dispersal. A similar bond cleavage in cyclopropane does not give two fragments, so the molecular ion is stronger than in propane, and is in fact responsible for the base peak. Loss of a hydrogen atom, either before or after ring opening, produces the stable allyl cation (m/z = 41). The third strongest ion in the spectrum has m/z = 39 (C_3H_3). Its structure is uncertain, but two possibilities are shown in Fig. 5.23. The small m/z = 39 ion in propane and the absence of m/z = 29 ion in cyclopropane are particularly significant in distinguishing these hydrocarbons.

5.2.5. Applications

1. **Isotope dating and tracking:** Mass spectrometer is to determine the $^{16}O/^{18}O$ and $^{12}C/^{13}C$ isotope ratio on biogenous carbonate. Mass

spectrometry is also used to determine the isotopic composition of elements within a sample. Differences in mass among isotopes of an element are very small and the less abundant isotopes of an element are typically very rare, so Mass spectrometer is used to study such kind of differences.

2. **Pharmacokinetics:** Pharmacokinetics is often studied using mass spectrometry because of the complex nature of the matrix often blood or urine, and the need for high sensitivity to observe low dose and long time point data.

3. **Protein characterization:** Mass spectrometry is an important emerging method for the characterization of proteins. The two primary methods for ionization of whole proteins are electrospray ionization (ESI) and matrix-assisted laser desorption/ionization (MALDI). In keeping with the performance and mass range of available mass spectrometers, two approaches are used for characterizing proteins. (1) In the first, intact proteins are ionized by either of the two techniques and then introduced to a mass analyser. This approach is referred to as top-down strategy of protein analysis. (2) In the second, proteins are enzymatically digested into smaller peptides using proteases such as trypsin or pepsin, either in solution or in gel after electrophoretic separation. Other proteolytic agents are also used. The collection of peptide products are then introduced to the mass analyser. When the characteristic pattern of peptides is used for the identification of the protein the method is called peptide mass fingerprinting (PMF), if the identification is performed using the sequence data determined in tandem MS analysis it is called denovo sequencing. These procedures of protein analysis are also referred to as the bottom-up approach.

4. **Space exploration:** Mass spectrometers are also widely used in space missions to measure the composition of plasma. For example, the Cassini spacecraft carries the Cassini Plasma Spectrometer (CAPS) which measures the mass of ions in Saturn's magnetosphere.

5. **Respired gas monitor:** Mass spectrometers are used in hospitals for respiratory gas. Mostly in the operating room, they were a part of a complex system in which respired gas samples from patients undergoing anesthesia were drawn into the instrument through a valve mechanism designed to sequentially connect up to the mass spectrometer. A computer directed all operations of the system. The

data collected from the mass spectrometer was delivered to the individual rooms for the anesthesiologist to use.

5.3. ELECTRON SPIN RESONANCE SPECTROSCOPY

5.3.1. Introduction

Electron paramagnetic resonance (EPR) or electron spin resonance (ESR) is defined as the form of spectroscopy concerned with microwave induced transitions between magnetic energy levels of electrons having a net spin and orbital angular momentum. When an atom or molecule with an unpaired electron is placed in a magnetic field, the spin of the unpaired electron can align either in the same direction or in the opposite direction as the field. These two electron alignments have different energies and application of a magnetic field to an unpaired electron lifts the degeneracy of the $\pm 1/2$ spins of the electron. Electron paramagnetic resonance (EPR) or Electron spin resonance (ESR) spectroscopy measures the absorption of microwave radiation by an unpaired electron when it is placed in a strong magnetic field. Species that contain unpaired electrons are free radicals, odd electron molecules, transition metal complexes, lanthanide ions and triplet state molecules.

5.3.2. Principle

The molecules of solid exhibit the paramagnetism as a result of unpaired electron spins, transitions can be induced between spin states by applying a magnetic field and then supplying electromagnetic energy, usually in the microwave range of frequencies. The resulting absorption spectra are described as electron spin resonance (ESR) or electron paramagnetic resonance (EPR). Electron spin resonance has been used as an investigative tool for the study of radicals formed in solid materials, since the radicals typically produce an unpaired spin on the molecule from which an electron is removed. Particularly fruitful has been the study of the ESR spectra of radicals produced as radiation damage from ionizing radiation. Study of the radicals produced by such radiation gives information about the locations and mechanisms of radiation damage.

The interaction of an external magnetic field with an electron spin depends upon the magnetic moment associated with the spin and the nature of an isolated electron spin is such that two orientations are possible. The application of the magnetic field then provides a magnetic potential energy which splits the spin states by an amount proportional to the magnetic field

(Zeeman Effect) and then radio frequency radiation of the appropriate frequency can cause a transition from one spin state to the other. The energy associated with the transition is expressed in terms of the applied magnetic field B, the electron spin g-factor g and the constant m_B which is called the Bohr magneton. If the radio frequency excitation was supplied by a klystron at 20 GHz, the magnetic field required for resonance would be 0.71 Tesla, a sizable magnetic field typically supplied by a large laboratory magnet.

For an electron of spin s = ½ the spin angular momentum quantum number will have values of $m_s = \pm\frac{1}{2}$. In the absence of magnetic field, the two values of m_s, i.e., $+\frac{1}{2}$ and $-\frac{1}{2}$ will give rise to a doubly degenerate spin energy state. If a magnetic field is applied, this degeneracy is removed and this leads to two non-degenerate energy levels. The low energy state will have the spin magnetic moment aligned with the field and corresponds to the quantum number $m_{s=-\frac{1}{2}}$. On other hand, the high energy state will have the spin magnetic moment opposed to the field and corresponds to the

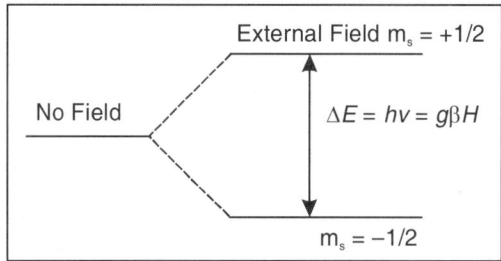

Fig. 5.24: Energy states

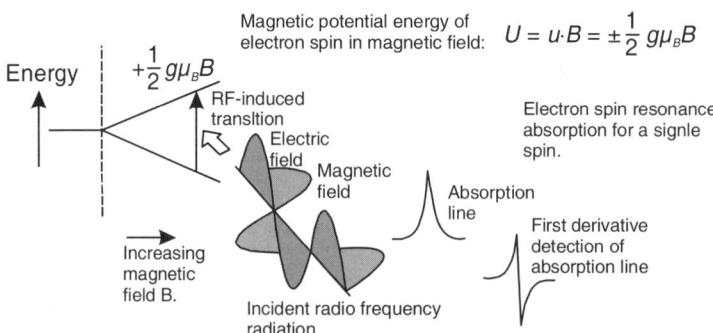

Fig. 5.25: Electron Spin Resonance absorption process

quantum number, $m_{s=+\frac{1}{2}}$. These energy stated are illustrated in Fig. 5.24 and 5.25. These two states will possess energies that are split up from original state with no applied magnetic field by the amount $+\mu_e H$ and $-\mu_e H$ for the low energy and high energy stated respectively; here H is the magnetic field acting on the unpaired electron and μ_e is the magnetic moment of the spinning electron. In ESR, a transition between the two different energy levels takes place by absorbing a quantum of radiation of frequency in the microwave region. Thus, the ESR spectrum of a free electron would consist of a single peak corresponding to a transition between these levels.

$$2\mu_e H = h\nu \qquad\qquad ... \; eq. \; 5.2$$

Where v is the frequency of the absorbed radiation in cycles per second. As the relation holds well for a free electron, the energy (ΔE) of transition is more accurately given by the following relation:

$$\Delta E = h = gH \qquad\qquad ... \; eq\text{-}5.2$$

Where h is the Planck's constant, H is the applied magnetic field, β is the Bohr's magneton which is a factor for converting angular momentum into magnetic moment, g is the proportionality factor which is a function of the electron's environment.

5.3.3. Instrumentation

The energies required to bring about a transition are different in ESR and NMR. In ESR, transitions occur at frequencies in the microwave region whereas in NMR, transitions occur in the radio frequency region. This reveals that the instrumentation for the NMR and ESR must be different.

The description of the various components of an ESR spectrometer is as follow:

5.3.3.1. Source – klystron

In most of ESR spectrometers the source is Klystron oscillator which is operation in the microwave band region of 3cm wavelength. A klystron is vacuum tube which can produce microwave oscillating centered on a small range of frequency. The frequency of the monochromatic radiation is determined by the voltage applied to klystron. A klystron oscillator is generally operated at 9500 Mc/sec.

5.3.3.2. Attenuator

Between the wave meter and circulator there is attenuator which adjusts the level of the microwave power incident upon the sample. It possesses an absorption element and corresponds to a neutral filter in light absorption measurements.

5.3.3.3 Circulator or magic – T

The microwave radiations produced by klystron oscillator are allowed to pass through the isolator, wavemeter and the attenuator, and finally enter the circulator through a wave guide by a loop of wire which couples with oscillating magnetic field and sets up a corresponding field in the wave guide. A wave guide is generally made up of hollow rectangular copper or brass tubing having silver or gold plating inside to produce a highly conducing flat surface. The operation of a four port circulator is indicated in the Fig. 5.26. The microwave radiations enter arm 1, and arm 2, is connected to resonant cavity and sample. Arm 3, generally having a terminating load, seems to absorb any power which might be reflected form the detector arm. Arm 4 is attached to the detector.

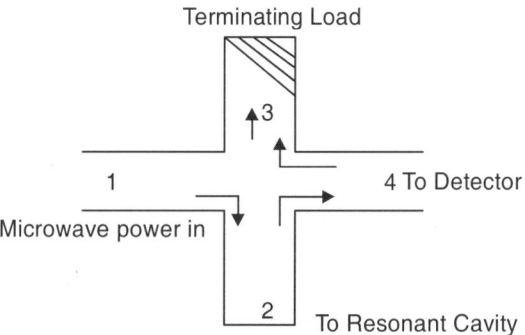

Fig. 5.26: A four-port microwave circulator showing the directions of microwave transmission among the several arms

5.3.3.4. Sample cavity

The heart of an ESR spectrometer is the resonant cavity containing the sample. The cavity system is constructed in such a way to maximize the applied magnetic field along the sample dimension. A sample volume of 0.15 to 0.5 ml can be used with samples which do not possess a high dielectric constant. Flat cells with a thickness of 0.25 mm and sample volume of 0.05 ml are generally used for such samples which possess high dielectric constant. For studying anisotropic effects in single crystals and in solid samples, rotatable cavities are generally utilized. In most of the ESR spectrometers, dual sample cavities are generally used. This is done for simultaneous observation of a sample and a reference material. By the use of a reference material, the sources of error are compensated by comparing relative signal heights.

5.3.3.5. Magnet system

The resonant cavity is placed between the pole pieces of an electromagnet. This provides a homogeneous magnetic field and can be a varied from zero to 500 gauss. The field should be stable and uniform over the sample volume. The stability of field is achieved by energizing the magnet with a highly regulated power supply. The stability of 1 part in 10^6 is satisfactory for resolution of ESR spectra of such samples whose g-factor ranges form 1.5 to 6. On the other hand, the stability might be as low as 1 part in 10^3 for paramagnetic ions and for free radicals in solid matrices. In order to sweep the magnetic field over a small range, the provision is made by varying the current in a pair of sweep coils.

5.3.3.6. Crystal detector

The most commonly used detector is a silicon crystal which acts as a microwave rectifier. This converts microwave power into a direct-current output.

5.3.3.7. Auto-amplifier and Phase sensitive detector

After detection by the crystal detector, the signal undergoes narrow band amplification. But the amplified signal contains a lot of noise. The reduction in noise is achieved by rejection of all noise components except those in a very narrow band by the operation of the phase sensitive detector.

5.3.3.8. Oscilloscope and Pen recorder

Finally, the signal from phase sensitive detector and sweep unit is recorded by the oscilloscope or pen recorder.

5.3.4. Applications

1. **Study of free radicals:** Even in very low concentration of sample ESR can study via free radicals. It is also applied in determination of structure of organic and inorganic free radicals. The intensity of ESR signal is directly proportional to the number of free radicals present. Hence using ESR we can measure relative concentration of free radicals.

2. **Investigation of molecules in the triple state:** A triple state molecule has a total spin S = 1 so that, its multiplicity can be given as 2S + 1 = 3. While free radicals with S = ½ has an odd number of unpaired electrons. A triple state molecule has an even number of electrons two of them unpaired. In triple state molecule

the unpaired electrons must interact whereas in diradical, the unpaired electrons do not interact for they are a great distance apart.

3. **Study of inorganic compounds:** ESR is very successful in the study of inorganic compounds. The ESR studies may be used in knowing the exact structures of solvated metal ions, in the study of catalysts and is used in the determination of oxidation state of metal. For example Copper is found to be divalent in copper protein complexes whereas it is found to be monovalent in some biologically active copper complexes. The information of unpaired electrons is very useful in various aspects in applications of ESR like spin labels, structural determination and reaction velocities and reaction mechanisms.

4. **Structural determination:** The ESR technique cannot be applied to determine molecular structure because the information obtained from the superfine structure is mostly about the extent of delocalization and Fermi contact interaction. It does not tell us about the arrangement of the atoms in the molecule although the symmetry of the molecule can be sometimes deduced from the sets of equivalent nuclei. In certain cases ESR is able to provide useful information about the shape of the radicals.

5. **Analytical applications:** Mn^{+2} ions can be measured and detected even when present in trace quantities. The method is very rapid and can be measured in aqueous solution over the range from 10-6 M to 0.1M. ESR method has proved to be a rapid and convenient method for determination of Vanadium in petroleum products. ESR can also be used to estimate Cu(II), Cr(II), Gadolinium (III), Fe(III) and Ti(III).

6. **Biological systems:** The ESR studies of variety of biological system such as, leaves, seeds and tissue preparation, it is found that a definite, correlation exists between the concentration of free radicals and the metabolic activity of the plant material. ESR has studied the presence of free radicals in healthy and diseased tissues. Most of the oxidative enzymes function via one electron redox reaction involving the production of either enzyme bound free radicals or by a change in the valence state of transition metal ion. This has been confirmed by ESR studies. Much of the ESR work on photosynthesis has been carried out with photosynthetic bacteria. The oxidation of bacteriochlorophyll in vitro produces an ESR signal.

7. **Modern biotechnology:** ESR being effectively used to reveal both structure and functional information it is very useful in modern biotechnology. There are three branches of modern biotechnology in which ESR is applied (a) Molecular biotechnology (b) Medical biotechnology (c) Classical biotechnology. Specific features of ESR in modern biotechnology are selectivity, specificity, non-invasiveness and sensitivity.

7.1. In molecular biotechnology

1. **ESR** is used to investigate the nucleotide-centered free radicals in DNA, either produced by irradiation or indirectly by other free radicals. ESR is applied to analysis of DNA hydration and the process of the hole or electron transfer from the hydration layer to DNA due to water ionization and to the analysis of DNA repair by DNA photolyase, by detection of flavin radical formation. ESR is useful in analysis of Reverse Transcriptase (RT) inhibition by polynucleotide.

2. **ESR** was employed to structure dependent molecular dynamics of Trans Activator Responsive (TAR) RNA of HIV-1. ESR is also used to determine the map of protein-RNA interactions between RNA and ribonuclease P from *E.coli.*

3. **Protein structure and dynamics:** The free radical damage of proteins in the field of research is still waiting for the complete exploration. For example, the ESR investigations of the interactions between ligands and target protein is the study on the ion siderophore complex and it's binding to site directed spin labeled ferric enterobactin receptor responsible for iron uptake by *Enterobacteria.*

   a. **Activity of enzymes:** ESR can effectively screen potential inhibitors interaction with the enzymes with high speed. Now a day, ESR is used in the analysis of enzymatic activity of nitric oxide synthetase (NOS), the main enzymes delivering nitric oxide (NO) in biological systems.

4. **Membranes:** The existence of phospholipid bilayers in biological systems is confusing from the point of view of evolutionary biology. The model of Fluid Mosaic appears

too simple to satisfactorily represent the details of membrane structure and the respective functions. The common view of the architecture of membrane has changed by the recent ESR evidence of the existence of structural domains stabilized by membrane proteins in the form of rafts.

5. **Glycobiology:** Spin labeled sugars, sugar residues, and spin labeled components interacting with sugar applied in two basic fields of carbohydrate research: Sugar metabolism (degradation and transport), Structural biochemistry of glycoproteins and membranes.

6. ESR is employed to analyze the process of sugar transport in bacteria.

7. ESR was applied to the analysis of the influence of diabetes on the properties of erythrocytes showing the decrease in erythrocyte deformability due to the non-enzymatic glycation of hemoglobin.

7.2. In medical biotechnology

Activation and transport of drugs: ESR is useful in several pharmacological investigations like interactions between DNA binding drugs and DNA. ESR may be used to characterize some herb derived products which act by increasing the level of free radicals and other reactive species produced during light induced oxidative stress of the cell. *In vivo* ESR experiments revealed that multimellar liposomes enhance the topical delivery of hydrophilic compound, drugs used to be more effective when applied in liposomes then in solutions. ESR imaging is a valuable tool for spatially resolved redox mapping of living tissues. Redox status of tumor tissues is significant for understanding tumor physiology and for determining the effects of chemotherapy and radiation.

7.3 In classical biotechnology

1. **Plant biotechnology:** ESR is helpful even at developing artificial photosynthesis, which is biggest biotechnological challenge for the mankind.

2. **Food production and storage:** Commercially ESR is used to analyze shelf life of beer and wine. It is based on free radicals generated in beer or wine due to the action of light

or spontaneously during the process of storage, contributes to the degradation and flavor changes of product. The level of free radical would depend on antioxidants presents in the solutions. Therefore antioxidant capacity of beer or wine helps to predict stability. Similar approach is applied to other food products, such as oils or milk. ESR measurement revealed also photosensitizing action of the important milk ingredient, vitamin-B_2, which may affect quality of the product. ESR also used in food science and hydration, water diffusion, and small molecule mobility in food systems.

5.4. SUMMARY

Infrared light lies between the Visible and Microwave portions of the electromagnetic spectrum. Infra red covers the range of the electromagnetic spectrum between 0.78 and 1000 mm. The most useful I.R. region lies between $4000 - 670 cm^{-1}$.

Near infrared light is closest in wavelength to visible light and far infrared is closer to the microwave region of the electromagnetic spectrum. Far infrared waves are thermal. Shorter, near infrared waves are not hot at all.

Interaction between vibrations can occur (*coupling*) if the vibrating bonds are joined to a single, central atom. Vibrational coupling is influenced by a number of factors - Strong coupling of stretching vibrations occurs when there is a common atom between the two vibrating bonds, Coupling of bending vibrations occurs when there is a common bond between vibrating groups ,Coupling between a stretching vibration and a bending vibration occurs if the stretching bond is one side of an angle varied by bending vibration, Coupling is greatest when the coupled groups have approximately equal energies and No coupling is seen between groups separated by two or more bonds.

FTIR Radiation Sources - Incandescent lamp, Nernst glower and Globar source. IR radiation which must be Intense enough for detection, Steady and Extend over the desired wavelengths.

Interferometer - The IR radiation produced by the source is directed to a mirror which sends the beam into the interferometer. Inside the interferometer, the beam is immediately split into two paths by the beam splitter. Each path leads to a mirror, which simply sends the beam back to the beam splitter, where it is recombined. This configuration causes the beams to interfere with each other constantly as the beams are travels through the interferometer.

Interferrogram - An interferogram can be generated regard less of whether a sample is present. The most commonly used interferometer is a Michelson interferometer. It consists of three active components: a moving mirror, a fixed mirror, and a beam splitter. The two mirrors are perpendicular to each other.

Mass spectrometry is an analytical tool used for measuring the molecular mass of a sample. The MS principle consists of ionizing chemical compounds to generate charged molecules or molecule fragments and measurement of their mass-to-charge ratios. The mass-to-charge ratio of the particles is then calculated from the motion of the ions as they transit through electromagnetic fields.

Chemical ionization is a lower energy alternative to EI for volatile analytes. In CI, there is a reagent gas (user ammonia or methane) in the ion chamber. In CI, ionization is due to proton transfer and is therefore a much lower energy process. The only way to analyze most small, biologically important molecules (sugars, amino acids, lipids etc.).

The technique of FAB is concept and design, in involve the bombardment of a solid spot of the analyte/matrix mixture on the end of a sample probe by a fast particle beam. The matrix (a small organic species like glycerol or 3-nitro benzylalcohol) is used to keep a homogenous sample surface. In FAB, the particle beam is a neutral inert gas (Ar or Xe) at 4-10 keV.

Electron Paramagnetic Resonance (EPR) and/or Electron Spin Resonance (ESR) is defined as the form of spectroscopy concerned with microwave-induced transitions between magnetic energy levels of electrons having a net spin and orbital angular momentum.

Electron-paramagnetic-resonance (EPR) or electron-spin-resonance (ESR) spectroscopy measures the absorption of microwave radiation by an unpaired electron when it is placed in a strong magnetic field. Species that contain unpaired electrons - free radicals, odd electron molecules, transition-metal complexes, lanthanide ions and triplet-state molecules.

The interaction of an external magnetic field with an electron spin depends upon the magnetic moment associated with the spin, and the nature of an isolated electron spin is such that two and only two orientations are possible. The application of the magnetic field then provides a magnetic potential energy which splits the spin states by an amount proportional to the magnetic field (Zeeman effect), and then radio frequency radiation of the appropriate frequency can cause a transition from one spin state to the other. The energy associated with the transition is expressed in terms of the applied magnetic field B, the electron spin g-factor g, and the constant m_B which is called the Bohr magneton.

ESR Source - the source is Klystron oscillator which is operation in the microwave band region of 3cm wavelength. A klystron is vacuum tube which can produce microwave oscillating centered on a small range of frequency. The frequency of the monochromatic radiation is determined by the voltage applied to klystron. A klystron oscillator is generally operated at 9500 Mc/ sec.

Attenuator - Between the wave meter and circulator there is attenuator which adjusts the level of the microwave power incident upon the sample. It possesses an absorption element and corresponds to a neutral filter in light absorption measurements.

ESR in Modern biotechnology - ESR being effectively used to revealed both structure and functional information. It is very useful in modern biotechnology. There are tree branches of modern biotechnology in which ESR is applied, a) Molecular biotechnology b) Medical biotechnology and c) Classical biotechnology. Specific features of ESR in modern biotechnology are - selectivity, specificity, non-invasiveness and sensitivity.

Chapter 6
ONLINE MONITORING
AND CONTROL DEVICES

6.0. pH

The concept of pH was first introduced by Sorensen in 1909. He recognized that hydrogen ion concentrations, while studying enzymatic reactions, he found it convenient to define a symbol which could represent the concentration of hydrogen ions and called this symbol as pH. It is defined by the following equation:

$$pH = -\log_{10}C_H \qquad \text{... eq. 6.1}$$

Where C_H is the hydrogen ion concentration.

$$C_H = 10^{-pH} \qquad \text{... eq. 6.2}$$

Pure water is known to be a weak electrolyte and it dissociated to form hydrogen ions and hydroxyl ions as follows:

$$H_2O = H^+ + OH^- \qquad \text{... eq. 6.3}$$

Assuming that activity coefficients are unit, the dissociation constant K_w of pure water is given by:

$$K_w = C_H^+ \times C_{OH}^- \qquad \text{... eq. 6.4}$$

The product of hydrogen and hydroxyl ions in water at 25°C is 1.008×10^{-14} moles^2 liters^{-2} and the concentrations of hydrogen and hydroxyl ions will of necessity be equal. Since the positive and negative electric charges in the solution must balance, each of these concentrations is given by:

$$C_H^+ = C_{OH}^- = 1.004 \times 10^{-7} \qquad \text{... eq. 6.5}$$

$$C_H^+ = 10^{-7} \qquad \text{... eq. 6.6}$$

Therefore pH of pure water is 7 it is obvious that the neutral point at which the hydrogen and hydroxyl ions are present in equal concentrations is located at pH. The lower case letter p in pH stands for the negative common (base ten) logarithm, while the upper case letter H stands for the element hydrogen. The pH is a logarithmic measurement of the number of moles of hydrogen ions (H^+) per liter of solution. The pH of an acidic solution at 25°C will be less than 7 and that of an alkaline solution greater than 7. The practical range of the pH scale is from −1 to 15 at room temperature, although most of the commercial instruments are designed to measure pH from 0 to 14.

The pH of any solution is a function of its temperature. Voltage output from the electrode changes linearly in relationship to changes in pH, and the temperature of the solution determines the slope of the graph. One pH unit corresponds to 59.16 mV at 25° C, the standard voltage and temperature to which all calibrations are referenced. The electrode voltage is decreases to 54.20 mV/pH units at 0.0°C and increases to 74.04 mV/pH units at 100.0°C.

Since pH values are temperature dependent, pH applications require some form of temperature compensation to ensure standardized pH values. Meters and controllers with automatic temperature compensation (ATC) receive a continuous signal from a temperature element and automatically correct the pH value based on the temperature of the solution. Manual temperature compensation requires the user to enter the temperature of the solution in order to correct pH readings for temperature. ATC is considered to be more practical for most pH applications.

6.0.1. pH Measurement

Measuring pH involves comparing the potential of solutions with unknown $[H^+]$ to a known reference potential. pH meters convert the voltage ratio between a reference half-cell and a sensing half-cell to pH values. In acidic or alkaline solutions, the voltage on the outer membrane surface changes proportionally to changes in $[H^+]$. The pH meter detects the change in potential and determines $[H^+]$ of the unknown by the Nernst equation:

$$E = E^o + (2.3RT)/nF \ \log \ \{unknown \ [H^+]/internal \ [H^+]\} \quad ...eq.6.7$$

Where: E = total potential difference (measured in mV); E^o = reference potential; R = gas constant; T = temperature in Kelvin; n = number of electrons; F = Faraday's constant; $[H^+]$ = hydrogen ion concentration.

pH is a measure of the acidity or alkalinity of a solution. The pH value states the relative quantity of hydrogen ions (H^+) contained in a solution. Greater the concentration of H^+ more acidic the solution and the pH is lower. In this relationship, pH is defined as the negative logarithm of hydrogen activity. A standard pH measuring system consists of three elements: (1) pH electrode (2) temperature compensation element (3) pH meter or controller. pH measurement is used in a wide variety of applications: agriculture, wastewater treatment, industrial processes, environmental monitoring, and in research and development.

The pH value of a solution is the negative log of its hydrogen ion activity (á), which is the product of hydrogen ion concentration [H$^+$] and the activity coefficient of hydrogen (gH$^+$) at that concentration. pH = –log α = –log γH$^+$[H$^+$] In pure water and in dilute solutions, the H$^+$ activity can be considered the same as the H$^+$ concentration.

$$pH = -\log \gamma H^+[H^+] = -\log [H^+] \qquad \textit{... eq. 6.8}$$

The pH of a solution measures the degree of acidity or alkalinity relative to the ionization of water. Pure water dissociates to yield 10^{-7} M of [H$^+$] and [OH$^-$] at 25°C; thus, the pH of water is 7, the point of neutrality. pH water = –log [H$^+$] = –log 10^{-7} = 7. Most pH readings range from 0 to 14. Solutions with a higher [H$^+$] than water (pH less than 7) are acidic; solutions with a lower [H$^+$] than water (pH greater than 7) are basic or alkaline.

Measuring pH involves comparing the potential of solutions with unknown [H$^+$] to a known reference potential. pH meters convert the voltage ratio between a reference half-cell and a sensing half-cell to pH values. Reference half-cells contains a conductor (usually silver with a silver chloride coating) immersed in a solution with known [H$^+$]. The potential between this internal conductor and the known solution is constant, providing a stable reference potential. Sensing half-cells (measuring half-cells) are made of a non-conducting glass or epoxy tube sealed to a conductive glass membrane. Like the reference half-cell, the sensing half-cell also contains a conductor immersed in a buffered electrolyte solution, ensuring constant voltages on the inner surface of the glass membrane and the sensing conductor. When the pH electrode is immersed in the solution to be measured, a potential is established on the surface of the sensing glass membrane. If the unknown solution is neutral, the sum of fixed voltages on the inner surface of the glass membrane and on the sensing conductor approximately balances the voltage on the outer surface of the glass membrane and the reference half-cell. This results in a total potential difference of 0 mV and a pH value of 7.

In acidic or alkaline solutions, the voltage on the outer membrane surface changes proportionally to changes in [H$^+$]. The pH meter detects the change in potential and determines [H⁺] of the unknown by the Nernst equation:

$$E = E° + 2.3RT \log \text{ unknown } [H^+] \quad ...eq.\ 6.9$$

where: E = total potential difference, T = temperature in Kelvin (measured in mV), n = number of electrons, E° = reference potential, F = Faraday's constant, R = gas constant, [H⁺] = hydrogen ion concentration.

The pH of a substance is an indication of how many hydrogen ions it forms in a certain volume of water. pH actually stands for power of hydrogen or potential of hydrogen. The proper definition of pH is that it's minus the logarithm of the hydrogen ion activity in a solution or the logarithm of the reciprocal of the hydrogen ion activity in a solution (Fig. 6.1).

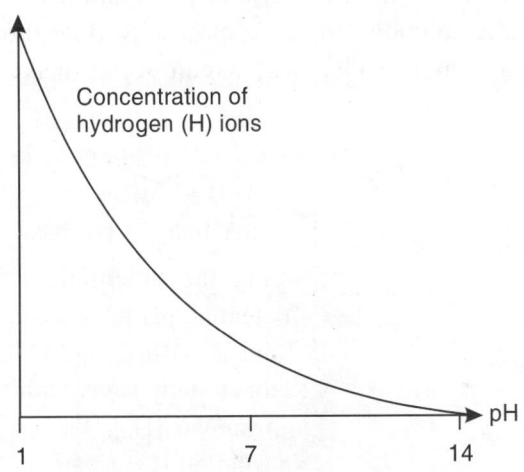

Fig. 6.1: The pH scale relates directly to the concentration of hydrogen ions in a solution, but not in a simple linear way. The relationship is what we call a negative exponential: the higher the pH (lower the acidity), the fewer the hydrogen ions but there are vastly fewer ions at high pH than at low pH.

6.0.2. pH system

pH reading is dependent upon all components of the system being operational. Problems with any one of the three: (a) Electrodes: A pH electrode consists of two half-cells; an indicating electrode and a reference electrode. Most applications today use a combination electrode with both half cells in one body. Over 90% of pH measurement problems are related to the improper use, storage or selection of electrodes. (b) Meters: A pH meter is a sophisticated volt meter capable of reading small milli-volt changes from the pH electrode system. The meter is seldom the source of problems for pH measurements. pH meters have temperature compensation either automatic or manual to correct for variations in slope caused by changes in temperature. Microprocessor technology has created many new convenience features for pH measurement. (c) Buffers: These solutions of known pH value allow the user to adjust the system to read accurate measurements. For best accuracy standardization should be performed with fresh buffer solutions, buffer used should frame the range of pH for the samples being tested and buffers should be at the same temperature as the samples.

A typical pH meter has two basic components (1) The meter itself which can be a moving-coil meter (one with a pointer that moves against a scale) or a digital meter (one with a numeric display) and (2) one or two probes that insert into the solution. To make electricity flow through something, have to create a complete electrical circuit so, to make electricity flow

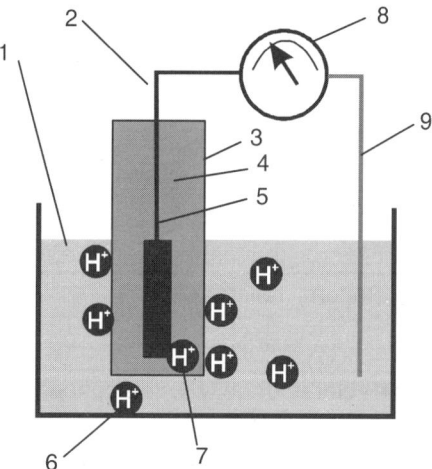

Fig. 6.2: **Key parts of a pH meter:** (1) Solution being tested; (2) Glass electrode, consisting of (3) a thin layer of silica glass containing metal salts, inside which there is a potassium chloride solution (4) and an internal electrode (5) made from silver/silver chloride. (6) Hydrogen ions formed in the test solution interact with the outer surface of the glass. (7) Hydrogen ions formed in the potassium chloride solution interact with the inside surface of the glass. (8) The meter measures the difference in voltage between the two sides of the glass and converts this potential difference into a pH reading. (9) Reference electrode acts as a baseline or reference for the measurement.

through the test solution have to put two electrodes or electrical terminals into it. If pH meter has two probes like the one in the photo at the top of this article, each one is a separate electrode; if have only one probe, both of the two electrodes are built inside it for simplicity and convenience.

The electrodes aren't like normal electrodes (metal wire); each one is a mini chemical set in its own right. The glass electrode, has a silver based electrical wire suspended in a solution of potassium chloride, contained inside a thin bulb or membrane made from a special glass containing metal salts typically compounds of sodium and calcium. The other electrode is called the reference electrode and has a potassium chloride wire suspended in a solution of potassium chloride (Fig. 6.2).

The potassium chloride inside the glass electrode is a neutral solution with a pH of 7, so it contains a certain amount of hydrogen ions (H^+). Suppose the unknown solution testing is much more acidic, so it contains a lot more hydrogen ions. What the glass electrode does is to measure the difference in pH between the orange solution and the blue solution by measuring the difference in the voltages their hydrogen ions produce.

6.0.2.1. Hydrogen electrode

Hydrogen electrode is a primary electrode. It is consists of an inert catalytically active metal surface made with platinum over which hydrogen is bubbled to achieve electrochemical equilibrium with the hydrogen ions in the solution. The following reaction takes place:

$$H^+ + e^- \rightarrow \frac{1}{2}H_2 \qquad \qquad ...\ eq.\ 6.10$$

The electrode is immersed in the solution under investigation and electrolytic hydrogen gas at 1 atm pressure is bubbled through the solution and over the electrode, in such a way that the electrode surface and the solution gets saturated with the gas at all times. Electrode life is 7-20 days. The potential set up at the hydrogen electrode by a given activity of hydrogen ions is governed by the Nernst equation. The hydrogen electrode is essentially a redox system and is affected by the presence of oxidizing and reducing agents it is therefore subject to a number of limitations in its applications.

6.0.2.2. Glass electrode

Glass electrode is based on the principle, when a thin membrane of glass is interposed between two solutions, a potential difference is observed across the glass membrane which depends on the ions present in the solutions.

The glass electrode consists of a thin walled bulb and sensitive glass sealed to a stem of non pH sensitive high resistance glass. The pH response is limited entirely to the area of the special glass membrane thus making the response independent of the depth of immersion. The membrane thickness of the order of 0.05–0.15 mm and bulbs are of the order of 10 mm in diameter (Fig. 6.3).

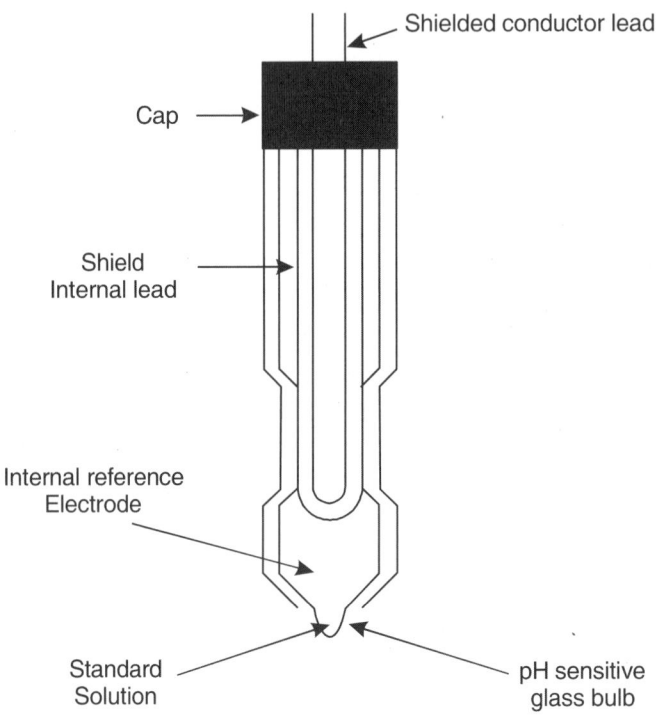

Fig. 6.3: A typical construction of a glass electrode

The glass pH electrodes are constructed of special glass to crate the ion selective barrier needed to screen out hydrogen ions from all the other ions floating around in the solution. The circuit path is contact from one electrode to other electrode through the glass barrier through the solution. On the inside of the membrane is a system of effectively constant pH. It is composed of silver or silver chloride dipped in HCl. Changes in electrical potential of the outer membrane surface is measured by means of an external reference electrode and its associated salt bridge. The complete pH cell is represented as follows:

The ideal pH response of a glass electrode behaving exactly in the same manner as a hydrogen electrode is given by:

$$E_2 - E_1 = 2.3026 \ RT/F \ (pH_2 - pH_1) \qquad ... \ eq. \ 6.11$$

Where E_1 and E_2 are the values of the electromotive force of cell 1 in test solutions of pH equal pH_1 and pH_2 respectively.

The useful pH range for a glass electrode generally lies between pH 1 and pH 11. Glass electrodes are covering temperature to about –10°C or up to about +120°C. Glass electrodes have two disadvantages (1) Measuring solutions containing matter can damage the glass membrane (2) The glass membrane is easily broken.

Care and maintenance of the glass electrode on a regular basis ensures the entire glass membrane must always be clean. Rinsing the membrane with distilled water. An alkaline hypochlorite solution can be used to clean electrode membranes subjected to solutions containing fat or proteins.

High temperature measurements, compounded by constant use in strong alkaline solutions or weak solutions of hydrofluoric acid will reduce the lifetime of the electrode since the glass membrane will slowly dissolve. Dry storage is recommended if the electrode will not be used for two weeks or more. Before use the electrode should be soaked well. New electrodes or those that have been stored dry should be conditioned or activated before use by soaking the bulb for a period of 12-24 hr. in 0.1 N hydrochloric acid.

6.0.2.3. Calomel electrode (or) reference electrode

The reference electrode is to provide a stable, reproducible voltage to which the working electrode potential may be referenced. The most common reference electrodes which meet these criteria are Mercury electrode and Silver/Silver chloride electrode

6.0.2.3.1. Mercury electrode

To measure the potential changes of the pH sensitive electrode directly, it is necessary that the pH cell be completed by means of a stable reference electrode, whose potential remains unaffected by changes in the composition of the cell solution. It consists of a metallic internal element, typical of mercury chloride, immersed in an electrolyte, which is usually a saturated solution of potassium chloride. The electrolyte solution forms a conductive salt bridge between the metallic element and the sample solution, in which the measuring and reference electrodes are emplaced. For a stable electrical connection between the internal metallic element and the sample solution, a small but constant flow of electrolyte solution is maintained through a liquid junction in the tip of the outer body of the reference electrode.

At a given temperature the activity of the mercurous chloride is constant and that of the mercury is unity, it is the chloride ion activity which is potential determining. When this is fixed, the electrode has a fixed potential at a fixed temperature. The most commonly used source of chloride is potassium chloride at saturated 3.8 M, 3.5 M or 0.1 M concentration. The useful lifetime of a reference electrode depends on the maintenance and care given to the electrode. This electrode should be stored in a small beaker containing the salt bridge solution for short term storage. For long term storage the electrode should be rinsed, dried and stored with the end cap on and the rubber band covering the filling hole in place (Fig. 6.4).

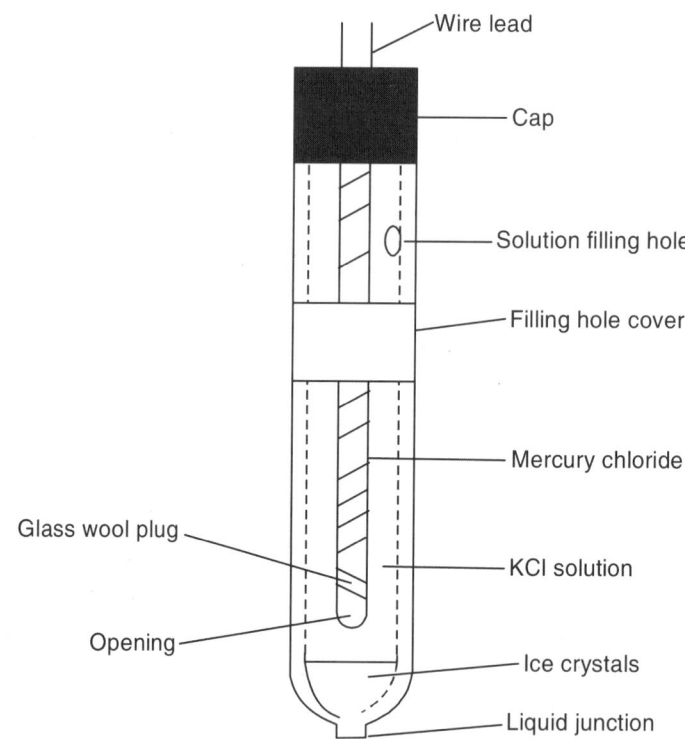

Fig. 6.4: Calomel electrode

6.0.2.3.2. Silver/silver chloride reference electrode

Fig. 6.5 shows the basic construction of Silver chloride electrode. A porous reference junction separated the filling solution in the electrode from the solution whose pH is to be measured. The filling solution's constant chloride ion concentration generates potential at a pure silver wire with silver chloride on it. The silver wire passes the signal from the solution being measured to the electrode's cable ore connector. This configuration of the electrode is called Single Junction Reference. If samples contain proteins sulfides, heavy metals or any other metals which interact with silver ions, they may react with the gel, causing a reduction in the reference output. This reaction can lead to reference signals or to precipitation at the reference junction leading to a short service life. A double junction reference electrode design as shown in figure-6.6, offers a barrier of protection to combat the above interactions. In this design, the inner chamber contains the slat concentration solutions so that stable outputs are generated. The outer chamber, which contacts the sample through the porous reference junctions, is filled with 0.1 M KCl. This lower ionic strength material more closely matches that of the sample and further reduces potentials. These electrodes are easily spoiled by drying. It is, therefore, advisable to keep the tips wetted at all times and store in 3M KCl when not in use. This helps to extend its life time. They usually last for 3-6 months.

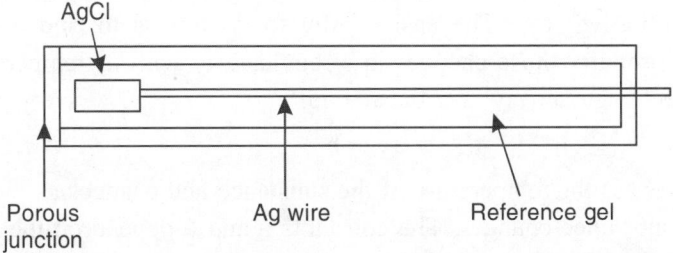

Fig. 6.5: Silver chloride reference electrode – Single Junction

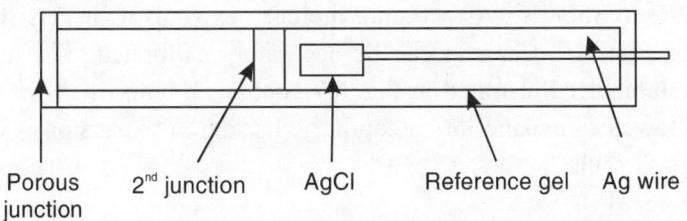

Fig. 6.6: Silver Chloride reference electrode – double junction

6.1. TEMPERATURE

Temperature is a physical property of a system that underlies the common notions of hot and cold, something that feels hotter generally has the greater temperature. Temperature is one of the principal parameters of Thermodynamics. It is easy to demonstrate that when two objects of the same material are placed together, the object with the higher temperature cools while the cooler object becomes warmer until a point is reached after which no more change occurs. When the thermal changes have stopped that the two objects are in thermal equilibrium. If three or more systems are in thermal contact with each other and all in equilibrium together, then any two taken separately are in equilibrium with one another. One of the three systems could be an instrument calibrated to measure the temperature i.e. a thermometer. When a calibrated thermometer is put in thermal contact with a system and reaches thermal equilibrium, then have a quantitative measure of the temperature of the system. For example mercury in glass clinical thermometer is put under the tongue of a patient and allowed to reach thermal equilibrium in the patient's mouth then see by how much the silvery mercury has expanded in the stem and read the scale of the thermometer to find the patient's temperature.

6.1.1. Thermometer

A thermometer is an instrument that measures the temperature of a system in a quantitative way. The easiest way to do this is to find a substance having a property those changes in a regular way with its temperature. The most direct regular way is a linear one:

$$t(x) = ax + b \qquad\qquad ...\ eq.\ 6.12$$

Where, t is the temperature of the substance and changes as the property x of the substance changes. The constants a and b depend on the substance used and may be evaluated by specifying two temperature points on the scale, such as 32°C for the freezing point of water and 212°C for its boiling point For example, the element mercury is liquid in the temperature range of –38.9°C to 356.7°C. As a liquid, mercury expands as it gets warmer its expansion rate is linear and can be accurately calibrated. The mercury in glass thermometer illustrated in Fig. 6.7 contains a bulb filled with mercury that is allowed to expand into a capillary. Its rate of expansion is calibrated on the glass scale.

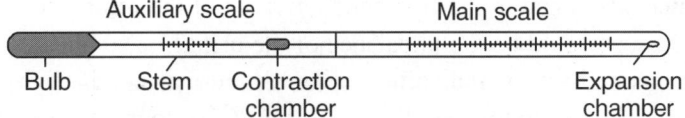

Fig. 6.7: Mercury in glass thermometer

6.1.2. Thermistors

The changes in temperature can be detected by using either platinum filament (hot wire) or thermistor. Fig. 6.8 shows the relative thermal conductivity of a series of gases of interest for analysis. In a typical hot wire cell thermal conductivity analyzer, four platinum filaments are employed as heat sensing elements. They are arranged in a constant current bridge circuit and each of them is place in a separate steel block and the block acts as a heat sink. The material used for construction of filaments is tungsten or platinum. Two filaments connected in opposite arms of the bridge act as reference arms, the other two filaments are connected in the gas stream, which act as measuring arm. The use of a four cell arrangement serves to compensate for temperature and power supply (Fig. 6.8).

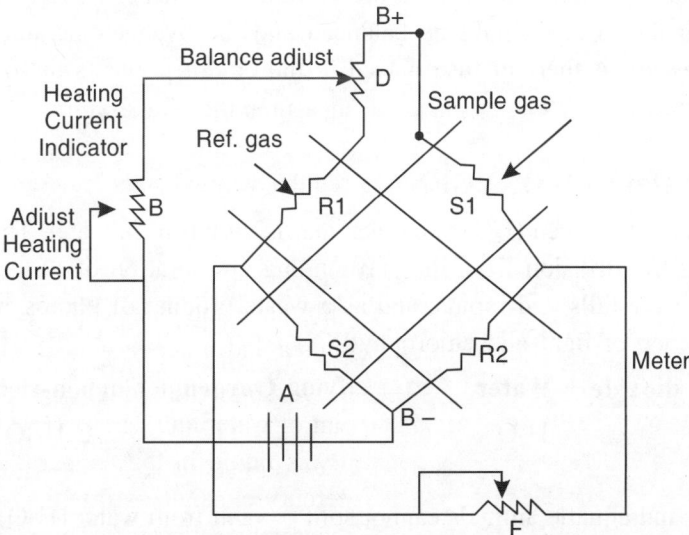

Fig. 6.8: Schematic diagram of a hot wire thermal conductivity analyzer

Initially, reference gas is made to flow through all the cells and the bridge is balanced with the help of potentiometer D (Fig. 6.8). When the gas stream passes through the measuring pair of filaments, the wires are cooled and there is a corresponding change in the resistance of the filaments.

The greater difference in thermal conductivities of the reference and sample gas, the greater would be the unbalance of the bridge. The unbalance current can be measured on an indicating meter. A thermal conductivity analyzer can be used in respiratory studies to follow CO_2 concentration change in the individual breaths of a patient.

6.1.3. Temperature controller

The temperature of the oven may be controlled by using a proportional temperature controller with a platinum resistance thermometer as a sensing element. The oven is thermally insulated so that heat loss to the atmosphere is minimized. The temperature sensing is done by the platinum resistance R1, which is placed in the oven. The temperature setting is done by adjusting the potentiometer VR1, calibrated in terms of temperature. This control is provided on the front panel of the instrument. When a setting is made, the bridge gets unbalanced and the amplifier, the synchronous rectifier and the UJT oscillator are actuated to open the gat of SCR. Thus, the current is applied to the oven hater and oven temperature begins to rise. The bridge approached nearer the balanced state and the hater current decreases. When the oven temperature reaches the preset value, the heater current would not flow and the bridge would be balanced and the oven temperature is thus kept constant. A thermal fuse placed in the circuit prevents the oven from overheating.

6.2. DISSOLVED OXYGEN

The dissolved oxygen (DO) is oxygen that is dissolved in water. The oxygen dissolves by diffusion from the surrounding air, aeration of water that has tumbled over falls and rapids, and as a waste product of photosynthesis in the presence of light and chlorophyll.

Carbon dioxide + Water $\longrightarrow$ Oxygen + Carbon-rich foods

$\quad\quad CO_2 \quad\quad\quad\quad H_2O \quad\quad\quad\quad\quad\quad O_2 \quad\quad\quad\quad C_6H_{12}O_6$

$$\dots eq.\ 6.13$$

Fish and aquatic animals cannot split oxygen from water (H_2O) or other oxygen containing compounds. Only green plants and some bacteria can do that through photosynthesis and similar processes. Virtually all the oxygen we breathe is manufactured by green plants. A total of three-fourths of the Earth's oxygen supply is produced by phytoplankton in the oceans.

6.2.1. Temperature effect

If water is too warm, there may not be enough oxygen in it. When there are too many bacteria or aquatic animal in the area they may overpopulate using DO in great amounts. Oxygen levels also can be reduced through over fertilization of water plants by run-off from farm fields containing phosphates and nitrates (ingredients in fertilizers). Under these conditions, the numbers and size of water plants increase (Fig. 6.9). Then, if the weather becomes cloudy for several days, respiring plants will use much of the available DO. When these plants die, they become food for bacteria, which in turn multiply and use large amounts of oxygen and this depleting all the oxygen. How much DO an aquatic organism needs depends upon its species, its physical state, water temperature, pollutants present and more. Numerous scientific studies suggest that 4-5 ppm of DO is the minimum amount that will support a large, diverse fish population. The DO level in good fishing waters generally averages about 9.0 ppm.

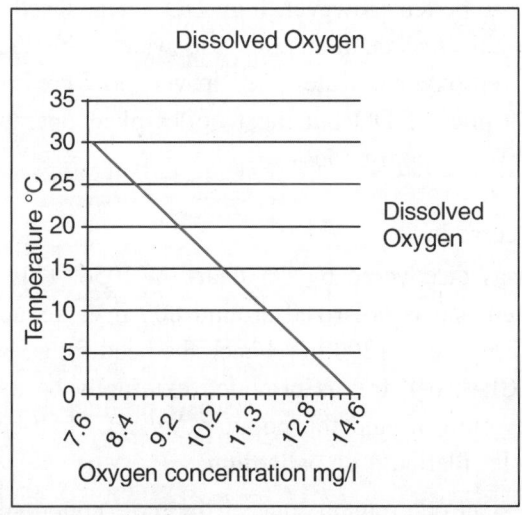

Fig. 6.9: The effect of the temperature in the DO

6.2.2. Environmental impact

Total dissolved gas concentrations in water should not exceed 110%. Concentrations above this level can be harmful to aquatic life. Fish in water containing excessive dissolved gases may suffer from gas bubble disease however this is a very rare occurrence. The bubbles or emboli block the flow of blood through blood vessels causing death. External bubbles (emphysema) can also occur and be seen on fins, on skin and on other tissue. Aquatic invertebrates are also affected by gas bubble disease but at levels higher than those lethal to fish. Adequate dissolved oxygen is necessary for good water quality. Oxygen is a necessary element to all forms of life. Natural stream purification processes require adequate oxygen levels in order to provide for aerobic life forms. As dissolved oxygen levels in water drop below 5.0 mg/l, aquatic life is put under stress. Oxygen levels that remain below 1-2 mg/l for a few hours can result in large fish kills. Dissolved oxygen is absolutely essential for the survival of all aquatic organisms not only fish but also invertebrates such as crabs, clams and zooplankton. Moreover, oxygen affects a vast number of other water indicators, not only biochemical but esthetic ones like the odor, clarity and taste. Consequently, oxygen is perhaps the most well established indicator of water quality.

6.2.3. Dissolved oxygen affects in water supplies

A high DO level in a community water supply is good because it makes drinking water taste better. However, high DO levels speed up corrosion in water pipes. For this reason, industries use water with the least possible amount of dissolved oxygen. Water used in very low pressure boilers have no more than 2.0 ppm of DO but most boiler plant operators try to keep oxygen levels to 0.007 ppm or less.

6.2.4. Oxygen electrode

The Clark cell was discovered by Dr. Clark in 1956. This is basically an ampherometric cell that is polarized around 800 mV. Reduction of oxygen is achieved between 400 to 1200 mV hence the need for a voltage of around 800 mV. In the Clark cell this is provided externally by a battery source. The Clark cell is built around the popular Ag/AgCl half-cell and a noble metal such as gold, platinum or palladium.

The Oxygen electrode remains one of the most commonly used devices for measuring the partial pressure of oxygen (oxygen tension) in the gas phase or dissolved solution. The Oxygen electrode finds application in a wide variety of diverse subject areas including areas such as:

1.   Environmental studies Example: O_2 levels in natural waters.
2.   Sewage treatment: Vital in monitoring the progress of bacterial attack.
3.   Alcohol production: O_2 levels in fermentation tasks need to be continuously monitored and controlled.

4.   Medicine-invasive and non-invasive monitoring of a key phy-
     siological analyte.

The typical range of detection of O_2 this device is from 10 - 4 atm
(0.01%) to 1atm (100%). The key to continuing supremacy of the Oxygen
electrode over other electrochemical devices for O_2 detection is the utilization
of a gas-permeable and ion-impermeable membrane that separates the test
system from the sensing electrode (platinum cathode). This membrane
prevents many problems of electrode passivation or poisoning that arise
when the sensing electrode is placed in direct contact with the system usually
an aqueous solution under test.

6.2.4.1. The cell

The Oxygen Electrode comprises two electrodes. The first is a small typically
2mm in diameter central platinum disc working electrode this is the cathode
and at this electrode, the O_2 diffusing through the membrane is reduced.
Set in a well surrounding, this is a silver ring counter and reference electrode
about ten times larger in surface area than the Pt cathode. Conduction
between the two electrodes is achieved using a 3M potassium chloride
solution to saturate the paper tissue covering the two electrodes. On top of
this is placed the key gas permeable membrane, usually 12.7mm thick Teflon,
sealed from the test sample in the incubation chamber by a silicone rubber

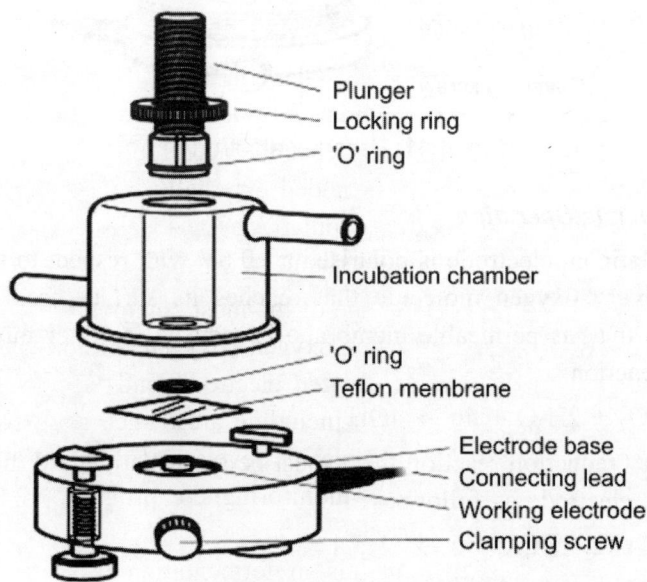

Plunger
Locking ring
'O' ring

Incubation chamber

'O' ring
Teflon membrane

Electrode base
Connecting lead
Working electrode
Clamping screw

Fig. 6.10: Glass electrode

'O' ring (Fig. 6.10 and 6.11). The controller is used to apply a voltage to the central platinum electrode that is sufficiently negative, with respect to the silver electrode, that all the oxygen diffusing through the membrane and reaching this electrode is reduced. The resultant current which flows between the two electrodes is proportional to the oxygen partial pressure in the test system, $P(O_2)$. The controller converts this current directly into a voltage and depending on the model will display this in units of percentage saturation.

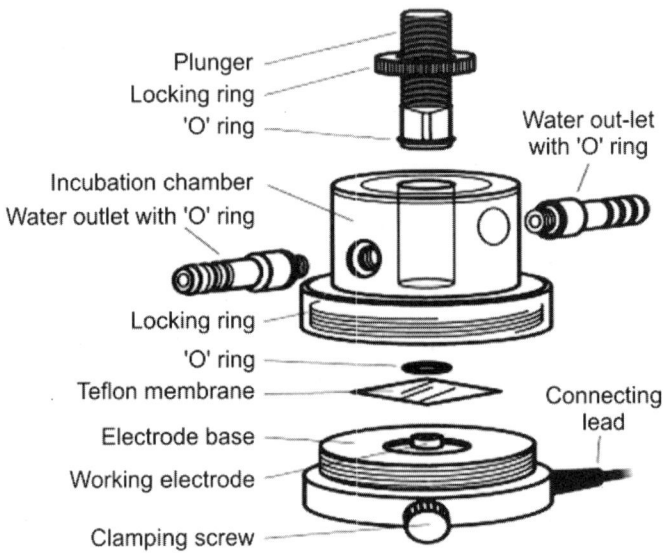

Fig. 6.11: Perspex electrode

6.2.4.2. *Principle/Operation*

When the platinum electrode is polarised at -0.6V with respect to the silver electrode, every oxygen molecule that reaches its surface from the test medium via the gas permeable membrane is reduced to water through the following reaction:

$$O_2 + 2H_2O + 4e \rightarrow 4OH^- \qquad \textit{... eq. 6.14}$$

For every reduction reaction there must be an oxidation and this occurs at the silver electrode as follows:

$$4Ag + 4Cl^- -4e \rightarrow AgCl \qquad \textit{... eq. 6.15}$$

Thus the overall electrochemical process that occurs in an Oxygen electrode is as follows ...

$$4Ag + O_2 + 2H_2O + 4Cl^- \rightarrow 4AgCl + 4OH \qquad \textit{... eq. 6.16}$$

As the oxygen electrode is repeatedly used the bright silver ring electrode rapidly becomes tarnished. Eventually an even coat of brown silver chloride forms on the silver electrode. The presence of this silver chloride layer is desirable it stabilizes the overall behavior of the electrode and should not be removed except if it grows very thick.

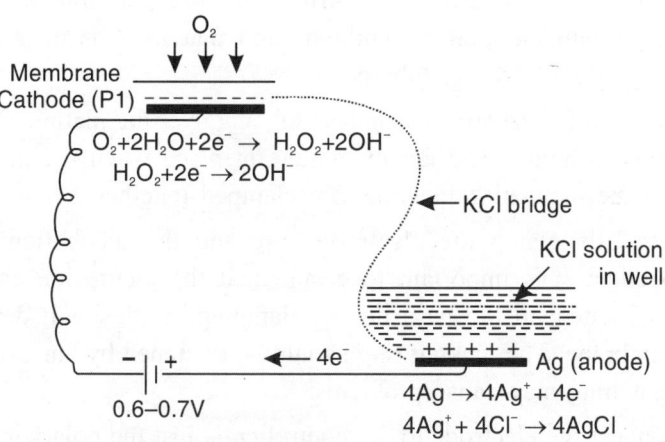

Fig. 6.12: The oxygen electrode reactions.

The Fig. 6.12, above represents the oxygen electrode reactions. When a potentiating voltage is applied across the two electrodes, the platinum (Pt) becomes negative i.e. becomes the cathode, and the silver (Ag) becomes positive (anode). Oxygen diffuses through the membrane and is reduced at the cathode surface so that a current flows through the circuit which is completed by a thin layer of KCl solution or other electrolyte. The silver is oxidized and silver chloride deposited on the anode. The current which is generated bears a direct, stoichiometric, relationship to the oxygen reduced and is converted to a digital signal and recorded by an electrode control unit.

6.2.4.3. Setting up the oxygen electrode

The electrode need to have the items small pair of sharp scissors, 3M potassium chloride solution, teat pipette, teflon membrane and tissue paper.

1. Using the teat pipette wet both electrodes and fills the small well containing the silver electrode with the potassium chloride solution.

2. Cut a 1.5cm square piece of tissue paper with a 2mm hole in its

centre and float this on the potassium chloride in the well ensuring that the hole is central above the platinum electrode.

3. Touch the empty teat pipette against the tissue paper and use it to suck off the excess electrolyte so that the paper is wet (but not very wet) and clings to the surface of the electrode.

4. Cut a 1.5cm square piece of teflon membrane and place it so that it covers both electrodes, ensuring that the platinum electrode is underneath the centre membrane and that there is no air bubbles trapped under the membrane.

5. Gently push the silicone rubber 'O' ring over the platinum electrode so as to hold the teflon membrane in place when the plastic base and the incubation chamber are clamped together.

6. Carefully clamp the electrode base and the incubation chamber together. It is important to ensure that the incubation chamber is not rotated on the base during clamping, as this will damage the membrane. The locking ring should be tightened by hand only. Over tightening may cause problems.

7. Connect the electrode to the controller, adjust the polarizing voltage to 0.6V and adjust the stirring speed to a suitable level. Connect the water jacket of the incubation chamber to a constant temperature water bath and allow the sample temperature to stabilize. The electrode is now ready for calibration.

6.2.4.4. Maintenance

When the electrode is not in use for a few hours (overnight) it is best dismantled and the electrodes left to soak in distilled water. If the electrode assembly must be left intact but non-operational for a few hours it is best if the electrode is left on, but with the stirrer switched off. The platinum electrode needs to have a mirror finish any surface damage will affect the response of the electrode, and it will thus need to be cleaned approximately once every 5-7 days of use or when it has lost its shine. A suitable polish can be made by mixing thick slurry of 0.3mm polishing alumina in distilled water. A piece of cotton wool can then is used to polish the platinum electrode until it is smooth, bright and clean (few minutes). The silver electrode will need to have the layer of silver chloride removed and the surface polished every 2-3 months of use. A 10% ammonia solution on cotton wool can be used to remove the silver chloride layer.

6.3. AGITATION

Mixing is a process where two or more substances enter a chamber and they are combined. The chamber can be a tank, hopper, length of pipe or even an extruder barrel. Most often stirring or agitation is the method to combine compounds. The stirring mechanism is a rotating arm that is often powered with a drive/motor. For a certain category of mixers, the chamber itself is rotated to achieve mixing. Different substances will require different levels of mixing, stirring speeds and elapsed times because of these variables drives can play a key role in the mixing process. Anything can be mixed and the type of mixer used depends on the nature of the substances to be mixed. Liquid-gas, liquid-liquid and liquid-solid systems will generally be mixed in a vessel or tank with an impeller to provide the mixing energy. The end product is almost always a liquid, suspension or slurry. A chemical reaction can often take place and the tank can have accommodations for the addition or removal of the heat of mixing or heat of reaction.

A wide range of equipment can be used for the mixing of pastes, aggregates, viscous materials or other solids. This is due to the wider variation in mixing properties for these different materials than for liquids. Typically, the agitating mechanism must be designed to extend very close to the walls of the vessel in these types of systems. The method used to keep chemicals in contact with photographic materials during processing. The main purposes of agitation have been include elimination of solution stagnation and dispersal of reaction products, increase of deposition rates by mass transfer enhancement, dissipation of heat at electrode/solution interfaces, incorporation of particles in the deposit and modification and control of deposit properties including thickness distribution.

It is possible to define the parameters of practical air agitation – air flow rate, air pressure and pipe length, air bubble size, bubble distribution pattern, bubble density in the tank. Fig. 6.13 showing the slow rise of spherical cap bubbles which induce eddying circulation in the tank and generate mist or fume when they break through the meniscus into the atmosphere. The stirring is cheap and simple but a rapid experiment with the stirrer altering its position and angle of insertion into the tank will reveal its need of optimization. One such approach has been to design more effective stirrer blades and a large number can be found in the design.

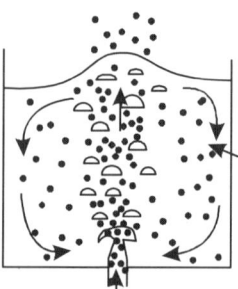

Fig. 6.13: Air agitation

6.3.1. Mixing features and properties
6.3.1.1. Liquids

The key to effective mixing for liquids is to create multiple flow patterns in the fluid being mixed. This motion is imparted to a fluid pocket as it contacts the blade on the rotating agitator. The momentum of this pocket will keep it in motion until it either contacts the wall of the vessel or runs into another moving pocket. Three components for the motion are (a) radial flow (outward from an agitator) (b) axial flow (parallel to the agitator arm) (c) angular flow (parallel to the vessel wall in a horizontal plane). The first two types are essential for effective mixing. Angular flow is undesirable, because it means that the fluid is orbiting around the agitator (pockets moving in parallel), but little mixing is occurring. Liquid flow can further be defined as either laminar or turbulent. The flow is predictable and many formulas and studies for fluid mechanics are based on laminar flow.

6.3.1.2. Solids, pastes and highly viscous materials

Solids, Pastes and other highly viscous materials cannot rely on internal flow momentum to help the mixing process, unlike liquids. Motion and therefore mixing must rely directly on contact with the agitating device. The designs of these will differ greatly from the propeller type agitators used for liquids and more energy per unit is required to mix solids over liquids as well. Other factors and properties of pastes and solids affect the mixing. Bulk densities, particle shapes and flow characteristics can affect energy requirements. In pure solids mixtures, particle size and abilities to hold static electricity charges must also be considered. A major objective of mixing solids can also be size reduction of particles and the breaking up of lumps.

Friction is an extremely important design consideration in the mixing of non-liquids. The heat produced by friction can either help a process as in an extruder or is considered waste energy that must be removed. Excessive temperatures from friction can also cause product degradation and the creation of unwanted side products or off gases.

6.3.2. Mixing equipments

6.3.2.1. Liquids

Major equipment for this is the mixing vessel, agitator, and baffles.

1. **Mixing vessels:** Mixing vessels come in a variety of styles and sizes. They can be used for batch processing or on a steady state flow basis where starting fluids enter the top and a mixed product is drawn from the bottom. Tanks typically have rounded bottoms to avoid stagnant areas that sharper corners could produce. A surrounding jacketed chamber holds steam or process water that might heat or cool the process depending on needs. Tanks could also be covered to prevent material losses due to volatiles or heat losses that waste energy such as in a mixing tank reactor.

2. **Agitators:** An agitator is a shaft with a propeller attached and the two components that provide for good mixing are radial and axial flow. A propeller with pitched blades promotes this behavior. The propeller speeds range from 350 to 1750 rpm with suspended matter being agitated at lower speeds and for rapidly dissolved solids or chemical reactions at the higher speeds. Propellers typically work best at 3 HP or below and in tanks whose diameters do not exceed six feet.

3. **Baffles:** Baffles are obstacles or barriers that are positioned parallel to the agitator arm at the edge of a tank wall. They inhibit tangential flow without affecting the radial component.

6.3.2.2. Solids, pastes and viscous materials

Mixing of non-liquids considerations to a wider choice of equipment and many more challenges due to the varying nature of the types of substances. Some of the typical ones are (i) Heating, cooling, or drying operations (ii) Size reduction (iii) Homogeneity of aggregates (iv) Coating of particles (v) Kneading. They all have in common is that the agitator must be designed so as to physically contact all of the substances in the mixer.

Mixing occurs due to the shearing action of the blades as no momentum from flow can be depended upon liquids. Shearing, is defined as the application of force via the mixing blade. The energy required per unit volume is greater for solids and pastes than for liquids. One notable exception is that shearing action is a tumbler-type mixer. Here, the vessel itself is either rotated or oscillated back and forth. A cement mixer is a good example of this. Tumbler types are useful when the objective is to break up clumps or agglomerates of material. They can have baffles to enhance mixing and are used for batch processing. They are generally low maintenance; low wear devices that are suited for dry solids and low adhesion pastes. One variation is to add inert ceramic spheres in the mixer. These serve the purpose of helping to break up chunks. The tumbler becomes a ball mill in these cases. One example where this is useful is in mixing certain types of paint.

Ribbon mixers consist of helical or spiral mixing blades that sweep across nearly the entire surface of the vessel. There are many variations to handle a wide variety of compounds. When pastes adhere to vessel walls, the tank may be aligned vertically so that the blades vertically lift the material as it is sheared. Heat transfer operations use this type of design. The main design criteria for ribbon mixers are the blade thickness, number of spirals, wall clearances and spiral pattern.

Twin rotor mixers use two parallel rotors with intermeshed flights along their axes. There is a tight clearance to the vessel walls, so high shearing action between the rotors and between rotors and wall occurs. This design is similar to a twin-screw extruder, except that the product is not forced through a die, nor is it normally melted. These, as well as single rotor mixers are useful for continuous mixing processes, and will produce a high degree of mixing for homogeneity.

6.4. SENSOR

A sensor is a device that measures a physical quantity and converts it into a signal which can be read by an observer or by an instrument. For example, a mercury thermometer converts the measured temperature into expansion and contraction of a liquid which can be read on a calibrated glass tube. A thermocouple converts temperature to an output voltage which can be read by a voltmeter. For accuracy, all sensors need to be calibrated against known standards. Sensor's sensitivity indicates how much the sensor's output changes when the measured quantity changes. For instance, if the mercury in a thermometer moves 1 cm when the temperature changes by 1°C, the sensitivity is 1cm/°C. Sensors that measure very small changes must have very high sensitivities.

In the laboratory, one of the best known types of sensor is the litmus paper test for acids and alkalis, which gives a qualitative indication by means of a colour reaction of the presence or absence of an acid. A more precise method of indicating the degree of acidity is the measurement of pH either by the more extended use of colour reactions in special indicator solutions or even by simple pH papers. However, the best method of measuring acidity is the use of the pH meter which is an electrochemical device giving an electrical response which can be read by a needle on a scale or on a digital read out device or input to a microprocessor. In such methods the sensor that responds to the degree of acidity is either a chemical the dye litmus or a more complex mixture of chemical dyes in pH indicator solutions or the glass membrane electrode in the pH meter. The chemical or electrical response then has to be converted into a signal that we can observe usually with our eyes. With litmus, this is easy, a colour change is observed because of the change in the absorbance of visible light by the chemical itself, this is immediately detected by our eyes in a lightened room. In the case of the pH meter the electrical response (a voltage change) has to be converted into an observable response movement of a meter needle or a digital display. The part of the device which carries out this conversion is called a transducer. Sensors can be divided into three types.

1. **Physical sensors:** Concerned with measuring physical quantities such as length, weight, temperature, pressure, and electricity. The response of a sensor is usually in the form of a physical response.

2. **Chemical sensors:** A sensor is concerned with detecting and measuring a specific chemical substance or set of chemicals. It is a device which responds to a particular analyte in a selective way through a chemical reaction and can be used for the qualitative or quantitative determination of the analyte.

3. **Biosensors:** A biosensor can be defined as a device incorporating a biological sensing element connected to a transducer. This sensor detects and measures may be purely chemical (inorganic), although biological components maybe the target analyte. The key difference is that the recognition element is biological in nature.

All of these devices have to be connected to a transducer so that a visibly observable response occurs. Chemical sensors and biosensors are generally concerned with sensing and measuring particular chemicals which may or may not be biological themselves. Fig. 6.14 shows schematically the general arrangement of a sensor.

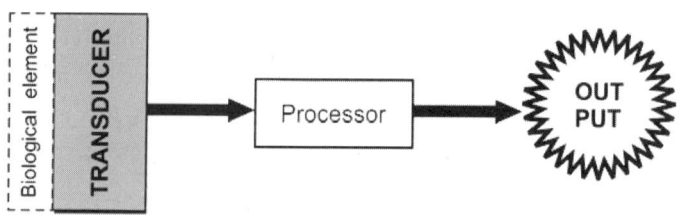

Fig. 6.14: Schematic layout of a Sensor

Sensors are changing one form of energy into another; hence they can be classified according to the type of energy transfer that they detect.

1. **Thermal sensors:** Thermometers, thermocouples, temperature sensitive resistors (Thermistors and resistance temperature detectors), bi-metal thermometers and thermostats. Heat sensors: bolometer, calorimeter, heat flux sensor.

2. **Mechanical sensors:** Pressure sensors: altimeter, barometer, barograph, pressure gauge, air speed indicator, rate-of-climb indicator. Gas and liquid flow sensors: flow sensor, anemometer, flow meter, gas meter, water meter, mass flow sensor. Gas and liquid viscosity and density: viscometer, hydrometer, oscillating U-tube. Geochemical sensors: acceleration sensor, position sensor, switch, strain gauge. Humidity sensors: hygrometer.

3. **Chemical sensors:** Chemical proportion sensors: oxygen sensors, ion-selective electrodes, pH glass electrodes, redox electrodes, and carbon monoxide detectors. Odor sensors: Tin-oxide gas sensors, and Quartz crystal microbalance (QCM) sensors. Gas sensors are often combined into an electronic nose.

4. **Light sensors:** Light sensors or photo detectors, including semi-conductor devices such as photocells, photodiodes, phototransistors, CCDs, and Image sensors; vacuum tube devices like photo-electric tubes, photomultiplier tubes; and mechanical instruments such. Scanning laser- A narrow beam of laser light is scanned over the scene by a mirror. A photocell sensor located at an offset responds when the beam is reflected from an object to the sensor, whence the distance is calculated by triangulation.

5. **Radiation sensors:** Geiger counter, dosimeter, Scintillation counter, Neutron detection Subatomic particle sensors: Particle detector, scintillator, Wire chamber, cloud chamber, bubble chamber.

6.5. X-RAYS

X-rays are electromagnetic radiation with wavelengths between about 0.02 Å and 100 Å (1Å $= 10^{-10}$ meters) (Fig. 6.15). They are part of the electromagnetic spectrum that includes wavelengths of electromagnetic radiation. X-rays have wavelengths similar to the size of atoms, they are useful to explore within crystals.

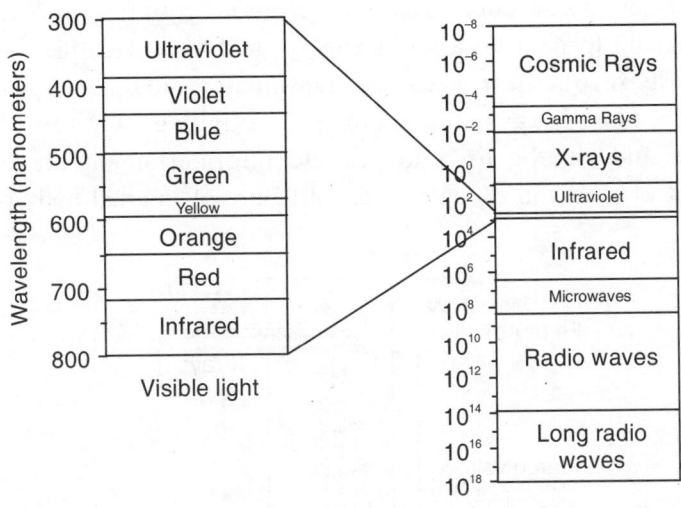

Fig. 6.15: Wavelengths of Electromagnetic Spectrum

The energy of X-rays like all electromagnetic radiation is inversely proportional to their wavelength as given by the *Einstein equation*:

$$E = hn = hc/l \qquad \qquad ... \textit{eq. 6.17}$$

Where E = energy, h = Planck's constant, 6.62517×10^{-27} erg sec, n = frequency, c = velocity of light = 2.99793×10^{10} cm/sec and l = wavelength

Thus, X-rays have a smaller wavelength than visible light and they have higher energy. With their higher energy, X-rays can penetrate matter more easily than can visible light. Their ability to penetrate matter depends on the density of the matter thus X-rays provide a powerful tool in medicine for mapping internal structures of the human body (bones have higher density than tissue, and thus are harder for X-rays to penetrate, fractures in bones have a different density than the bone, thus fractures can be seen in X-ray pictures).

6.5.1. X- rays source

X-rays are produced in a device called an X-ray tube (Fig. 6.16). It consists of an evacuated chamber with a tungsten filament at one end of the tube, called the cathode and a metal target at the other end called an anode. Electrical current is run through the tungsten filament causing it to glow and emit electrons. A large voltage difference measured in kilovolts is placed between the cathode and the anode, causing the electrons to move at high velocity from the filament to anode target. Upon striking the atoms in the target, the electrons dislodge inner shell electrons resulting in outer shell electrons having to jump to a lower energy shell to replace the dislodged electrons (Fig. 6.16). These electronic transitions results in the generation of X-rays. The X-rays then move through a window in the X-ray tube and can be used to provide information on the internal arrangement of atoms in crystals or the structure of internal body parts.

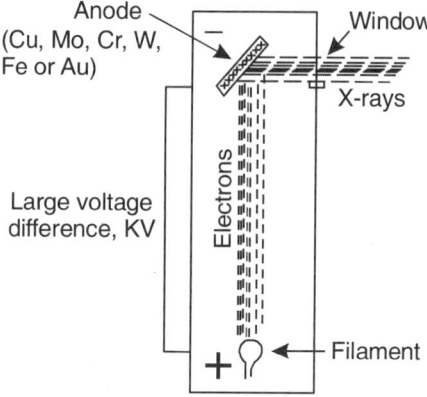

Fig. 6.16: X-Rays produced tube

6.5.2. Diffractometer

X-rays are generally filtered to a single wavelength (monochromatic) and collimated to a single direction before they are allowed to strike the crystal. The filtering not only simplifies the data analysis but also removes radiation that degrades the crystal without contributing useful information. Collimation is done either with a collimator (long tube) or with a clever arrangement of gently curved mirrors. Mirror systems are preferred for small crystals (0.3 mm) or with large unit cells (150 Å).

6.5.3. Electron density

Crystallographic experiment is not really a picture of the atoms but a map of the distribution of electrons in the molecule, *i.e.* an electron density map. However, the electrons are tightly localized around the nuclei the electron density map gives a pretty good picture of the molecule. This is because electromagnetic radiation including X-rays interacts with matter through its fluctuating electric field which accelerates charged particles. The electrons fluctuating in position and through their accelerations emitting electromagnetic radiation in turn because electrons have a much higher charge to mass ratio than atomic nuclei or even protons, they are much more efficient in this process. Intensity of scattered radiation is proportional to the square of the charge/mass ratio and the proton is about 2000 times as massive as the electron.

6.5.4. Crystals

X-ray scattering from a single molecule would be incredibly weak and extremely difficult to detect the noise level which would include scattering from air and water. A crystal arranges huge numbers of molecules in the same orientation, so that scattered waves can add up in phase and raise the signal to a measurable level (Fig. 6.17). In a sense, a crystal acts as an amplifier. If the waves add up in phase in some directions, they have to cancel out in a lot of other directions. That is why the diffraction pattern from a crystal is an array of spots (Fig. 6.18).

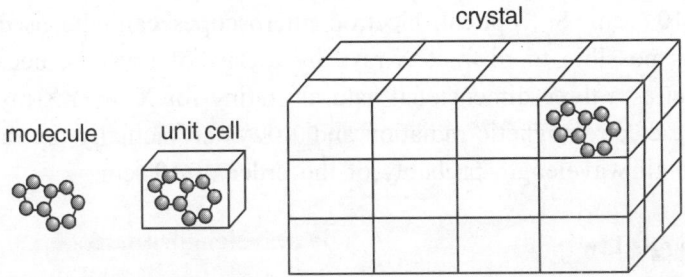

Fig. 6.17: A crystal arranges huge numbers of molecules in the orientation

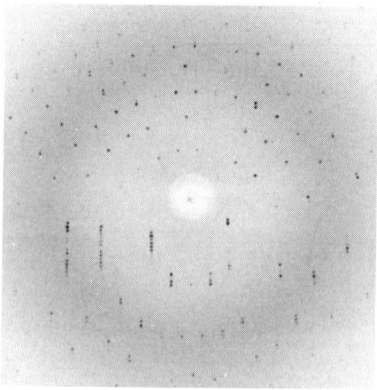

Fig. 6.18: Diffraction pattern from a crystal is an array of spots

There are a number of potential bottlenecks in determining a crystal structure but growing a useful crystal can be the most serious one.

6.5.5. X–ray diffraction

X–Rays are now possible to ascertain the special arrangement of the structural units of a substance in the crystalline state hence X-rays can be employed in investigating the interior of a crystal. The most important feature of the crystals of a substance is that although the shapes and sizes of individual crystals differ considerably, the interfacial angles remain constant. The distances between the atoms in crystals have been found to be roughly equal to 10^{-8} cm. So, optical, electron microscopes can't be used in this field. It is possible to diffract X-rays by means of crystals, because the crystals act as a three dimensional natural grating for X-rays, X-rays act as part of the electromagnetic radiation and X-rays are actually the radiations of very small wavelength probably of the order of 10^{-8}cm.

6.5.6. Bragg's law

A beam of X-rays consists of a bundle of separate waves; the waves can interact with one another such interaction is termed as interference. If all the waves in the bundle are in phase, that is their crests and troughs occur at exactly the same position, the same as being an integer number of wavelengths out of phase, $n\lambda$, $n = 1, 2, 3, 4$, etc., the waves will interfere with one another and their amplitudes will add together to produce a resultant wave that is has a higher amplitude the sum of all the waves that are in phase (Fig. 6.19).

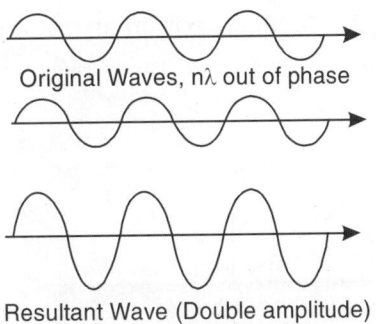

Fig. 6.19: X-ray waves that are in phase

If the waves are out of phase, being off by a non-integer number of wavelengths then destructive interference will occur and the amplitude of the waves will be reduced. In an extreme case, if the waves are out of phase by a multiple of $1/2\lambda$ $(n/2\lambda)$, the resultant wave will have no amplitude and thus be completely destroyed (Fig. 6.20).

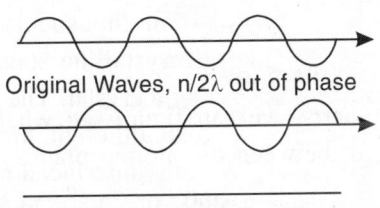

Resultant Wave-completely destroyed

Fig. 6.20: X-ray waves that are out of phase

The atoms in crystals interact with X-ray waves in such a way as to produce interference. The interaction can be thought of as if the atoms in a crystal structure reflect the waves. But, because a crystal structure consists of an orderly arrangement of atoms, the reflections occur from what appears to be planes of atoms. A beam of X-rays entering a crystal with one of these planes of atoms oriented at an angle of θ to the incoming beam of monochromatic X-rays.

Two such X-rays are shown in Fig. 6.21 where the spacing between the atomic planes occurs over the distance, d. Ray 1 reflects off of the upper atomic plane at an angle θ equal to its angle of incidence. Similarly, Ray 2 reflects off the lower atomic plane at the same angle θ. While Ray 2 is in the crystal, however, it travels a distance of 2a farther than Ray 1. If this distance 2a is equal to an integral number of wavelengths (nλ), then Rays 1 and 2 will be in phase on their exit from the crystal and constructive interference will occur.

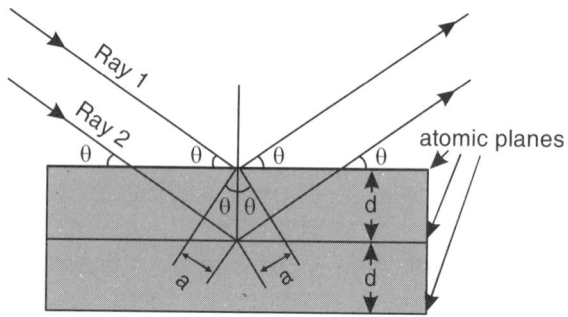

Fig. 6.21: The atoms in crystals interact with X-ray waves

If the distance 2a is not an integral number of wavelengths, then destructive interference will occur and the waves will not be as strong as when they entered the crystal. Thus, the condition for constructive interference to occur is:

$$n\lambda = 2a \qquad \text{... } eq.\ 6.17$$

But, from trigonometry, we can figure out what the distance 2a is in terms of the spacing, d, between the atomic planes ...

$$a = d\ \sin\theta \quad \text{or} \quad 2a = 2\ d\ \sin\theta \qquad \text{... } eq.\ 6.18$$

Thus, $$n\lambda = 2d\ \sin\theta \qquad \text{... } eq.\ 6.19$$

This is known as Bragg's Law for X-ray diffraction. If the wavelength of the X-rays going in to the crystal and can measure the angle θ of the diffracted X-rays coming out of the crystal, then know the spacing referred to as d-spacing between the atomic planes.

$$d = n\lambda/2\ \sin\theta \qquad \text{... } eq.\ 6.20$$

Again it is important to point out that this diffraction will only occur if the rays are in phase when they emerge, and this will only occur at the appropriate value of n (1, 2, 3, etc.) and θ.

6.5.7. Fourier theory

The diffraction pattern is related to the object diffracting the waves through a mathematical operation called the Fourier transform. Both the amplitude and the phase of the diffracted waves to compute the inverse Fourier transform. The number of photons gives the intensity, which turns out to be proportional to the square of the amplitude (peak height) of the diffracted wave. X-Ray diffraction crystallography has become a major tool for chemistry and biochemistry. It can easily provide detailed structures of molecules of a few atoms or with more difficulty molecular systems of a hundred thousand atoms. It is a tool used in physics and materials science to examine structural features of simple materials in exquisite detail. X-Ray diffraction is an optical technique in the sense that electromagnetic radiation is used to create an image of an object (Fig. 6.22).

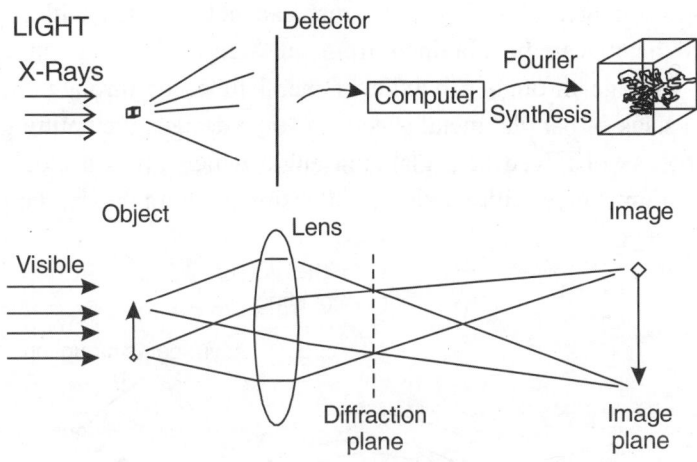

Fig. 6.22: The relation between x-ray crystallography and simple optics can be visualized.

In the X-ray case the objects are submicroscopic and the X-ray light has a wavelength very much shorter than visible light. The interactions between the light and the object to be imaged are similar in both visible and X-ray cases light is scattered from the electrons in the object. The similarities end when the scattered light is to be combined to form the image; the X-ray does not have the lenses or sharply curved mirrors possessed by the optician working with visible light to reconstruct the image. Instead the phase of the scattered rays must be deduced and then the image must be calculated by Fourier synthesis.

In the case of lens optics, one can think of rays of light scattering from points on an object and being collected and directed by the lens to form a magnified image of that point. Alternately, one can realize that all rays scattering in each direction are steered by the lens to meet at a point on a plane (the diffraction plane) where they will interfere. These diffracted rays then diverge and continue to the image plane, where they interfere again with all rays coming from a single point on the object but from different directions, to give the final image. In X-ray diffraction there is no lens, so the scattered rays must be measured directly. This measurement destroys the phase information, and this lost information must be recovered by some other method.

6.5.8. Production and detection of X-rays

Equipment for X-ray diffraction is essentially comparable to an optical grating spectrometer. Lenses and mirrors can not be used with X-rays. A collimated beam can be obtained from an X-ray tube with an extended target by passage through a bundle of metal tubes or through the spaces between a stack of parallel metal sheets. In some designs the emitting surface of the target is observed at a glancing angle which gives a close approxi mation to align source with maximum intensity. Such a tube can be mounted

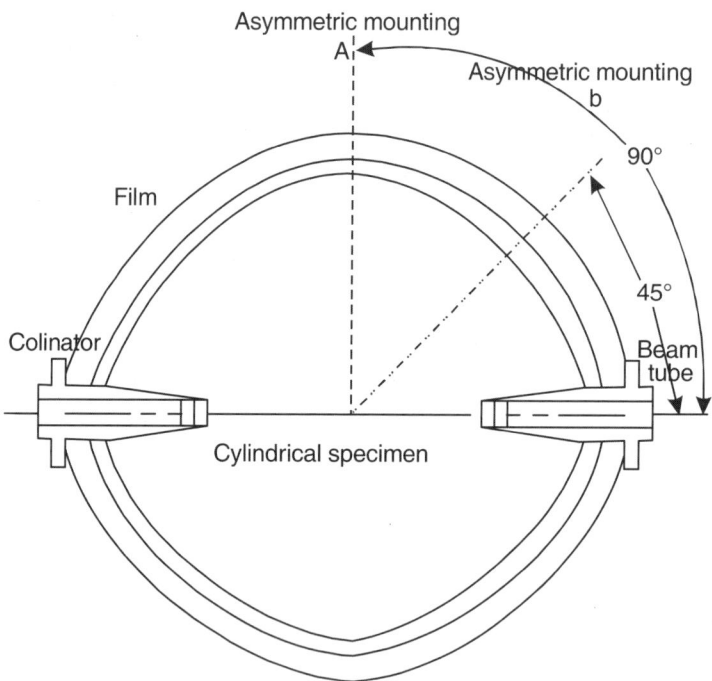

Fig. 6.23: The Debye Scherrer circular powder camera which uses a thin cylindrical powder specimen set on the axis of the film cylinder

vertically anode up and beams can be taken through each of several in different horizontal directions (Fig. 6.23).

X-ray diffracted beam can be detected photographically or by a Geiger or proportional counter and by a scintillation counter of the cadmium iodide type. The most common instrument for photographic recording of diffraction patterns in Debye-Scherer powder camera in which a thin cylindrical specimen is set on the film cylinder axis and can be rotated to increase the number of diffracting particles. The sample is prepared in the form of a fine homogenous powder, and a thin layer of it is inserted in path of the X-rays.

The beam form the Coolidge tube is filtered to produce a nearly monochromatic beam, which is collimated by passing through a narrow tube. The undiffracted radiation then passes out of the camera via a narrow exit tube. The power can be mounted on any non-crystalline supporting material, such as paper, with an organic mucilage or glue as adhesive.

6.5.9. Limitations

Two limiting cases of X-ray crystallography small-molecule and macromolecular crystallography are often discerned. Small molecule crystallography typically involves crystals with fewer than 100 atoms in their asymmetric unit; such crystal structures are usually so well resolved that the atoms can be discerned as isolated blobs of electron density. Macromolecular crystallography often involves tens of thousands of atoms in the unit cell. Such crystal structures are generally less well-resolved; the atoms and chemical bonds appear as tubes of electron density, rather than as isolated atoms.

6.5.10. Applications

1. X-ray diffraction leads primarily to the identification of crystalline compounds. Elements as such will be observed only if in the free crystalline state in which the response is to the element present. For example each of the oxides of iron gives its own pattern, and the appearance of a certain pattern proves the presence of that particular compound in the material being examined.

2. Quantitative determination of the relative amounts of the constituents of a mixture of solids is also possible because the power of diffracted beam depends on the quantity of the corresponding crystalline material in the sample.

3. Its discovery in 1912 by Von Lave, X-ray diffraction has provided of important information to science and industry. The arrangement and the spacing of atoms in crystalline materials have been directly deduced from diffraction studies. These studies proved important in understanding the physical properties of metals, polymeric materials and other solids. The X-ray diffraction patterns of a crystal structure provide sufficient information in determine the dimensions of the unit cell of the crystal lattice and the atomic arrangement within the cell.

4. X-ray diffraction provides a convenient and practical means for the qualitative identification of crystalline compounds. This application is based on the fact that the X-ray diffraction pattern is unique for each crystalline substance and so chemical identity can be assumed.

6.6. SUMMARY

The concept of pH was introduced by Sorensen in 1909. He recognized that hydrogen ion concentrations, while studying enzymatic reactions, he found it convenient to define a symbol which could represent the concentration of hydrogen ions and called this symbol as pH. It is defined by the following equation ...pH $= -\log_{10} C_H$

Nernst equation - The potential of the measuring electrode may be written by means of the Nernst equation ...

$$E = E_o + 2.3026 \ RT/F \ \log C_H$$

$$E = E_o - 2.3026 \ RT/F \ pH_c$$

The hydrogen electrode is the primary electrode. The hydrogen electrode consists of an inert but catalytically active metal surface, made with platinum, over which hydrogen is bubbled to achieve electrochemical equilibrium with the hydrogen ions in the solution. The potential set up at the hydrogen electrode by a given activity of hydrogen ions is governed by the Nernst equation.

Glass Electrode - Glass electrode is based on the principle, that when a thin membrane of glass is interposed between two solutions, a potential difference is observed across the glass membrane, which depends on the ions present in the solutions. The glass pH electrodes are constructed of special glass to crate the ion selective barrier needed to screen out hydrogen ions from all the other ions floating around in the solution.

Reference Electrode - The most common reference electrodes which meet these criteria are - Mercury/Mercurous electrode and Silver/Silver chloride electrode.

Temperature is a physical property of a system that underlies the common notions of hot and cold, something that feels hotter generally has the greater temperature. Temperature is one of the principal parameters of Thermo-dynamics. A thermometer is an instrument that measures the temperature of a system in a quantitative way. The easiest way to do this is to find a substance having a property those changes in a regular way with its temperature. The changes in temperature can be detected by using either platinum filament (hot wire) or thermistor.

The dissolved oxygen (DO) is oxygen that is dissolved in water. The oxygen dissolves by diffusion from the surrounding air, aeration of water that has tumbled over falls and rapids, and as a waste product of photosynthesis.

Mixing is a process where two or more substances enter a chamber where they are combined. The chamber can be a tank, hopper, length of pipe, or even an extruder barrel. Most often, stirring or agitation is the method to combine compounds. The main purposes of agitation - Elimination of solution stagnation and dispersal of reaction products, Increase of deposition rates by mass transfer enhancement, Dissipation of heat at electrode/solution inter-faces, Incorporation of particles in the deposit and Modification and control of deposit properties including thickness distribution.

A sensor is a device that measures a physical quantity and converts it into a signal which can be read by an observer or by an instrument. Laboratory sensors - In the laboratory, one of the best known types of sensor is the litmus paper test for acids and alkalis, which gives a qualitative indication, by means of a color reaction, of the presence or absence of an acid.

Biosensor - A biosensor can be defined as a device incorporating a biological sensing element connected to a transducer. The analyte in this sensor detects and measures may be purely chemical (even inorganic), although biological components maybe the target analyte. The key difference is that the recognition element is biological in nature.

X-rays are electromagnetic radiation with wavelengths between about 0.02 Å and 100 Å ($1Å = 10^{-10}$ meters). They are part of the electromagnetic spectrum that includes wavelengths of electromagnetic radiation called visible light which our eyes are sensitive to different wavelengths of visible light appear to us as different colors.

X-rays are produced in a device called an X-ray tube. It consists of an evacuated chamber with a tungsten filament at one end of the tube, called the cathode, and a metal target at the other end, called an anode. Electrical current is run through the tungsten filament, causing it to glow and emit electrons.

A crystal arranges huge numbers of molecules in the same orientation. The distances between the atoms in crystals have been found to be roughly equal to 10^{-8}cm. So, optical, electron microscopes can't be used in this field. It is possible to diffract X-rays by means of crystals, because - The crystals act as a three dimensional natural grating for X-rays, X-rays act as part of the electromagnetic radiation and X-rays are actually the radiations of very small wavelength probably of the order of 10^{-8}cm.

X- Ray Crystallography - Two limiting cases of X-ray crystallography small-molecule and macromolecular crystallography are often discerned. Small-molecule crystallography typically involves crystals with fewer than 100 atoms in their asymmetric unit; such crystal structures are usually so well resolved that the atoms can be discerned as isolated "blobs" of electron density. Macromolecular crystallography often involves tens of thousands of atoms in the unit cell. Such crystal structures are generally less well-resolved; the atoms and chemical bonds appear as tubes of electron density, rather than as isolated atoms.

Applications of X-Ray Diffraction - to the identification of crystalline compounds and determination of the relative amounts of the constituents of a mixture of solids is also possible because the power of diffracted beam depends on the quantity of the corresponding crystalline material in the sample.

Chapter 7
SEPARATION TECHNIQUES

7.0. HIGH PERFORMANCE LIQUID CHROMATOGRAPHY (HPLC)

7.0.1. Introduction

During 1970's, most chemical separations were carried out using a variety of techniques including open column chromatography, paper chromatography and thin-layer chromatography. However, these chromatographic techniques were inadequate for quantification of compounds and resolution between similar compounds. During this time, pressure liquid chromatography began to be used to decrease flow through time thus reducing purification times of compounds being isolated by column chromatography. However, flow rates were inconsistent and the question of whether it was better to have constant flow rate or constant pressure was debated. High pressure liquid chromatography was developed in the mid 1970's and quickly improved with the development of column packing materials and the additional convenience of on-line detectors. By the 1980's HPLC was commonly used for the separation of chemical compounds. Computers and automation added to the convenience of HPLC. Improvements in type of columns thus reproducibility was made as such terms as micro-column, affinity columns and fast HPLC began to immerge.

The past decade has seen a vast undertaking in the development of the micro-columns and other specialized columns. The dimensions of the typical HPLC column are XXX mm in length with an internal diameter between 3-5mm. The usual diameter of micro-columns or capillary columns ranges from 3μm to 200μm. proposed high pressure systems capable of operating at pressures up to 2.07×10^7 Nm^{-2} (3000 psi). In recent years a high pressure pump capable of operating at 1000 – 6000 psi is also used because of wide applications of HPLC in organic and inorganic chemistry in recent years.

Fast HPLC utilizes a column that is shorter than the typical column with a length of about 3 mm long and they are packed with smaller particles. The technique of high performance liquid chromatography is also known as high speed or high pressure liquid chromatography. The HPLC is a method of separation in which the stationary phase is contained in a column one end of which is attached to a source of pressurized liquid eluent/mobile phase.

7.0.2. Principle

In HPLC, eluent from the solvent reservoir is filtered, pressurized and pumped through the chromatographic column. A mixture of solutes injected at the top of the column is separated into components on traveling down the column and the individual solutes are monitored by the detector and recorded as peaks on a chart recorder. Hence the main components of a HPLC are a high pressure pump, a column/injector system and a detector. In addition components such as solvent reservoir, inline filters, pressure gauges, recorders, integrators and minor components are required. The essential features of a modern HPLC are shown in Fig. 7.1 and comprise the following basic components:

1. A solvent delivery system including pump
2. Sample injection system
3. A chromatographic column
4. A detector and recording system.

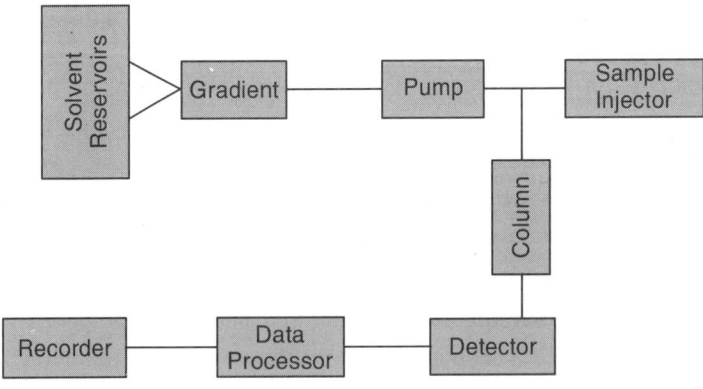

Fig. 7.1: Block diagram of modern HPLC

7.0.2.1. Solvent delivery system

The mobile phase is pumped under pressure from one or several reservoirs and flows through the column at a constant rate. It is advisable to use deaerated mobile phase solvent mixture using a vacuum pump means of deaeration that has no effect on the composition of the mixture. The choice of mobile phase is very important in HPLC and the eluting power of the mobile phase is determined by its overall polarity, the polarity of the stationary phase and the nature of the sample components. For normal phase separations eluting power increases with increasing polarity of the solvent, but for reversed phase separations, eluting power decreases with increasing solvent polarity. Optimum separating conditions can be achieved by making use of a mixture of two solvents and gradient elution is used where sample components vary widely in polarity.

Gradient elution system can be classified as low pressure and high pressure systems. In low pressure gradient elution system, the eluent components are mixed in proportion varying with time at low pressure and the mixture is pumped in order to be delivered at high pressure to the column. A single pump is used with low pressure gradient systems. The system must have as small as possible a volume, in order to limit the delay between the time of formation of a specific composition and the time it enters the column. In high pressure gradient elution system, components of fixed composition are each pumped by separate pump, and then mixed a high pressure in a ratio varying with time. In this system 2 or 3 pumps are used, each one pumping a different component of the eluent at a rate which depends on the composition of the eluent. The sum of the flow rated delivered by each pump is maintained constant during the run.

Solvents properties are separation of boiling point, viscosity of the solvent increases with increasing pressure, detector compatibility; flammability–many of the solvent used in HPLC are flammable as well as toxic, toxicity and compressibility for examples, a solvent generally decreases with increasing pressure.

7.0.2.2. Pumps

Another most important component of HPLC is the pump because its performance directly affects the retention time, reproducibility and detector sensitivity. In HPLC used pumps are reciprocating piston pumps/constant flow pumps, syringe type pumps, constant pressure pumps/non-reciprocating pumps and pneumatic pumps.

7.0.2.2.1. Reciprocating piston pumps

Reciprocating piston pumps consist of a small motor driven piston which moves rapidly back and forth in a hydraulic chamber that may vary from 35-400µl in volume. On the back stroke, the separation column valve is closed and the piston pulls in solvent from the mobile phase reservoir. On the forward stroke the pump pushes solvent out to the column from the reservoir. A wide range of flow rates can be attained by altering the piston stroke volume during each cycle or by altering the stroke frequency. Dual and triple head pumps consist of identical piston-chamber units which operate at 180 or 120 degrees out of phase. This type of pump system is significantly smoother because one pump is filling while the other is in the delivery cycle. Single head mechanical reciprocating pumps have been developed because of the following important advantage.

    a.   These pumps are the most convenient to use.

    b.   These are relatively easy to flush and rinse.

    c.   These are cheaper than multiple head pumps at similar levels of sophistication.

    d.   They are more reliable because most maintenance problems occur in connection with check valve.

7.0.2.2.2. Syringe type pumps

Syringe type pumps are most suitable because these pumps deliver only a finite volume of mobile phase before it has to be refilled. These pumps have a volume between 250 to 500 ml. The pump operates by a motorized lead screw that delivers mobile phase to the column at a constant rate. The rate of solvent delivery is controlled by changing the voltage on the motor.

7.0.2.2.3. Constant pressure pumps

In constant pressure pumps the mobile phase is driven through the column with the use of pressure from a gas cylinder. A low-pressure gas source is needed to generate high liquid pressures. The valving arrangement allows the rapid refill of the solvent chamber whose capacity is about 70 ml. This provides continuous mobile phase flow rates.

7.0.2.3. Sample injection system

The amounts of the sample to be injected into the column after the preparation can be determined by the components of interest have to be detected with accuracy after separation on the column. This amount depends on the two main factors (1) The sensitivity of the detector for these components (2) The extent of dilution undergone in the column. The injection process is depends on the following factors:

   a. Type of the injection system used.

   b. Connection pattern between the injector and the column.

   c. Injected volume as well as the injection time.

The injection system should be convenient to use, able to operate at high pressures, chemically inert with the eluent and the sample, and reproducible. The following three important ways are used to introduce the sample into the injection port.

7.0.2.3.1. Fixed volume valve injection

Fixed volume valve injection in which a fixed volume is introduced by making use of a fixed volume loop injector. The sample storage line is loaded by means of a load syringe. Excess sample is used in order to ensure that the storage line is completely filled. This line may either be an external loop with adequate volume which may be replace when injection volume is to be changed of fixed volume at drilled into the rotor. These devices deliver highly reproducible sample amounts.

7.0.2.3.2. Variable injection valve

Variable injection valve which a variable volume is introduced by making use of an injection valve. This type of injectors contains a needle port which can be closed at high pressure to insert the syringe. The required sample amount is injected into the loop which becomes partially filled with sample. Then solvent line switching causes the sample to be injected into the column. The sample is loaded at almost atmospheric pressure by means of syringe in a non-flushed flow line while the eluent flows directly from the pumping system to the column.

7.0.2.3.3. On column injection

On column injection is a variable volume is introduced by means of a syringe through a septum. With syringe injectors the needle of the syringe which contains the desired amount of the sample to be injected is introduced onto the top of the pressurized column through a self sealing elastomeric membrane and its content is discharged into the eluent stream. This is the simplest form of sample introduction.

7.0.2.4. Column packing

Depending on particle size two methods are used for column packing (1) For relatively large particles (>15 – 20 μm) a dry packing technique has been used which involves adding the dry column packing material slowly either continuously or in small portion while bouncing, tapping and rotating the column or by a modified tap fill method. (2) In the case of small particles, above technique is not suitable so a slurry packing technique is used. This technique involves coupling the column to be packed to a reservoir filling the column with supporting liquid and the reservoir with slurry of the column packing material in a suitable supporting liquid. The reservoir is connected to pump and a pressure of $3000 - 10,000$ psi is applied to the system in order to force the slurry into the analytical column. During slurry packing particle fractionation should be avoided by packing the slurry into the column at high pressure by making use of correct slurry liquid and by agitating the slurry in an ultrasonic bath before packing.

Silica and Alumina have been packed using balance density slurries i.e. with liquid such as methyl iodide and 2 dibromoethene. High viscosity liquids have also been used. Carbon tetrachloride slurries have been used successfully. The packing used in modern HPLC consist of small and rigid particles having a narrow particle size distribution. There are 3 main types of column packing in HPLC. (a) Porous, polymeric beds based on styrene-divinyl benzene co-polymers. These are used mainly for ion exchange and size exclusion chromatography. These have been now replaced by silica based pickings which are more efficient and mechanically more stable. (b)Porous layer beds (diameter 30-55 µm) consisting of a thin shell (1-3µm) of silica on a spherical inert core. After the development of totally porous micro particulate packing, these have not been used much in HPLC. (c) Totally porous silica particles with narrow particle size range (diameter <10µm). These packing have been used for analytical HPLC in recent years.

7.0.2.5. Detector

The detector for an HPLC is the component that emits a response due to the eluting sample compound and subsequently signals a peak on the chromatogram. It is positioned immediately posterior to the stationary phase in order to detect the compounds as they elute from the column. The detector for HPLC consists of a photometric detector fitted with a low volume flow cell. HPLC detectors are usually two types (1) Bulk property detectors, which compare an over all changes in a physical property of the mobile phase with and without an eluting solute. Examples of such detectors are refractive index and conductivity detectors. Refractive Index (RI) detectors measure the ability of sample molecules to bend or refract light. This property for each molecule or compound is called its refractive index. For most RI detectors, light proceeds through a bi-modular flow-cell to a photo detector. One channel of the flow-cell directs the mobile phase passing through the column while the other directs only the mobile phase. Detection occurs when the light is bent due to samples eluting from the column, and this is read as a disparity between the two channels. (2) Solute property detectors, which respond to a physical property of the solute which is not exhibited by the pure mobile phase. Such type of detectors is about 1000 times more sensitive, giving a detectable signal for a few nanograms of sample. Ultraviolet, visible adsorption, fluorescence and electrochemical detectors have achieved popularity in this category of detectors.

Radiochemical detection involves the use of radio labeled material, usually tritium (^3H) or carbon-14 (^{14}C). It operates by detection of fluorescence associated with beta-particle ionization, and it is most popular in metabolite research. (a) Homogeneous- Where addition of scintillation fluid to column effluent causes fluorescence. (b) Heterogeneous - Where lithium silicate and fluorescence caused by beta-particle emission interact with the detector cell, and has sensitivity limit up to 10^{-9} to 10^{-10} gm/ml.

7.0.3. Applications

1. The wide applicability, speed and sensitivity of HPLC have resulted in it becoming the most popular form of chromatography and all types of biological molecules have been purified.

2. Reverse phase HPLC is useful for the separation of polar compounds such as drugs and their metabolites, peptides, vitamins, polyphenols and steroids.

3. This technique is particularly widely used in clinical and pharmaceutical work as it is possible to apply biological fluids such as serum and urine directly to the column.

4. The separation of highly polar compounds such as amino acids, organic acids and the catecholamines.

5. HPLC has the biggest impact on the separation of oligopeptides and proteins. Proteins purified by this technique include aldolase, transferring, cytochrome-C and thyroglobulin.

7.1. GAS CHROMATOGRAPHY

7.1.1. Introduction

In gas chromatography, the sample is vaporized and injected onto the head of a chromatographic column. Elution is brought about by the flow of an inert gaseous mobile phase. In most other types of chromatography, the mobile phase does not interact with molecules of the analyte; its only function is the transport the analyte through the column. Two types of gas chromatography are encountered (1) Gas-solid chromatography (GSC): Gas solid chromatography is based upon a solid stationary phase in which retention of analytes is the consequence of physical adsorption. GSC has limited application owing to semi permanent retention of active polar molecules and severe tailing of elution peaks. Thus this technique has not found wide application except for the separation of certain low molecular weight gaseous species. (2) Gas-liquid chromatography (GLC): The concept of GLC was first developed in 1941 by Martin and Synge. In 1955, the first commercial apparatus for gas liquid chromatography appeared on the market. Gas liquid chromatography finds widespread use in all fields of science where its name is usually to gas chromatography (GC). Gas liquid chromatography is based upon the partition of the analyte between a gaseous mobile phase and a liquid phase immobilized on the surface of an inert solid.

7.1.2. Principle

The principle of gas liquid chromatography is almost similar to that of liquid partition chromatography except that the mobile phase in gas liquid chromatography is a gas rather than a liquid.The sample is introduces as a gas at the head of the column. As a result, components having finite solubility in the stationary liquid phase distribute themselves between this phase and the mobile gas phase in accordance with the equilibrium law. Elution is then carried out by forcing an inert gas such as He or N_2 through the column. The rate of movement of the various components along the column depends upon their tendency to dissolve in the stationary liquid phase. Components having a negligible solubility in the stationary phase move rapidly through the column, while those components whose distribution

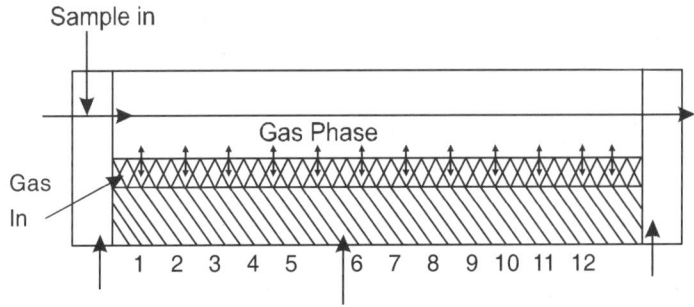

Fig. 7.2: Schematic representation of the chromatographic process

coefficient favors the solvent liquid phase, move with a low rate through the column. Ideally bell shaped elution curves are obtained. Qualitative identification of the components is based upon the time required for the peak to appear at the end of the column. Quantitative data are obtained from the evaluation of peak areas. A schematic representation of the gas liquid chromatography process is shown in the Fig. 7.2.

The mobile gas phase is continuously moving over the stationary liquid phase. When a sample of solute vapor is introduced into the gas stream at the head of the column, it is swept into the column and undergoes distribution between the gas and liquid phases in amore or less stepwise manner indicated by the arrows (Fig. 7.2). The greater total amount of liquid phase in the column, the more solute will dissolve in it and the greater will be the partition ratio. Similarly, an increase in the amount of liquid phase results in increased residence time in that phase, and the partition ratio increases. It should be noted that the gas phase simply serves to move the solute down the column between excursions into the liquid phase. All solutes spend the same time in the gas phase in any particular column. The partition ratio is simply the ratio of the amount of solute in the stationary phase to the amount of solute in the mobile phase.

$$K = \frac{\text{Amount of solute in the liquid phase}}{\text{Amount of solute in the gas phase}}$$

$$= \frac{\text{Time in the liuqid phase}}{\text{Time in the gas phase}} \qquad \textit{... eq. 7.1}$$

The partition ratio depends on the nature of solute, the nature of liquid phase, the amount of liquid phase and the temperature.

7.1.3. Apparatus

The Gas chromatography consists of the following main components.

1. A tank of carrier gas

2. An injection port of the sample

3. The column

4. Detector with appropriate readout

The essential components of the apparatus used for Gas Liquid Chromatography are given in the Fig. 7.3.

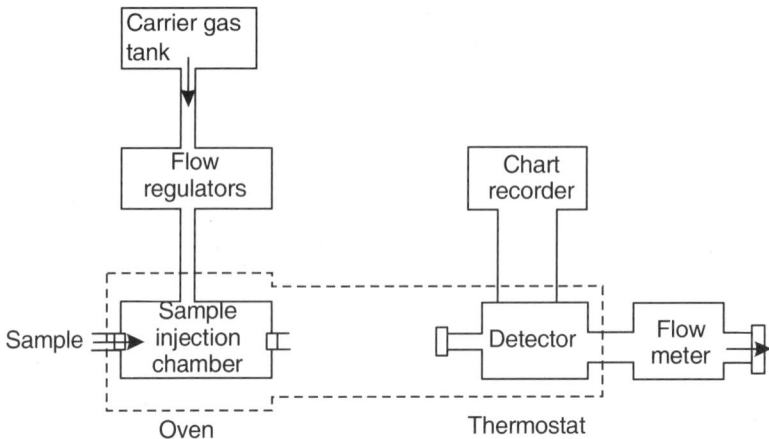

Fig. 7.3: Block diagram of a gas chromatographic apparatus

7.1.3.1. Carrier gas

The mobile phase in Gas Liquid Chromatography is usually Helium or Nitrogen, although Carbon dioxide and Hydrogen from tank sources have also been tried. The most important requirements of a carrier gas are (i) It should be inert. (ii) It should be available at low cost, because large quantities are used. (iii) It should allow the detector to respond in an adequate manner.

Helium or Nitrogen gases are most commonly used as carrier gas because there are two main reasons (a) Detectors depend on the thermal conductivity of the gas, a property that is much greater for hydrogen and helium than for other gases. (b) Other advantage also shared by hydrogen, is that because of its low density greater flow rates can be employed this reduces the time required for a separation.

The detector undergoes a marked increase in temperature Helium has a very high thermal conductivity by virtue of its molecular weight and size. When Nitrogen or Carbon dioxide is used as carrier gas detection by thermal conductivity is less sensitive. Disadvantages of hydrogen are its fire and explosion hazards, and its reactivity toward reducible or unsaturated sample compounds. Hydrogen has a specific advantage, however, in that it can be generated electrolytically.

A high pressure gas cylinder is used as a carrier gas reservoir. The cylinder is also attached with a pressure regulator to reduce and control the gas flow through the separation column.

7.1.3.2. Injection port

The most common injection method is micro syringe which is used to inject sample through a rubber septum into a flash vapouriser port at the head of the column. The temperature of the sample port is usually about 50°C higher than the boiling point of the least volatile component of the sample. For packed columns, sample size ranges from tenths of a micro liter up to 20μl. Capillary columns on the other hand need much less sample typically around 10^{-3}μl. For capillary GC, split/split less injection is used (Fig. 7.4).

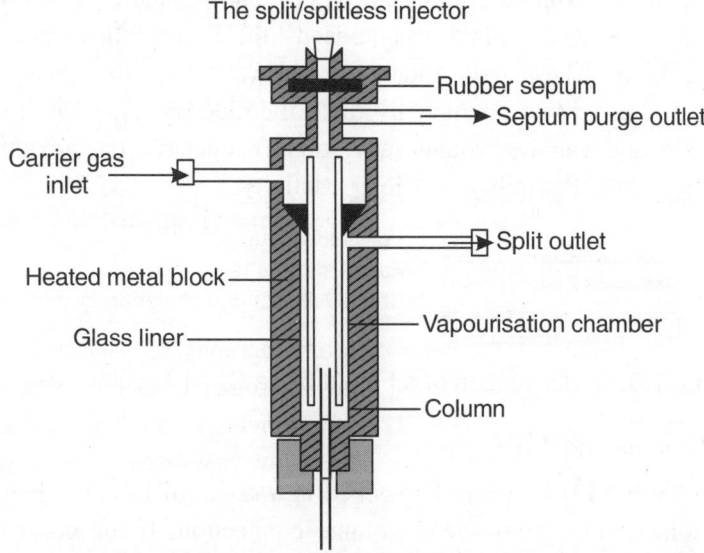

Fig. 7.4: The diagram of a split/split less injector

The injector can be used in one of two modes split or split less. The injector contains a heated chamber containing a glass liner into which the sample is injected through the septum. The carrier gas enters the chamber and can leave by three routes when the injector is in split mode. The sample vapourises to form a mixture of carrier gas, vapourised solvent and vapourised solutes. A proportion of this mixture passes onto the column, but most exits through the split outlet. Solid, liquid and gas samples are conveniently injected into the sample port by a syringe.

7.1.3.3. Columns

In Gas chromatography two types of columns are used (1) Packed columns contain a finely divided, inert and solid support material commonly based on diatomaceous earth coated with liquid stationary phase. Most packed columns are 1.5-10m in length and have an internal diameter of 2-4mm.

(2) Capillary columns have an internal diameter of a few tenths of a millimeter. These columns are classified into two types (a) Wall-coated open tubular (WCOT): Wall-coated columns consist of a capillary tube whose walls are coated with liquid stationary phase (Fig. 7.5). (b) Support-coated open tubular (SCOT): In support-coated columns, the inner wall of the capillary is lined with a thin layer of support material such as diatomaceous earth, onto which the stationary phase has been adsorbed.

SCOT columns are generally less efficient than WCOT columns. Both types of capillary column are more efficient than packed columns. In 1979, a new type of WCOT column was devised - the Fused Silica Open Tubular (FSOT) column. These have much thinner walls than the glass capillary columns, and are given strength by the polyimide coating. These columns are flexible and can be wound into coils. They have the advantages of physical strength, flexibility and low reactivity.

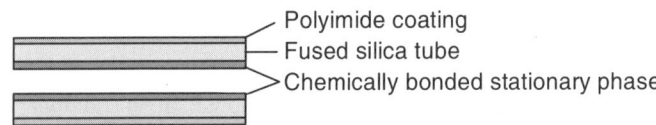

Fig. 7.5: Cross section of a Fused Silica Open Tubular Column

7.1.3.3.1 Column packing

Columns up to 50 ft long can be packed in a series of U's and then each U column joined with a low dead volume connection. If the columns were glass they were usually filled through an opening at the top of each U which was terminated in a plug of quartz wool and sealed-off in a blow-pipe flame. These long packed columns could be operated at a maximum of 200 psi. and could provide efficiencies of up to 50,00 theoretical plates. Straight columns are clumsy to use and occupy a large amount of space which is often difficult to thermostat. The coiled column although more difficult to pack has been readily accepted due to the compact nature of their design. To obtain adequate efficiencies, however, a special packing procedure had to be developed.

The packing is placed in a reservoir attached to a gas supply that forces the packing through the column. The column exit is connected to a vacuum pump. A wad of quartz wool is placed at the end of the column, constrained by a small restriction, which prevents the wad from being sucked into the pump. The vacuum and gas flow are turned on simultaneously and the packing is swept rapidly through the column (Fig. 7.6). This causes the material to be slightly compacted along the total length of the column and to produce well packed columns. The difficulties involved preparing packed columns have also contributed to the preferential popularity of the open tubular columns. The production of capillary columns can be largely automated and several columns can be prepared simultaneously.

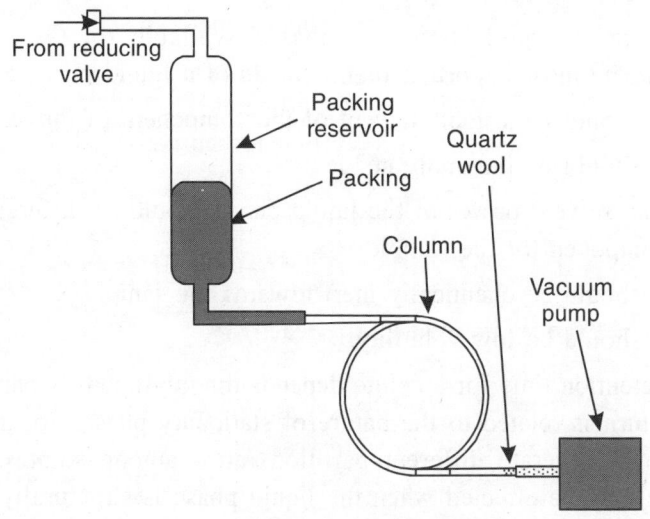

Fig. 7.6: Column packing apparatus

7.1.3.3.2. Column temperature

For precise work, column temperature must be controlled to within tenths of a degree. The optimum column temperature is dependant upon the boiling point of the sample. Temperature slightly above the average boiling point of the sample results in an elution time of 2-30 minutes. Minimal temperatures give good resolution but increase elution times.

7.1.3.4. Solid inert support

The main purpose of the solid support is to provide support to the thin and uniform film of liquid phase. The most important requirements of a solid support are:

   i. It should be porous hence has a large surface area.

   ii. It should be capable of providing a good mechanical strength.

   iii. It should consist of small, uniform and spherical particles.

   iv. It should be chemically inert at elevated temperatures.

   v. It should be readily wetted by the liquid phase to give a uniform coating.

The solid support should have no effect on the chromatographic process but it serves as a mechanical matrix for the liquid phase. Various supports have been formed from powdered Teflon, Alumina, Caborundum and Micro glass beads.

7.1.3.5. Stationary liquid phase

The number of liquid stationary phases available for GLC is almost unlimited. The most important requirements of a liquid phase are:

i. It should be a good solvent of the component of the sample.

ii. It should be thermally stable.

iii. The solvent power of the liquid phase should be different for each component of the sample.

iv. It should be chemically inert towards the sample.

v. It should be low volatility.

The retention time for a solute depends directly upon its partition ratio which in turn is related to the nature of stationary phase. So, in GLC the solvent must generate different partition ratios among solutes. The best separation may be effected when the liquid phase is structurally similar to the compounds being separated. For example, water and methyl alcohol, a semi polar liquid phase is most suitable. Similarly, hydrocarbons can be separated very effectively by using a non polar liquid phase. For aromatic hydrocarbons, benzyldiphenyl is good solvent (Table 7.1).

Table 7.1: Stationary phases and their application

S.No	Stationary phase	Max. Temp.	Applications
1.	Silicone oil	180 – 220	A wide variety of compound types
2.	Silicon rubber gum	300 – 350	Alcohols, Aromatics, Urinary compounds, Drugs and Alkaloids, Fatty acids, Sugars, Vitamins and Pesticides
3.	Cabowax 20M	200 – 250	Alcohols, Aromatics, Gases, Pesticides
4.	Cabowax 200	150	Aldehydes and Ketones

7.1.3.6. Detector

Detection devices for a gas liquid chromatograph must respond rapidly and reproducibility to the low concentrations of the solutes emitted from the column. Detector is used to respond to concentrations that are smaller by one or two orders of magnitude. In general detectors can also be classified into two groups (1) Those detectors that respond to the concentration (in mole fraction) of solute in the carrier gas (2) Those detectors that respond to the mass flow rate of the solute in moles per unit time.

The general requirements of a gas chromatographic detector are:

a. High sensitivity.

b. Physically suitable.

c. Capable of operable up to a maximum column temperature.

d. Ease of operation.

e. No response to undesired compounds.

f. Response to compounds for which analysis is required.

g. An output signal which is a linear function of the concentration of sample in the detector.

h. Linear response extending to high concentration.

The effluent from the column is mixed with hydrogen and air, and ignited. Organic compounds burning in the flame produce ions and electrons which can conduct electricity through the flame. A large electrical potential is applied at the burner tip, and a collector electrode is located above the flame. The current resulting from the pyrolysis of any organic compounds is measured. FIDs are mass sensitive rather than concentration sensitive; this gives the advantage that changes in mobile phase flow rate do not affect the detector's response. The FID is a useful general detector for the

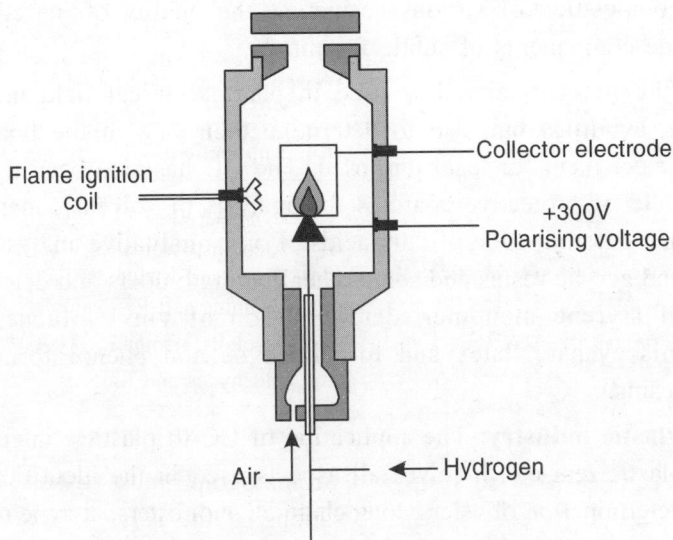

Fig. 7.7: The Flame Ionization Detector

analysis of organic compounds. It has high sensitivity, a large linear response range, and low noise. It is also robust and easy to use (Fig. 7.7).

7.1.4. Applications

1. **Food industry:** It has been used for determination of the colour and flavor of the foods. The technique is also been used for the determination of residual solvents in spice, oleoresins and pesticides in foods.

2. **Biochemical and clinical fields:** The technique is especially useful for applications involving body components of all types. Blood gases, estrogens, hydroxyl corticosteroids and vanilimnadelic acid have been determined and analysed in clinical medicine (areas of fatty acid and steroid analysis).

3. **Herbicides and pesticides:** To measure the small quantities of herbicides and pesticides control agents present on the surface of crops and animal tissues. Most effective pesticides are halogenated compounds which can be measured with excellent quantitative results by microcoulometric methods after pyrolysis of the gas chromatographic column effluents. Analysis of pesticide residues and determination of organophosphates in pesticides have successfully been carried out by using gas chromatography.

4. **Cosmetics and perfumes:** It is helpful in determining the composition of various cosmetics, the quality of ingredients and the components of subtle fragrances.

5. **Pharmaceuticals:** It is used in pharmaceutical field not only to assay drugs but also to determine their fate in the body and to detect them for legal purposed. The GC has also been used in the field of protective coatings for analysis of solvents, identification and determination of fatty acids of oils, qualitative analysis of alkyl and acrylic resins and some other coating binders and determination of styrene monomer, determination of vinyl toluene, toluene diisocyanate, latex and formaldehyde and phenol formaldehyde resins.

6. **Plastic industry:** The application of GC to plastics, intermediates, plasticizers etc. involves all aspects used in the identification and determination of esters, long chain alcohol esters, styrene monomers in styrene plastics and vinyl acetate in its copolymers.

7.2. ION-EXCHANGE CHROMATOGRAPHY

7.2.1. Introduction

Ion exchange can be defined as a reversible process in which ions are exchanged between liquid and solid, a highly insoluble body in contact with it. The solid known as an ion exchanger and no substantial organic polymer has been used as ion exchanger. The most common properties of all ion exchangers which have been used in analysis are:

1. They are almost insoluble in water and organic solvents, like benzene, carbon tetrachloride and ether.

2. They are complex in nature and infact they are polymeric.

3. They contain active or counter ions that will exchange reversibility with other ions in a surrounding solution without any substantial change in the material.

Exchange of ions with the electrolyte solution is known as Ion exchangers. Ion exchangers are porous solids swelling in water without dissolving in it. According to the composition of the main skeleton which binds the ionogenic groups into a whole system, the ion exchange materials may be classified as mineral and organic. The ion exchange materials of mineral origin include silicates, alumino silicates, aluminium hydroxide, zirconium phosphate etc. The ion exchange materials of organic origin are the products of chemical treatment of coal or lignin or manmade high molecular organic compounds containing the ionogenic groups. Natural or artificial ion exchangers can be used for water treatment. Natural materials are glauconites and humic coals. Manmade ion exchange materials are sulphonated coals and synthetic ion exchange resins. Glauconites are amorphous ferroalumino silicates, Humic coals are used to prepare cation exchangers, contain 15% of humic acids and sulphonated coals are prepared by treating coals extracted at the coal basins with conc. suphuric acid.

Ion exchange resins are three dimensional network polymers insoluble in water and containing ionogenic groups i.e. groups capable of ion exchange. Ion exchange resin as a special type of a polyelectrolyte and consists of three dimensional polymeric hydrocarbon network to which are bonded a large number of electrically charged groups, such as SO_3^- (sulphonated group) or $N(CH_3)_4^+$ (quaternary ammonium group).

7.2.2. Cation exchange resins

A cation exchange resin may be defined as a high molecular weight, cross-linked polymer containing sulphonic, carboxylic, phenolic etc. groups as an integral part of the resin and equivalent amount of the cations. Ion exchange resins having sulphonic groups as the exchange sites are usually known as strongly acidic cation exchanger resins. The identity of the cation which is held by each sulphonated is determined by chemical equilibrium with the surrounding solution. When such a resin is treated with a strong acid (5% HCl) all the sulphonated present will be converted to the acid form i.e. the hydrogen form. The resin in this form acts as an insoluble strong acid and can be represented as $(RSO^-_3) H^+$, Where R represents the resin network. Now a solution of NaCl is passed through the hydrogen form of the resin, the H^+ ions will be replaced by Na^+ ions, because the sulphonated group has less attraction for the hydrogen ion than for the sodium ion. The equilibrium can be represented as:

$$(RSO^-_3) H^+ + Na^+ \rightarrow (RSO^-_3) Na^+ + H^+ \qquad \textit{... eq. 7.2}$$

From the above equilibrium for each equivalent of sodium ion, one equivalent of hydrogen ion is freed. The equilibrium position is dependent mainly upon concentration of Na^+ ions in the solution in contact with resin and acid strength of the resin. For a strong acidic cation exchange resin, the exchange affinity for cations depends on the charge of the cation.

7.2.3. Anion exchange resins

An anion exchange resin is a polymer containing anion or quaternary ammonium groups as integral parts of the resin and equivalent amount of anions, such as Cl^-, SO^{2-}_4, OH^- ions etc. The representing an anion exchange resin also includes $R - N (CH_3) H$ and $R-N (CH_3)_2$. When an anion exchange resin is treated with hydrochloric acid, the substituted ammonium cations are obtained.

$$R - NH_2 + H^+ + Cl^- \rightarrow R - NH^+_3 Cl^- \qquad \textit{... eq. 7.3}$$

When treated with solutions of any inoised material, exchange takes place as.

$$R - NH^+_3 Cl^- + Na^+ + OH^- \rightarrow R - NH^+_3 OH^- + Na^+ + Cl^- \textit{... eq. 7.4}$$

$$2R - NH^+_3 Cl^- + 2H^+ + SO^{2-}_4 \rightarrow (R - NH^+_3)_2 SO^{2-}_4 + 2HCl \textit{... eq. 7.5}$$

7.2.4. Types of ionisable groups

Ionisable groups are two types which retain their useful properties and are attached to the hydrocarbon skeleton by covalent bonds in a resin. (a) Acidic groups which give rise to cation exchange resins. (b) Basic groups which give rise to anion exchanger resins. Both these groups can be subdivided according to the strength of their groups. (i) Strong acid resin: Such types of resins are useful for amino acids. The resin is effective over the entire range of pH in both the acid and the salt forms. Resins prepared by nuclear sulphonated of cross linked polystyrene belong to this class. (ii) Weak acid resign: These resins are based on polymers of methancrylic acid and possess carboxyl groups. These resins are available with several degrees of cross linking. (iii) Strong base resins: Resins with positively charged quaternary ammonium groups attached to cross linked polystyrene frame work belong to this class. Trimethyl ammonium groups are used for this resin. These groups $[C_6H_4CH_2N (CH_3)_3]^+$ are ionised over the wide range of pH. (iv) Weak base resins: Tertiary amine resins and polyamine resins, having a mixture of primary, secondary and tertiary amine groups on the polystyrene net work are well known examples of weak base resins. The tertiary amine resins are effective between 0-9 pH, while polyamine resins are effective only between 0-7 pH.

7.2.5. Ion exchange equilibrium

Ion exchange is a reversible process, in which the replacement of the exchangeable ions A_x in the resin by ions of like sing $B\gamma$ from a solution takes place. Thus, the in exchange process can be represented as:

$$A_x + B\gamma \rightarrow B_x + A\gamma \qquad \qquad ...\ eq.\ 7.6$$

The various factors determining the distribution of ions between an exchange resin and a solution are:

1. Nature of exchanging ion

a. In has been observed that at ordinary temperature the extent of exchange increases with the increasing valence of the exchanging ion, provided the concentration is low.

   Example: $Na^+ < Ca^{2+} < Al^{3+} < Th^{4+}$

b. Under similar conditions and constant valence, for univalent ions, the extent of exchange increases with decrease in size of the hydrated cation. But for divalent ions, both ionic size and the incomplete dissociation of salts of bivalent metals are important.

c. For strongly basic anion exchange resins, the behavior of univalent anions is similar to univalent cations.

   $(F^- < OH^- < HCO^-_3 < Cl^- < HSO^-_3 < CN^- < Br^- < NO^{2-} < I^-)$

d. When a cation in solution is being exchanged for an ion of different valency the relative affinity of the ion having higher valence increases in direct proportion to the dilution. Thus the exchange will be favored by high dilution, if the lower valence ion is present in the exchanger and the higher valence ion is in solution, whereas the exchanger will be favored by increase in concentration. If lower valence ion is in the solution and higher valence ion is in the exchanger.

2. Nature of ion exchange resin

The adsorption of ions depends upon the nature of the functional groups in the resin and upon the degree of cross linking. The resin becomes more selective towards ions of different sizes as the degree of cross linking increases. Under such conditions ions with smaller hydrated volume will be absorbed preferentially.

7.2.6. Basis for separations

Separation of mixture is based on one of the three different principles (1) At low concentration, the extent of exchange increases with increasing valence of the exchanging ion. (2) At low concentration and constant valence, the extent of exchange increases with increasing atomic number. (3) The extent of exchange is strongly influenced by the formation of complexes.

7.2.7. Principle

The principle of gas chromatography is the attraction between oppositely charged particles. Many biological materials like amino acids and proteins have ionisable groups and the fact that they may carry a net positive or negative charge can be utilized in separating mixtures of such compounds.

Ion-exchange separations are carried out in columns packed with an ion exchanger. There are two types of ion-exchanger namely cation and anion exchangers. Cation exchangers possess negatively charged groups and these will attract positively charged molecules. These exchangers also called acidic ion exchange materials since their negative charges result from the protolysis of acidic groups. Anion exchangers have positively charged groups which will attract negatively charged molecules. The term basic ion exchange materials are also used to describe exchangers since positive charges generally result from the association of protons with basic groups.

The actual ion-exchange mechanism is composed of 5 steps.

    i. Diffusion of the ion to the exchanger surface this occurs very quickly in homogeneous solutions.

ii. Diffusion of the ion through the matrix structure of the exchanger to the exchange site. This is dependent upon the degree of cross-linkage of the exchanger and the concentration of the solution. This process is thought to be the feature which controls the rate of the whole ion-exchange process.

iii. Exchange of ions at the exchange site. This is thought to occur instantaneously and is an equilibrium process.

Cation exchanger ...

$$RSO^-_3 ... Na^+ + N^+H_3R' \rightarrow RSO^-_3\ N^+H_3\ R' + Na^+ \qquad ...\ eq.\ 7.7$$
Excha- Counter   bound   exchanged
Nger   ion   molecular ion   ion

Anion exchanger ...

$$(R)_4\ N^+ ... Cl^- + {}^-OOCR \rightarrow (R)_4\ N^+ ... OOCR' + Cl^- \qquad ...eq.7.8$$

The more highly charged the molecule to be exchanged, the tighter it binds to the exchanger and the less readily it is displaced by other ions.

iv. Diffusion of the exchanged ion through the exchanger to the surface.

v. Selective desorption by the eluent and diffusion of the molecule into the external solution. The selective desorption of the bound molecule is achieved by changes in pH and ionic concentration or by affinity elution, in which case an ion which has greater affinity for the exchanger than has the bound molecule is introduced into the system.

In the separation of amino acids various variables which are to be taken into consideration are choice of ion exchanger, pH, and concentration of ionic species in the eluting agent and the type of side chain on the amino acid. The amino acid mixture is quantitatively applied to the cation resin in the column and the eluting agent is pumped through the column. The effluent is mixed with ninhydrin in reaction chamber and passed through a spectrophotometer. When an amino acid emerges from the column, the ninhydrin reaction takes place and absorption is observed.

In the column the mixture is heated to 105°C to develop the color, the intensity of which is then determined at 570 nm to monitor the majority of the amino acids and a second set at 440 nm to monitor the colour produced by proline and hydroxyproline. Many amino acid analyzers are using two separate columns, the second one containing an anion exchanger to separate the basic amino acids and ammonia faster and more effectively.

7.2.8. Applications

1. Ion exchange resins have been used in the laboratory in many ways in the removal of interfering ions, for group separations, for concentration of samples, in establishing the charges on ions, in the purification of samples, in the preparation of standard solutions, in determining the formation constants of complexes and the resins can be used as acid and base catalysts.

2. Demineralization water - Ion exchange resins have been used for the complete demineralization of water for chemical use. The resulting water is as pure as distilled water and is far less expensive. Water purification by this method in comparison to distillation has two main advantages the ion exchange procedure is cheaper as well as faster than distillation in the laboratory and it satisfies the purity requirements of a scientific laboratory.

3. In removing carbonate from a solution of sodium hydroxide.

4. Total cation content of a sample solution can be determined by passing the sample through a hydrogen cation exchanger.

5. Preparation of standard solutions.

6. Useful application of ion exchange resin is in determining the concentration of traces of an ion from a very dilute solution. Cation exchange resins are used to collect traces of metallic elements from larger volumes of natural water.

7. Separation of isotopes–Isotopes of boron, beryllium, calcium, cobalt and uranium are separated on ion exchange columns.

8. Recently ion exchange finds a large number of applications in metallurgy, like separation of rare earths, isolation of transuranic elements, production of Uranium and separation of Zirconium and Hafnium.

7.3. GEL FILTRATION CHROMATOGRAPHY

7.3.1. Introduction

Gel filtration chromatography is a separation technique based on size and it is also called as molecular exclusion or gel permeation chromatography. In gel filtration chromatography, the stationary phase consists of porous beads with a well defined range of pore sizes. The stationary phase for gel filtration is said to have a fractionation range, meaning that molecules within that molecular weight range can be separated.

Proteins that are small and can fit inside all the pores in the beads and are said to be included. These small proteins have access to the mobile phase inside the beads as well as the mobile phase between beads and elute last in a gel filtration separation. Proteins that are too large to fit inside any of the pores are said to be excluded. They have access only to the mobile phase between the beads and, therefore, elute first. Proteins of intermediate size are partially included, they can fit inside some but not all of the pores in the beads. These proteins will then elute between the large (excluded) and small (totally included) proteins.

The separation of a mixture of glutamate dehydrogenase (MW 290,000), lactate dehydrogenase (MW 140,000), serum albumin (MW 67,000), ovalbumin (MW 43,000) and cytochrome c (MW 12,400) on a gel filtration column packed with Bio-Gel P-150 (fractionation range 15,000 - 150,000). When the protein mixture is applied to the column, glutamate dehydrogenase would elute first because it is above the upper fractionation limit. Therefore it is totally excluded from the inside of the porous stationary phase and would elute with the void volume (V_0). Cytochrome c is below the lower fractionation limit and would be completely included, eluting last. The other proteins would be partially included and elute in order of decreasing molecular weight.

These separations can be described by the equation

$$Vr = V_0 + KV_i \qquad \qquad \textit{... eq. 7.9}$$

$$V_r = V_0 + KV_i \qquad \qquad \textit{... eq. 7.10}$$

Where V_r is the retention volume of the protein, V_0 is the volume of mobile phase between the beads of the stationary phase inside the column (void volume), V_i is the volume of mobile phase inside the porous beads (included volume) and K is the partition coefficient (the extent to which the protein can penetrate the pores in the stationary phase, with values ranging between 0 and 1). The partition coefficient (K) for glutamate dehydrogenase would be 0 (totally excluded), $K = 1$ for cytochrome c (totally included) and K would be between 0 and 1 for the other proteins, which are within the fractionation range for the column.

Gel filtration can be used to separate proteins by molecular weight at any point in a purification of a protein. It can also be used for buffer exchange - a protein dissolved in a sodium acetate buffer, pH 4.8, can be applied to a gel filtration column that has been equilibrated with tris buffer pH 8.0. Using the tris buffer pH 8.0 as the mobile phase the protein moves into the tris mobile phase as it travels down the column while the much smaller sodium acetate buffer molecules are totally included in the porous beads and travels much more slowly than the protein.

7.3.2. Principle

In gel filtration or molecular exclusion chromatography, molecules in solution are separated by size as they pass through a column of cross linked beads that form a three dimensional network. These polymer beads are made of dextran, agarose, or acrylamide comprise the stationary phase. The liquid phase is the solvent that is found both around the beads and in the pores of the stationary phase matrix.

Sample passes through the column there are two routes that molecules can take through the column depending on their size and the size of the pores in the beads. Molecules that are larger than the pores will not enter the stationary phase staying in the solution (mobile phase) surrounding the beads hence they elute first from the column. Smaller molecules will enter the pores in the beads and so move more slowly through the column (Fig.-7.8). Molecules of intermediate size will enter the stationary phase to some extent, but will not spend as much time there as do the smaller molecules. Larger molecules will elute from the column first, and the smallest molecules will elute last, with intermediate molecules strung out in between.

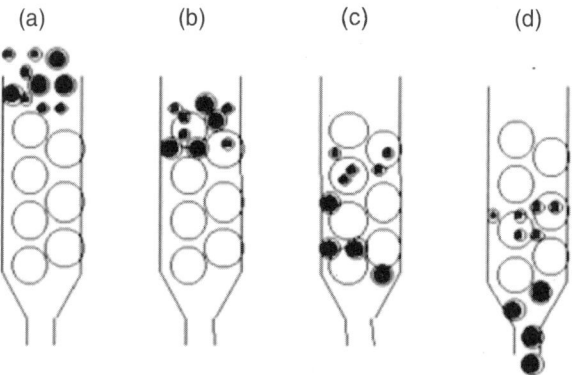

Fig. 7.8:   Gel filtration chromatography. Open circles represent the gel filtration beads, large filled circles are large molecules, and small filled circles are small molecules.

As the samples elute from the bottom of the column, they are collected in tubes as fractions. Usually fractions of a particular volume are collected. In most cases, the eluted fractions must be tested to determine both what fractions contain the samples (usually proteins) and how much of the sample (protein) is in the fraction. Several methods that are commonly used are spectrophotometric examination of the fractions, SDS-PAGE of the fractions and assaying the fractions for a particular enzymatic activity.

Once the sample concentrations in the fractions have been determined, the concentrations can be graphed against the elution volume to create an elution profile (Fig. 7.9). The first volume that elutes from the column before any fractions is called the void volume (V_0) of the column. Any proteins or other molecules that are too large to enter the pores of the beads will elute immediately after the void volume.

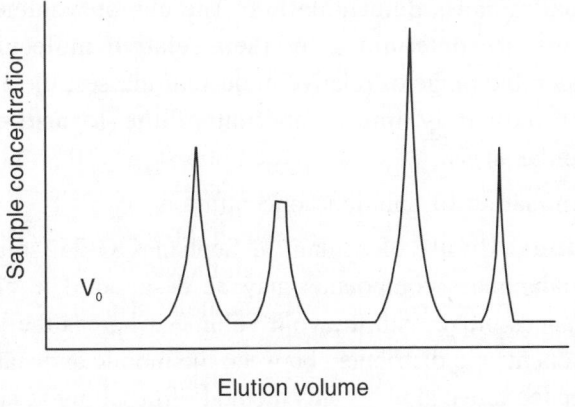

Fig. 7.9: The concentrations graphed against the elution volume to create an elution profile.

If the molecular weights of the proteins are already known, then these data can be used to create a standard curve to calculate the molecular weight

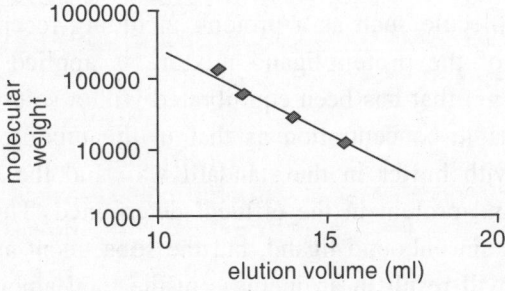

Fig. 7.10: The elution volume is plotted against the known molecular weights of the proteins on semi log graph paper.

of an unknown protein. The elution volume is plotted against the known molecular weights of the proteins on semi log graph paper (Fig. 7.10) and a line derived. From this line and the elution volume of an unknown protein, the molecular weight of the unknown can be estimated.

7.3.3. Applications

1. **Purification:** The main application of exclusion chromatography is in the purification of biological macromolecules by facilitating their separation from larger and smaller molecules. Viruses, enzymes, hormones, antibodies, nucleic acids and polysaccharides have all been separated and purified.

2. **Molecular mass determination:** The elution volumes of globular proteins are determined by their relative molecular mass. A considerable range of relative molecular masses, the elution volume is approximately linear function of the logarithm of relative molecular mass.

3. Determination of solution concentration

4. **Desalting:** By use of column of Sephadex G-25, solutions of high molecular mass compounds may be desalted. The high molecular substances move with the void volume, whereas the low molecular components are distributed between the mobile and stationary phases and hence move slowly. This method of desalting is faster and more efficient than dialysis. Applications include removal of phenol from nucleic acid preparations, ammonium sulphate from protein preparations and salt from samples eluted from ion-exchange chromatography columns.

5. **Protein-binding studies:** Exclusion chromatography is one of method used to study the reversible binding of ligand to a macromolecule such as a protein, including receptor proteins. A sample of the protein/ligand mixture is applied to a column of suitable gel that has been equilibrated with a solution of the ligand of the same concentration as that in the mixture. The sample is eluted with buffer in the standard way and the concentration of ligand and protein in the effluent determined. The early fractions will contain unbound ligand, but the subsequent appearance of the protein will result in an increase in the total amount of ligand.

7.4. AFFINITY CHROMATOGRAPHY

7.4.1. Introduction

Affinity chromatography is a term which now covers a variety of methods of enzyme purification, the common factor of which is the more or less specific interaction between the enzyme and the immobilized ligand. In its most specific form the immobilized ligand is a substrate or competitive inhibitor of the enzyme. Ideally it should be possible to purify an enzyme from a complex mixture in a single step and indeed purification factors of up to several thousand-folds have been achieved. An alternative, equally specific approach is to use an antibody to the enzyme as the ligand. Such specific matrices are very expensive and cannot be generally employed on a large scale. Additionally, they often do not perform as well as might be expected due to non-specific binding effects. In general, affinity chromatography achieves a higher purification factor than ion-exchange chromatography.

A specific approach suitable for many enzymes is to use analogues of coenzymes such as NAD^+, as the ligand. This method has been used successfully but has now been superceded by the employment of a series of water soluble dyes as ligands. These are much cheaper and usually by trial and error have been found to have surprising degrees of specificity for a wide range of enzymes. This dye-affinity chromatography was purportedly discovered by accident, certain enzymes being found to bind to the bluedyed dextran used as a molecular weight standard to calibrate gel exclusion columns. Another fortuitous discovery was hydrophobic interaction chromatography found when it was noted that certain proteins were unexpectedly retained on affinity columns containing hydrophobic spacer arms. Hydrophobic adsorbents now available include octyl or phenyl groups. Hydrophobic interactions are strong at high solution ionic strength so samples need not be desalted before application to the adsorbent. Elution is achieved by changing the pH or ionic strength or by modifying the dielectric constant of the eluent using for instance ethanediol. Cellulose has more hydroxyl groups hence this material (Whatman HB1) is designed to interact with proteins by hydrogen bonding. Samples are applied to the matrix in a concentrated 50% saturated > 2M solution of ammonium sulphate and proteins are eluted by diluting the ammonium sulphate. This introduces more water which competes with protein for the hydrogen bonding sites. The selectivity of both of these methods is similar to that of fractional precipitation using ammonium sulphate but their resolution may be somewhat improved by their use in chromatographic columns rather than batch wise.

Careful choice of matrices for affinity chromatography is necessary. Particles should retain good flow and porosity properties after attachment of the ligands and should not be capable of the non-specific adsorption of proteins. Agarose beads fulfill these criteria and are readily available as ligand supports. Affinity chromatography is not used extensively in the large scale manufacture of enzymes primarily because of cost.

7.4.2. Matrix

The most common matrixes are the cross linked dextrans (Sephacryl S), agarose (Sepharose, Bio-Gel A), Polyacrylamide gels (Bio-Gel P), cellulose, porous glass and silica. An ideal matrix for affinity chromatography must possess the following characteristics.

1. It must contain suitable and sufficient chemical groups to which the ligand may be covalently bind and must be stable under the conditions of the attachment.

2. It should exhibit good flow properties.

3. It must be stable during binding of the macromolecule and its subsequent elution.

4. It must interact only weakly with other macromolecules to minimize non-specific adsorption.

7.4.3. Ligand

The chemical nature of a ligand is determined by the prior knowledge of the biological specificity of the compound to be purified. It is possible to select a ligand that displays absolute specificity in that it will bind exclusively to one particular compound. It is possible to select a ligand that displays group selectivity in that it will bind to a closely related group of compounds that possess a similar in built chemical specificity (Table 7.2).

The most common method of attachment of the ligand to the matrix involves the preliminary treatment of the matrix with cyanogens bromide (CNBr). The reaction conditions and the relative proportion of the reagents will determine the number of ligand molecules that can be attached to each matrix particle. Alternative coupling procedures involve the use of bis-epoxides, N, N'-disubstituted carbodiimides, sulphonyl chloride, N-hydroxysuccinide ester and dichlorotriazines.

Table 7.2: Examples of group-specific ligands commonly used in affinity chromatography

S. No.	Ligand	Affinity
1.	5′AMP	NAD^+-dependent dehydrogenases
2.	2′5′-ADP	$NADP^+$-dependent dehydrogenases
3.	Calmodulin	Calmodulin-binding enzymes
4.	Avidin	Biotin-containing enzymes
5.	Fatty acids	Fatty-acid-binding enzymes
6.	Proteins A & G	Immunoglobulins
7.	Poly (A)	RNA containing poly (U) sequences, some RNA-specific proteins
8.	Lysine	rRNA

7.4.4. Principle

The goal of affinity chromatography is to separate all the molecules of a particular specificity from the whole choice of molecules in a mixture such as a blood serum. For example the antibodies in a serum sample specific for a particular antigenic determinant can be isolated by the use of affinity chromatography. In affinity chromatography the following steps are present.

Step–1: An immunoadsorbent is consists of a solid matrix to which the antigen has been coupled (covalently). Agarose, Sephadex, derivatives of cellulose, or other polymers can be used as the matrix (Fig. 7.11).

Step–2: The serum is passed over the immunoadsorbent. As long as the capacity of the column is not exceeded those antibodies in the mixture specific for the antigen will bind (non-covalently) and be retained. Antibodies of other specificities and other serum proteins will pass through unimpeded (Fig. 7.11).

Step–3: Elution: A reagent is passed into the column to release the antibodies from the immunoadsorbent. Buffers containing a high concentration of salts or low pH are often used to disrupt the noncovalent interactions between antibodies and antigen. A denaturing agent such as 8 M urea will also break the interaction by altering the configuration of the antigen binding site of the antibody molecule. Another approach is to elute with a soluble form of the antigen. These compete with the immunoadsorbent for the antigen-binding sites of the antibodies and release the antibodies to the fluid phase (Fig. 7.11).

Step–4: Dialysis: The eluate is then dialyzed for example buffered saline in order to remove the reagent used for elution (Fig. 7.11).

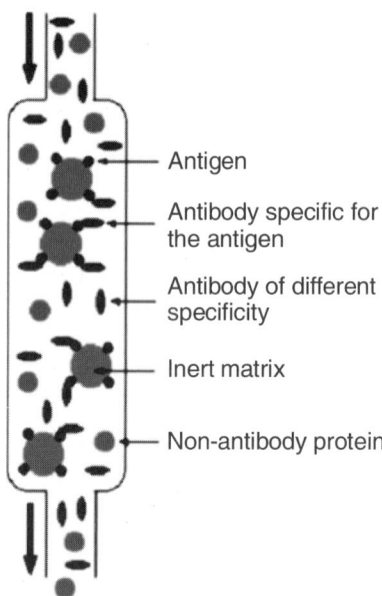

Fig. 7.11: Affinity Chromatography.

Affinity chromatography is designed to purify a particular protein from a mixed sample (Fig. 7.12 – 7.17).

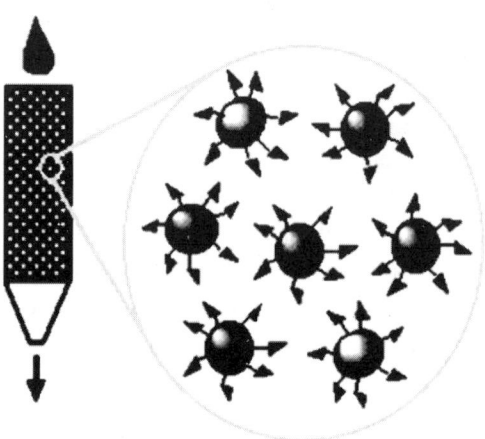

Fig. 7.12: Loading affinity column

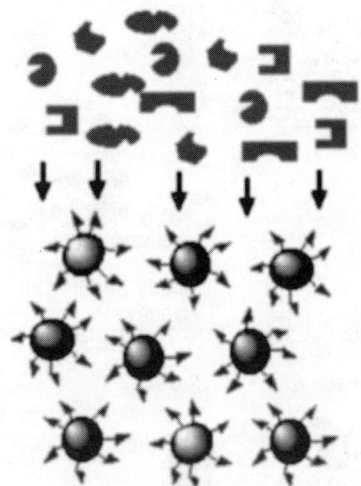

Fig. 7.13: Proteins sieve through matrix of affinity beads

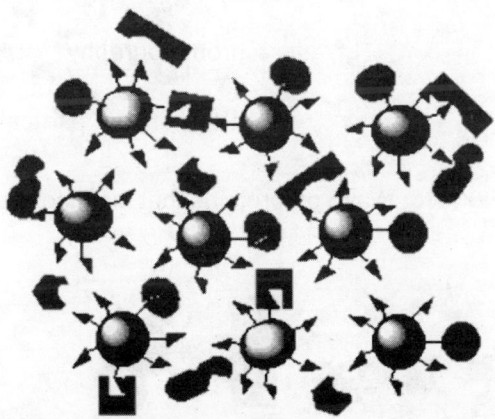

Fig. 7.14: Proteins interact with affinity ligand with some binding loosely and others tightly

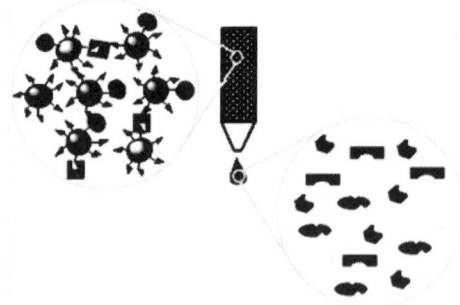

Fig. 7.15: Wash off proteins that do not bind

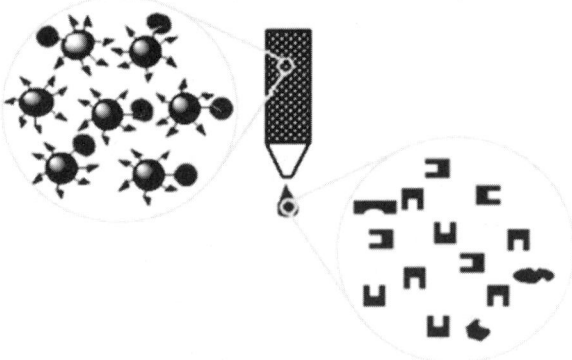

Fig.7.16: Wash off proteins that bind loosely

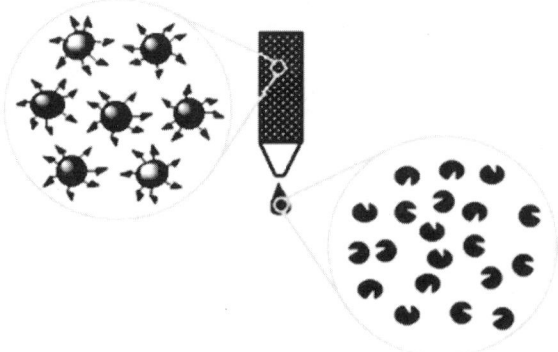

Fig. 7.17: Elute proteins that bind tightly to ligand and
collect purified protein of interest

7.4.5. Applications

1. A wide range of enzymes and other proteins including receptor proteins and immunoglobulins has been purified by affinity chromatography.

2. The application of the technique is limited only by the availability of immobilized ligands.

3. The principles have been extended to nucleic acids and have made a considerable contribution to developments in molecular biology. Messenger RNA for example is isolate by selective hybridization on poly (U) Sepharose 4B by exploiting its poly (A) tail.

4. Immobilized single-stranded DNA can be used to isolate complementary RNA and DNA.

5. Affinity chromatography is used for the separation of a mixture of cells into homogeneous populations.

6. The technique relies on the antigenic properties of the cell surface or the chemical nature of exposed carbohydrate residues on the cell surface or on a specific membrane receptor-ligand interaction.

7. The immobilized ligands used include protein-A, which binds to the F_c region of I_gG, a lectin or the specific ligand for a membrane receptor.

7.5. MEMBRANE SEPARATION

The water like solutions that pass through the membrane are referred to as permeate. The rate at which the permeate flows through the membrane is called the flux rate. Membrane processing is a technique that permits concentration and separation without the use of heat. Particles are separated on the basis of their molecular size and shape with the use of pressure and specially designed semi-permeable membranes. When a solution and water are separated by a semi-permeable membrane the water will move into the solution to equilibrate the system this is known as osmotic pressure. If a mechanical force is applied to exceed the osmotic pressure up to 700 psi, the water is forced to move down the concentration gradient i.e. from low to high concentration. **Permeate** designates the liquid passing through the membrane and **retentate** (concentrate) designates the fraction not passing through the membrane (Fig. 7.18 and 7.19).

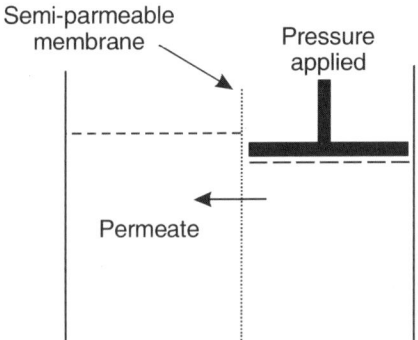

Fig. 7.18: Membrane processing

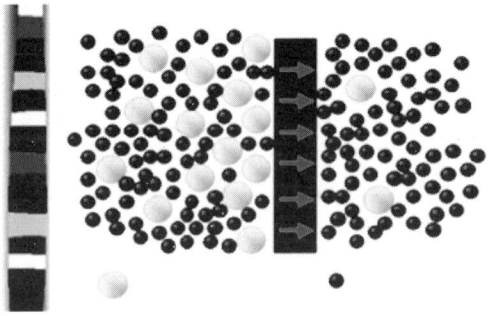

Fig. 7.19: Membrane separation.

In membrane separation spent metal removal fluids are pumped from a process tank at a moderate pressure typically 30 to 50 PSIG and rapid flow to a series of membranes. This flow is typically between 750 to 1,100 gallons per square foot of membrane per day and is referred to as the feed rate. Large molecules and virtually all petroleum products are blocked at the membrane surface. The compounds that do not   pass through the membrane are referred to as the reject.

Membranes can be configured in various ways and have varying life spans are (a) round tubes with approximately 0.5" or 1" internal diameter which can last from 3 to 8 years (b) hollow fibers with an approximate internal diameter of 0.030" which can last from 1 to 2 years (c) flat sheets wrapped in a spiral configuration lasting from 3 to 8 years and (d) flat sheets that are vibrated or turbulated with mechanical wipers lasting from 3 to 8 years. Membranes come in two basic sizes (i) **Micro filtration** rated at 0.1 to 1.0 micron (Fig. 7.22) (ii) **Ultra filtration** rated from 0.001 to 0.1 micron, the most typical membrane size rate at 0.005 micron (Fig. 7.21). The general flow patterns of the various membrane separation systems are illustrated from Fig. 7.20 to Fig. 7.22. Table 7.3 shows size of materials retained, driving force, and type of membrane for various membrane separation processes. Table 7.4 shows the applications of membrane separation processes and their alternative processes.

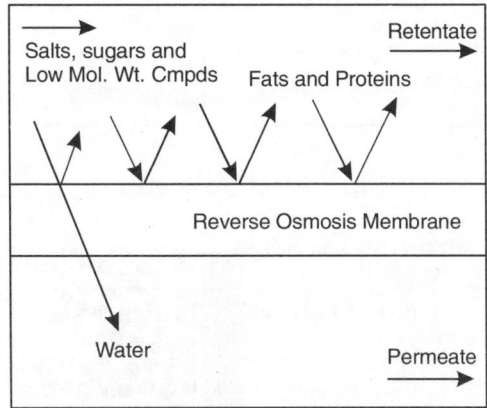

Fig. 7.20: Reverse Osmosis

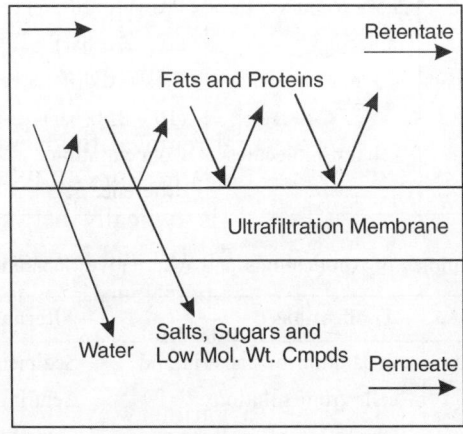

Fig. 7.21: Ultrafiltration

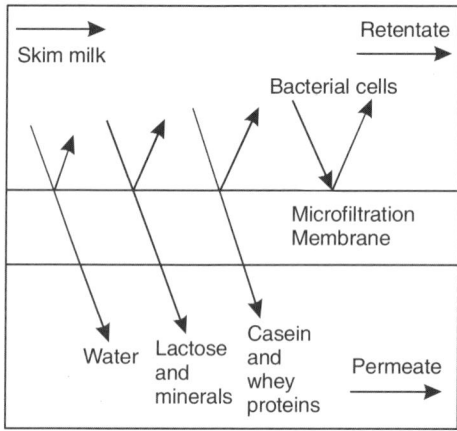

Fig.-7.22: Microfiltration

Table 7.3: Size of Materials Retained, Driving Force, and Type of Membrane

S.No.	Process	Size of materials retained	Driving force	Type of membrane
1.	Micro filtration	0.1 - 10 μm micro particles	Pressure difference (0.5 - 2 bar)	Porous
2.	Ultra filtration	1 - 100 nm macromolecules	Pressure difference (1 - 10 bar)	Micro porous
3.	Nano-filtration	0.5 - 5 nm molecules	Pressure difference (10 - 70 bar)	Micro porous
4.	Reverse Osmosis	< 1 nm molecules	Pressure difference (10 - 100 bar)	Nonporous
5.	Dialysis	< 1 nm molecules	Concentration difference	Nonporous (or) Micro porous

Table 7.4: Examples of Applications and Alternative Separation Processes

S.No.	Process	Applications	Alternative processes
1.	Micro filtration	Separation of bacteria and cells from solutions	Sedimentation, Centrifugation
2.	Ultra filtration	Separation of proteins and virus, concentration of oil-in-water emulsions	Centrifugation
3.	Nano-filtration	Separation of dye and sugar, water softening	Distillation, Evaporation

| 4. | Reverse Osmosis | Desalination of sea and brackish water, process water purification | Distillation, Evaporation, Dialysis |
| 5. | Dialysis | Purification of blood (artificial kidney) | Reverse osmosis |

Advantages

1. Membrane separation consistently separates a wide variety of emulsions, surfactants, chelating chemistries and various mixtures.
2. It requires no specific chemical knowledge.
3. Complex instrumentation is not required.
4. The method does not require constant attention.
5. The basic concept of membrane separation process is simple to understand.

Disadvantages

1. Membranes are expensive.
2. Certain solvents can quickly and permanently destroy the membrane.
3. Certain colloidal solids especially graphite and residues from vibratory deburring operations can permanently foul the membrane surface.
4. The energy cost is higher than chemical treatment although less than evaporation.
5. Oil emulsions are not chemically separated so secondary oil recovery can be difficult.
6. Synthetics are not effectively treated by this method.

7.5.1. Ultra filtration (UF)

Suspended materials and macromolecules can be separated from a waste stream using a membrane and pressure differential called Ultra-filtration. This method uses a lower pressure differential than reverse osmosis and doesn't rely on overcoming osmotic effects. It is useful for dilute solutions of large polymerized macromolecules where the separation is roughly proportional to the pore size in the membrane selected.

Ultra-filtration membranes are commercially fabricated in sheet, capillary and tubular forms. The liquid to be filtered is forced into the assemblage and dilute permeate passes perpendicularly through the membrane while concentrate passes out the end of the media. This may prove useful for the recovery and recycle of suspended solids and macromolecules (fig.-7.23). Excellent results have been achieved in textile finishing applications and other situations where neither entrained solids that could clog the filter nor dissolved ions that would pass through are present. Membrane life can also be affected by temperature, pH and fouling.

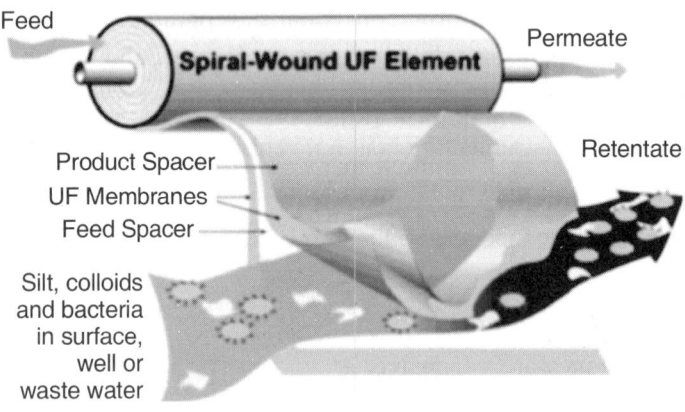

Fig. 7.23: Ultra filtration membrane process

Ultra-filtration is a selective fractionation process utilizing pressures up to 145 psi (10 bar). It concentrates suspended solids and solutes of molecular weight greater than 1,000, and permeate contains low molecular weight organic solutes and salts. UF is widely used in the fractionation of milk and whey, and also finds application in protein fractionation.

Ultra-filtration (UF) is a membrane process that uses moderate hydraulic pressure to transfer water and low molecular weight species through a membrane while retaining contaminants such as suspended solids, colloids and large organic molecules (Fig. 7.23). It is generally used for separations where particle sizes are larger than that of salt ions. UF membranes are frequently used in conjunction with EDR and RO in ultra-pure water and other applications.

Convection or ultra-filtration solute is carried in solution a fluid across a semi-permeable membrane in response to a trans-membrane pressure gradient. This mimics what actually happens in the normal human kidney. The rate of ultra-filtration depends upon the porosity of the membrane and the hydrostatic pressure of the blood, which depends upon blood flow. This is very effective in removal of fluid and middle sized molecules, which are thought to cause uremia. Moreover, most of the cytokines involved in sepsis are middle molecules.

7.5.1.1. Importance

1. Effective fluid management is a primary goal for end stage kidney disease patients on PD therapy, since the patient's own kidneys have limited if any ability to eliminate excess fluid from the bloodstream.

2. Effective fluid management improves the ability to control weight gain and edema-the abnormal accumulation of fluid in the tissue spaces, cavities or joint capsules of the body.

3. Hypertension is a common manifestation of too much fluid in the body contributing to worsened cardiovascular disease in renal patients.

4. Achieving proper fluid balance can optimize a patient's quality of life and decrease morbidity.

7.5.1.2. Applications

1. Oil emulsion waste treatment.
2. Treatment of whey in dairy industries.
3. It is used to study the concentration of biological macromolecules.
4. Electro coat paint recovery.
5. Concentration of textile sizing.
6. Used to study the concentration of heat sensitive proteins for food additives and gelatin.
7. Pulp mill waste treatment.
8. Production of ultra pure water for electronics industry.
9. Bioprocessing-separation and concentration of biologically active components.
10. Protein harvesting useful for grass proteins, algal/plankton proteins.

7.5.2. Reverse osmosis

Diffusion is the movement of molecules from a region of higher concentration to a region of lower concentration. Osmosis is a special case of diffusion in which the molecules are water and the concentration gradient occurs across a semi permeable membrane. The semi permeable membrane allows the passage of water, but not ions (Na^+, Ca^{2+}, Cl^-) or larger molecules (glucose, urea, bacteria). Diffusion and osmosis are thermodynamically favorable and will continue until equilibrium is reached. Osmosis can be slowed, stopped or even reversed if sufficient pressure is applied to the membrane from the 'concentrated' side of the membrane. If the liquid on the more concentrated side is maintained at a higher pressure however this process can be reversed. The solvent will flow from the concentrated side to the less concentrated side. Since the membrane blocks the passage of the dissolved waste constituents, the concentrated solution becomes even more concentrated this process is called reverse osmosis.

7.5.2.1. Components

A typical reverse osmosis system consist of the pre-filter, RO membrane unit, a pressurized storage tank for the treated water, a post-filter and a separate delivery tap for the treated water supply (Fig. 7.24). The water supply entering the RO unit should be bacteriologically safe. RO units will remove virtually all microorganisms but they are not recommended for that use because of the possibility of contamination through pinhole leaks or deterioration due to bacterial growth. Water softeners are commonly used in Minnesota and the Dakotas in advance of the RO system.

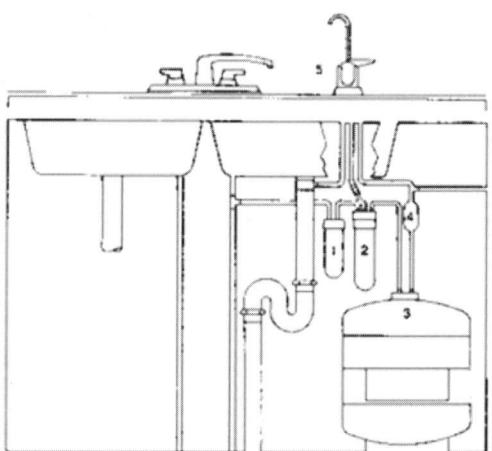

Fig. 7.24: A Typical RO System components. (1) Particle filter, (2) Reverse osmosis membrane unit (3) Pressurized treated-water storage container (4) Carbon adsorption post-filter (5) Separate treated-water tap.

7.5.2.1.1. Pre-filter

The pre-filter is sometimes referred to as a sediment filter. It removes small suspended particles to extend the life of the membrane. Some membrane units are damaged by chlorine and others by bacterial growth. Where chlorine is present, a carbon pre-filter may also be recommended.

7.5.2.1.2. RO membrane

Several kinds of reverse osmosis membranes are available. The most common materials are cellulose acetate or polyamide resins. Mixtures or variations of these materials are also used. Each product has certain advantages and limitations and these need to be considered carefully.

7.5.2.1.3. Storage tank

Most RO units supply treated water at very low rates so a storage tank of 2 to 5 gallons is used to provide a suitable supply. These units are pressurized to produce an adequate flow when the tap is open. Under sink storage requires minimum pressure to deliver water. Other locations may require increased delivery pressure which may reduce membrane performance.

7.5.2.1.4. Post-filter

The main reason for post filtration is to remove any undesirable taste and any residual organics from the treated water. Usually a carbon filter is used for this purpose. Where a carbon filter is used as a part of the pre-filtration step, post filtration is normally eliminated.

7.5.2.1.5. Delivery tap

A separate delivery tap for the treated water is used so that both treated and untreated water are available.

7.5.2.1.6. Others

No special controls are required on most systems since they operate by the use of pressure-sensitive switches, check valves, or flexible bladders. Shut-off valves are important to conserve water during low use periods. Monitoring gauges or servicing lights are becoming increasingly common and assist greatly in knowing whether the system is or isn't working.

7.5.2.2. Principle

Reverse osmosis is referred to as ultra filtration because it involves the movement of water through a membrane as shown in fig.-7.25. The membrane has microscopic openings that allow water molecules but not larger compounds to pass through. Some RO membranes also have an electrical charge that helps in rejecting some chemicals at the membrane surface. Proper maintenance is essential to retain effectiveness over time. Some units are equipped with automatic membrane flushing systems to clean the membrane.

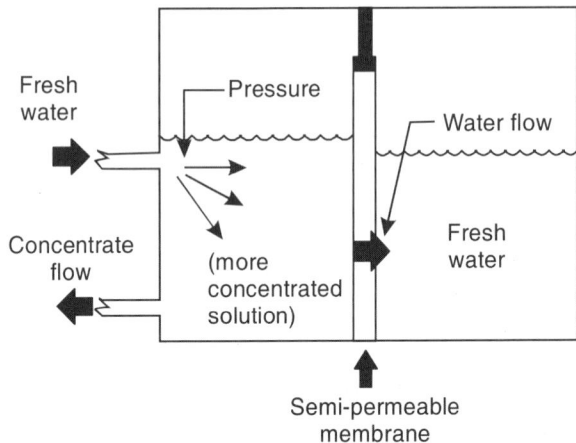

Fig.-7.26: Diffusion/Dialysis process.

The effectiveness of RO units is characterized by the rejection rate or rejection percentage. The rejection rate is the percent of a contaminant that does not move through or is rejected by the membrane. Some typical rejection rates for common contaminants are shown in Table 7.5 and these rejection rates are for single contaminants under design conditions.

Table 7.5: Typical rejection rates for common contaminants

S.No	Contaminant	Rejection Rate Range	
		Laboratory Tests	**Field Tests**
1.	Nitrates	83 - 92%	92%
2.	Total Dissolved Solids	95 - 99%	60% - 99%
3.	Sulfates	90 - 98%	60% - 98%
4.	Sodium	87 - 93%	60% - 93%

Where water contains more than one contaminant, the rejection rate for each contaminant may be reduced or one of the contaminants may be reduced in preference to the other contaminant. For example, cases have been reported where water supplies containing either high TDS levels or high sulfates in combination with nitrates show no decrease in nitrates after treatment. To determine the needed rejection rate, it is necessary to consider the initial concentration. For example, if a water supply contains nitrates at a concentration of 20 milligrams per liter (mg/l), an RO unit rejecting at a rate of 85 percent, which means 15 percent remaining, would reduce the level to 3 mg/l..

7.5.2.3. Disadvantages

RO units use a lot of water. They recover only 5 to 15 percent of the water entering the system. The remainder is discharged as waste water. Because waste water carries with it the rejected contaminants, methods to recover this water are not practical for household systems. Waste water is typically connected to the house drains and will add to the load on the household septic system. An RO unit delivering 5 gallons of treated water per day may discharge 40 to 90 gallons of waste water per day to the septic system.

7.5.2.4. Applications

Reverse osmosis (RO) has become a common method for the treatment of household drinking water supplies. Effectiveness of RO units depends on initial levels of contamination and water pressure. RO treatment may be used to reduce the levels of naturally occurring substances that cause water supplies to be unhealthy or unappealing (foul tastes, smells or colors) and Substances that have contaminated the water supply resulting in possible adverse health effects or decreased desirability.

RO systems are typically used to reduce the levels of total dissolved solids and suspended matter. The principal uses of reverse osmosis in Minnesota and the Dakotas are for the reduction of high levels of nitrate, sulfate, sodium and total dissolved solids.

RO units with carbon filters may also reduce the level of some SOCs (soluble organic compounds) like pesticides, dioxins and VOCs (volatile organic compounds like chloroform and petrochemicals). An RO unit alone may not be the best solution for these types of contaminants, but installing a properly designed RO unit to reduce the levels of other contaminants may provide a reduction in SOCs and VOCs.

7.5.3. Diffusion/Dialysis

The movement of solutes from a compartment in which they are in high concentration to one in which they are in lower concentration along an electrochemical gradient. An electrolyte solution runs countercurrent to blood flowing on the other side of a semi permeable (small pore) filter. Small molecules such as urea move along the concentration gradient into the dialysate fluid. Larger molecules are poorly removed by this process. Solute removal is directly proportional to the dialysate flow rate (Fig. 7.26).

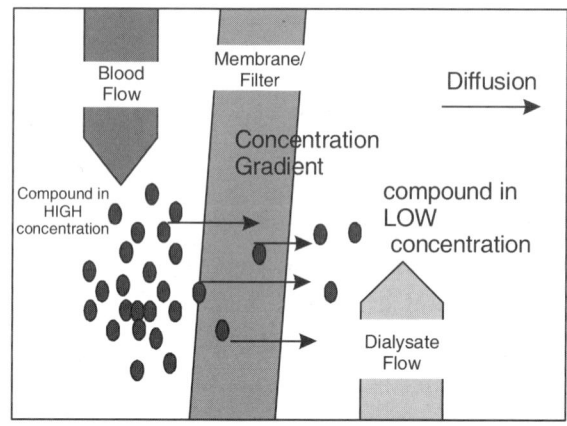

Fig. 7.26: Diffusion/Dialysis process

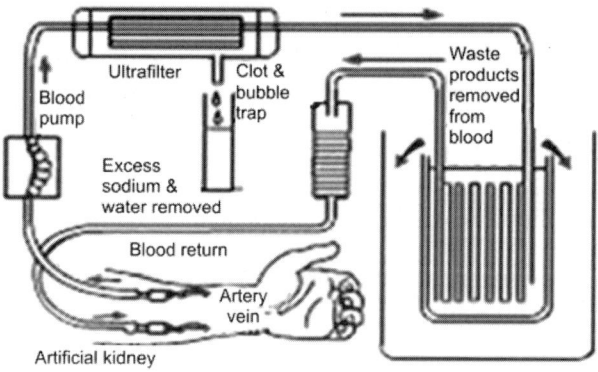

Fig. 7.27: Dialysis process: Waste products removed from blood

The kidneys filter and purify the blood throwing out the impurities through the urine. If the kidneys become diseased and cease functioning the blood gets poisoned and the patient dies. The only way to deal with kidney failure is to transplant a healthy kidney obtained from a living or just dead person or to remove the impurities from the blood artificially. The technique for removing waste products from the blood is called dialysis. In dialysis, blood from an artery in the arm (leg) is drawn into a dialysis machine and the purified blood is fed into a vein. The person requiring dialysis has to undergo an 8 hours session connected to the dialysis machine every two or three days. It is an exhausting and costly process (Fig. 7.27).

7.6. SUMMARY

The HPLC is a method of separation in which the stationary phase is contained in a column, one end of which is attached to a source of pressurized liquid eluent/mobile phase.

The packing used in modern HPLC consist of small, rigid particles having a narrow particle size distribution. There are 3 main types of column packing in HPLC. These are - Porous, polymeric beds based on styrene-divinyl benzene co-polymers. Porous layer beds (diameter 30-55 μm) consisting of a thin shell (1-3μm) of silica on a spherical inert core and totally porous silica particles with narrow particle size range (diameter <10μm).

HPLC Applications - The wide applicability, speed and sensitivity of HPLC have resulted in it becoming the most popular form of chromatography and all types of biological molecules have been purified, Reverse phase HPLC is useful for the separation of polar compounds such as drugs and their metabolites, peptides, vitamins, polyphenols, steroids used in clinical and pharmaceutical work as it is possible to apply biological fluids such as serum and urine directly to the column, separation of highly polar compounds, Proteins purification which includes aldolase, transferring, cytochrome c and thyroglobulin.

Gas solid chromatography is based upon a solid stationary phase in which retention of analytes is the consequence of physical adsorption. GSC has limited application owing to semi permanent retention of active polar molecules and severe tailing of elution peaks.

Gas liquid chromatography is based upon the partition of the analyte between a gaseous mobile phase and a liquid phase immobilized on the surface of an inert solid.

The partition ratio is simply the ratio of the amount of solute in the stationary phase to the amount of solute in the mobile phase.

$$K = \frac{\text{Amount of solute in the liquid phase}}{\text{Amount of solute in the gas phase}}$$

$$= \frac{\text{Time in the liuqid phase}}{\text{Time in the gas phase}}$$

The mobile phase in GLC is usually helium or nitrogen, although carbon dioxide and hydrogen from tank sources have also been tried. The most important requirements of a carrier gas - It should be inert It should be available at low cost and It should allow the detector to respond in an adequate manner.

Capillary columns have an internal diameter of a few tenths of a millimeter. They can be one of two types – Wall-coated open tubular (WCOT) Wall-coated columns consist of a capillary tube whose walls are coated with liquid stationary phase and Support-coated open tubular (SCOT). In support-coated columns, the inner wall of the capillary is lined with a thin layer of support material such as diatomaceous earth, onto which the stationary phase has been adsorbed.

GLC Liquid stationary phases requirements are - It should be a good solvent of the component of the sample, It should be thermally stable. The solvent power of the liquid phase should be different for each component of the sample, It should be of chemically inert towards the sample and It should be of low volatility.

Ion exchange can be defined as a reversible process in which ions of like sign are exchanged between liquid and solid, a highly insoluble body in contact with it. The solid known as an ion exchanger and no substantial organic polymer has been used as ion exchanger.

The most common properties of all ion exchangers which have been used in analysis are insoluble in water and organic solvents, like benzene, carbon tetrachloride and ether. A cation exchange resin may be defined as a high molecular weight, cross-linked polymer containing sulphonic, carboxylic, phenolic etc. groups as an integral part of the resin and equivalent amount of the cations.

An anion exchange resin is a polymer containing anion or quaternary ammonium groups as integral parts of the resin and equivalent amount of anions, such as Cl^-, So^{2-}_4, OH^- ions etc. The representing an anion exchange resin also includes $R - N (CH_3) H$ and $R-N (CH_3)_2$.

Ion exchange is a reversible process, in which the replacement of the exchangeable ions A_x in the resin by ions of like sing B γ from a solution takes place. Thus, the in exchange process can be represented as

$$A_x + B \gamma B_x + A \gamma$$

Ion exchange resins uses - the removal of interfering ions, For group separations, For concentration of samples, establishing the charges on ions, purification of samples, preparation of standard solutions, determining the formation constants of complexes and used as acid and base catalysts.

Gel filtration chromatography is a separation based on size. It is also called molecular exclusion or gel permeation chromatography. In gel filtration chromatography, the stationary phase consists of porous beads with a well-defined range of pore sizes. Gel filtration can be used to separate proteins by molecular weight at any point in a purification of a protein.

In gel filtration or molecular exclusion chromatography, molecules in solution are separated by size as they pass through a column of cross-linked beads that form a three-dimensional network. These polymer beads (frequently made of dextran, agarose, or acrylamide) comprise the stationary phase. The liquid phase is the solvent that is found both around the beads and in the pores of the stationary phase matrix.

Affinity chromatography is a term which now covers a variety of methods of enzyme purification, the common factor of which is the more or less specific interaction between the enzyme and the immobilized ligand. In general, affinity chromatography achieves a higher purification factor than ion-exchange chromatography.

Membrane processing is a technique that permits concentration and separation without the use of heat. Particles are separated on the basis of their molecular size and shape with the use of pressure and specially designed semi-permeable membranes.

Dialysis - The technique for removing waste products from the blood is called dialysis.

Suspended materials and macromolecules can be separated from a waste stream using a membrane and pressure differential, called Ultra-filtration. Ultra-filtration membranes are commercially fabricated in sheet, capillary and tubular forms. Ultra-filtration (UF) is a membrane process that uses moderate hydraulic pressure to transfer water and low molecular weight species through a membrane while retaining contaminants such as suspended solids, colloids and large organic molecules.

Reverse Osmosis - Reverse osmosis is sometimes referred to as ultra filtration because it involves the movement of water through a membrane. The membrane has microscopic openings that allow water molecules, but not larger compounds, to pass through.

Chapter 8
NUCLEAR MAGNETIC RESONANCE SPECTROSCOPY

8.1. INTRODUCTION

A radically different type of interaction between matter and electromagnetic forces can be observed by subjecting a substance simultaneously to two magnetic fields, one stationary and the other varying at some radio frequency. At a particular combination of fields energy is absorbed by the sample and the absorption can be observed as a change in the signal developed by a radio frequency detector and amplifier. This energy absorption can be related to the magnetic dipolar nature of spinning nuclei hence this technique known as Nuclear Magnetic Resonance (NMR).

Radio waves are considered to be the lowest form of electromagnetic radiation and this amount of energy is not sufficient to excite or vibrate or rotate an atom or molecule. The amount of energy available in radio frequency region is just sufficient to affect the nuclear spin of the atom in a molecule and hence constitute the most fundamental part of Nuclear Magnetic Resonance (NMR) Spectroscopy. The method of Nuclear Magnetic Resonance (NMR) was first developed by E.M.Purcell and Felix Bloch in 1946.

In case of all forms of spectroscopy may be described in terms of three important factors are frequency of spectral lines or bands, intensity of spectral lines or bands and shape of spectral lines or bands. In the case of NMR these molecular parameters are found to be shielding constant of nuclei, coupling constant of nuclei and lifetime of energy level, and all these parameters are of fundamental importance in NMR Spectroscopy.

8.2. NUCLEAR SPIN AND MAGNETIC MOMENT

All nuclei carry a charge, in some nuclei this charge spins on the nuclear axis and this circulation of nuclear charge generates a magnetic dipole along the axis. Nuclei of atoms are composed of protons and neutrons. Like electrons these particles also have property to spin on their own axis and each of them possesses angular momentum $1/s$ $(h/2\pi)$ in accordance with quantum theory. The net resultant of the angular momentum of all nuclear particles is called nuclear spin (Fig. 8.1).

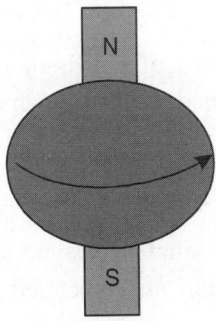

Fig. 8.1: Spinning charge in proton generates magnetic dipole

A nucleus having nuclear spin quantum number I there are $(2I + 1)$ spin states. Two properties of nuclear particles which are important in understanding of NMR spectroscopy are the net spin associated with the protons and neutrons, and the distribution of positive charge. The net spin number or spin quantum number I of a particular nucleus can be obtained by adding spin numbers of individual proton and neutron of ½ each assuming that neutrons cancel only neutrons and protons cancel only protons because of paring or spinning in opposite directions. The spin numbers I have values 0, 1/2, 1, 3/2, 5/2 and so forth (I = 0 represents no spin). There are three broad principles for the nuclear spin:

    a.   If the sum of protons and neutrons is even, I is zero or integral (0.1.2.3 …). If the spins of all the particles are paired there will be no net spin and the nuclear spin quantum number I will be zero. Ex: ^{12}C, ^{16}O, ^{18}O, ^{32}S nuclei have even number of neutrons and even number of protons and so will show zero spin. Particles in such nuclei are paired (spinning in opposite directions) and hence there is no resultant in opposite directions and so there is no resultant spin and consequently there will be no magnetic moment.

b. If the sum of the protons and neutrons is odd, I is half integral (1/2, 3/2, 5/2 etc). Ex: ^{1}H, ^{19}F, ^{13}C and ^{31}P have I = ½ and a uniform charge distribution. ^{11}B, ^{35}Cl, ^{79}Br and ^{81}Br are the nuclei having I = 3/2.

c. If both protons and neutrons are even numbered, I is zero. ^{12}C and ^{16}O fall in this category and give no NMR signal. The magnetic properties occur with those nuclei which have odd atomic number and odd mass number. Ex: ^{1}H, ^{15}N, ^{19}F, ^{31}P etc., odd atomic number and even mass number. Ex: ^{1}H, ^{14}N etc. and even atomic number and odd mass number. Ex: ^{13}C

Nuclei with a spin number I of 1 or higher have a non-spherical charge distribution. This asymmetry is described by an electrical quadrupole moment and affects the relaxation time consequently the coupling with neighboring nuclei. In those nuclei where the magnitude of the spin is not zero, the nuclear spin quantum number I may assume any of the values ½, 3/2 etc (table 8.1). So for study by nuclear magnetic resonance a nucleus must possess angular momentum and an associated magnetic moment.

Table 8.1: Spin quantum of various nuclei

No. of protons	No. of Neutrons	Spin quantum	Examples
Even	Even	0	^{12}C, ^{16}O, ^{32}S
Odd	Even	½	^{1}H, ^{19}F^{31},P
		3/2	^{11}B, ^{79}Br
Even	Odd	1/2	^{13}C
		3/2	^{127}I
Odd	Odd	1	^{2}H, ^{14}N

The NMR spectroscopy is most often concerned with I = 1/2 the best example of which is proton ^{1}H with a spin of 1/2 so NMR spectroscopy is also known as Proton Magnetic Resonance (PMR) Spectroscopy. A hydrogen nucleus consists of only one proton and consequently it can have spin of +1/2 or − ½, depending upon the direction of spin. The deuterium nuclei consists of one proton and one neutron, i.e. the sum is (1 + 1) = 2 which is even. Hence deuterium molecule may have a spin of 1 or 0 depending on whether the spins of the proton and neutron are parallel or anti-parallel. However the nucleus of deuterium has been found to have a spin of one which indicates that neutron and proton have parallel alignment each having spin quantum number I equal to 1/2. For the nuclei of helium and oxygen

I=0. The value of I for nitrogen nuclei is one similarly the values of I for sodium and chlorine are 3/2 and 5/2 respectively.

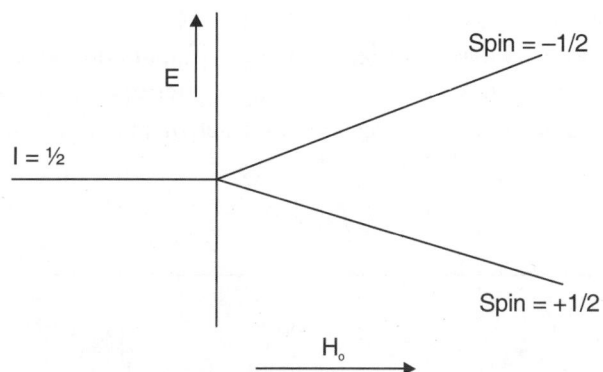

Fig. 8.2: Two energy levels for a proton in magnetic field H_0

Protons and neutrons possess magnetic properties they are associated with a magnetic moment µ for the nuclei. The relation between this magnetic moment and nuclear spin quantum number is given by:

$$\mu = k\sqrt{[I(I+1)]} \qquad \text{... eq. 8.1}$$

Where k is a constant and I is a spin quantum number.

The spin number I determines the number of orientations a nucleus may assume in an external uniform magnetic field in accordance with the formula 2I + 1. Consider the specific case of ^1H (Proton) nucleus, which ahs spin quantum number I = 1/2 so there are two spin states given by (2I + 1) = 1 × 1/2 + 1 = 2 which have the values +1/2 and −1/2. Thus there are two energy levels and a slight excess population in the lower energy state. In the absence of a magnetic field the energy levels corresponding to the two spin states are degenerate but in the presence of a magnetic field the two spin levels are separated by an energy difference E. This energy difference depends upon the strength of the magnetic field in which the nucleus is situated.

8.2.1. Spinning nuclei-magnetic moments

Some elements have isotopes with nuclei they were spinning about an axis and the spinning of a charged particle generates a magnetic field. Isotopes with nuclear spin thus have a magnetic field, the magnitude and directions which can be described by a vector know as magnetic movement (Fig. 8.3).

The spinning nuclei behave as though they were tiny bar magnets having a north pole and a south pole. A nucleus or an electron bears a charge its spin gives rise to a magnetic field that is analogous to the field produced when an electric current is passed though a coil of wire. The resulting magnetic dipole μ is oriented along the axis of spin and has a value that is characteristic for each kind of particle. The interrelation between particle spin and magnetic moment gives rise to a set of magnetic quantum states given by

$$m = I, I - 1, I - 2 \ldots - I$$

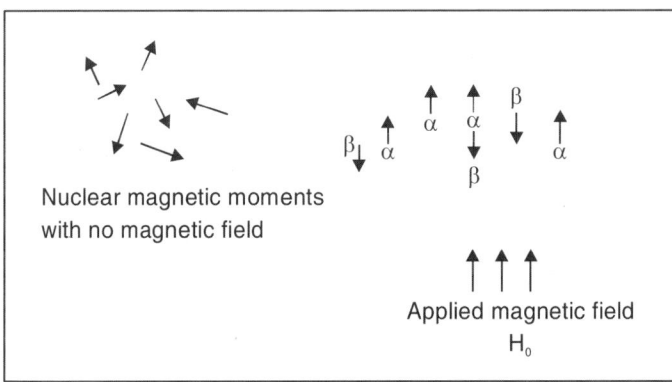

Nuclear magnetic moments with no magnetic field

Applied magnetic field
H_0

Fig. 8.3: Orientation of nuclear magnetic moments

8.2.2. Magnetic moment and magnetic field

A particle that possesses a magnetic moment when brought into the influence of an external field tends to become oriented in such manner that it's magnetic dipole μ and hence its spin axis is parallel to the field. These magnetic moments are oriented in random fashion in the absence of a magnetic field. The behavior of the particle in the presence of magnetic field is some what like that of a small bar magnet the potential energy of either being dependent upon the orientation of the dipole (Fig. 8.5). With the magnet this energy can assume an infinite number of values depending upon its alignment. The energy of the atomic particle however is limits to (2I + 1) discrete values. The potential energy in quantized or non quantized case is given be:

$$E = -\mu_H H_0 \qquad\qquad \textit{... eq. 8.2}$$

Where μ_H is the component of magnetic moment in the direction of the field, H_0 is the strength of external field.

The quantum character of atomic particles limits to a few number of possible energy levels. For a spin quantum number I and magnetic quantum number m, the energy of a quantum level is given by:

$$E = -\sqrt{\frac{m\mu}{I}}\beta H_0 \qquad \text{... eq. 8.3}$$

Where H_0 is the strength of the external field in gauss B is a constant known as nuclear magneton μ is magnetic moment of the particle expressed in units of nuclear magneton. The value of μ for proton is 2.7927 nuclear magneton while that for an electron it is − 1836. The nuclear spin can have only two alternative values associated with quantum numbers +½ (= α) and −1/2 (= β). When these nuclei are placed in a magnetic moments tend to align with the field corresponding to α spin or opposite to the field corresponding to β spin.

The proton has magnetic quantum numbers of +1/2 and −1/2. The energies of these two states in the magnetic field are given by:

$$M = +1/2 \qquad E = -\frac{1/2(\mu\beta Ho)}{1/2} = -\mu\beta Ho \qquad \text{... eq. 8.4}$$

$$M = -1/2 \qquad E = -\frac{-1/2(\mu\beta Ho)}{1/2} = +\mu\beta Ho \qquad \text{... eq. 8.5}$$

These two quantum levels correspond to the two possible orientations of the spin axis and the magnetic field shown in Fig. 8.5.

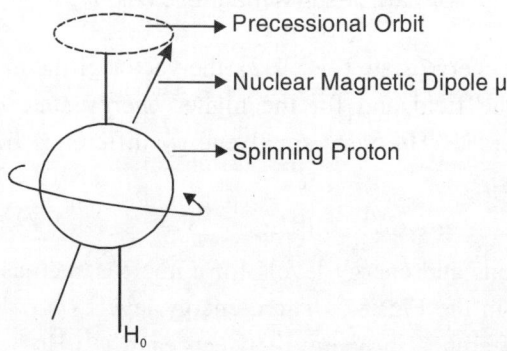

Fig. 8.4: Proton processing in a magnetic field Ho

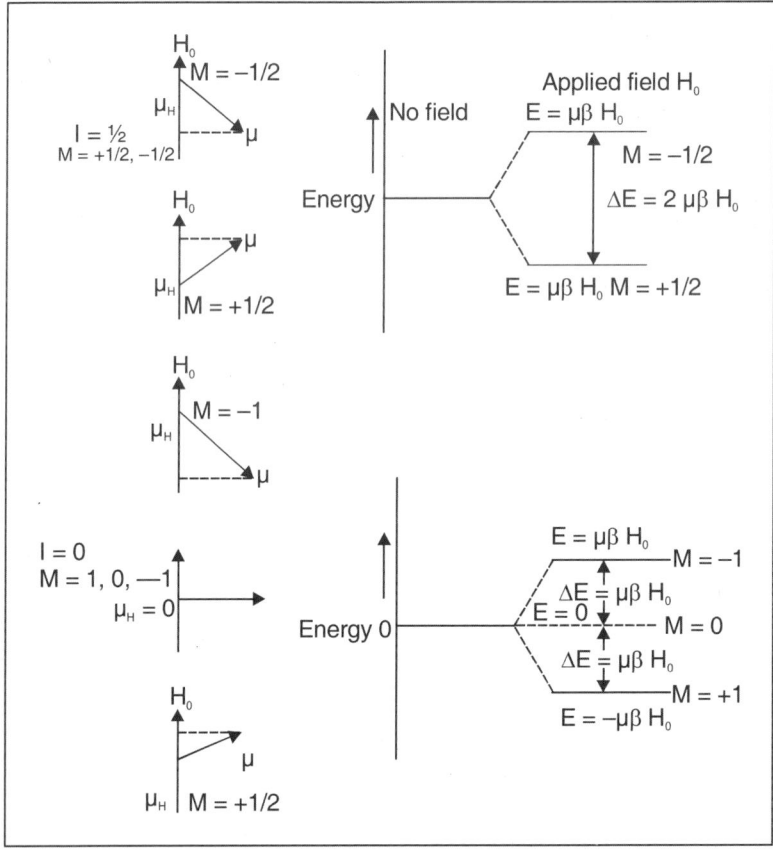

Fig. 8.5: Orientations of magnetic moments and energy levels
for particles in a magnetic field H_0.

For the lower energy state (m = 1/2), the vector of the magnetic moment is aligned with the field and for the higher energy state (m = −1/2) the alignment is reversed. The very small energy difference between the two levels is given by:

$$\Delta E = 2\mu\beta Ho \qquad\qquad ... \; eq. \; 8.6$$

The orientations and energy levels for a nucleus such as N ($I = 1$) have also been shown in the Fig. 8.5. Three energy levels ($m = 1$, 0 and −1) are found and the difference in energy between each is μHo.

$$E = \frac{\mu\beta Ho}{I} \qquad\qquad ... \; eq. \; 8.7$$

Excitation to a higher nuclear magnetic quantum level can be brought about by absorption of a photon with energy hv (= ΔE). Thus above equation can be written as:

$$hv = \frac{\mu \beta Ho}{I} \qquad \qquad \textit{... eq. 8.8}$$

8.3. TYPES

There are two types of NMR spectrometers (1) Single coil spectrometers, in which absorption is measured (2) Two coil instruments in which resonant radiation is measured. NMR spectrometers also been classified as low resolution instruments and high resolution instruments.

8.3.1. Low resolution NMR

A single coil NMR spectrometer containing a glass sample tube filled with distilled water. Now set the radio frequency oscillator at 5 MHz and vary the magnetic field at a constant rate from 0 to 1 T by monitoring the output of the detector. Effective quantitative and Quantitative measurements are also possible such type of experiments can be performed by making use of an instrument called wide line or low resolution NMR spectrometer.

8.3.2. High resolution NMR

The high resolution instruments are capable of resolving the fine structure that is associated with the absorption peak for a given kind of nucleus. By making use of chemical environment of the nucleus it is possible to determine the nature of this fine structure such details can not be detected however by making use of a wide line instrument. This type of instrument is useful for quantitative element analysis and physical environment of a nucleus. A high resolution spectrometer can evidence two distinct types of structures in NMR absorption due to proton resonance know as chemical shift and spin-spin Coupling.

8.3.2.1. Chemical shift

Every nucleus in a compound is surrounded by a cloud of electrons which are in constant motion. Under the influence of the applied magnetic force these electrons are caused to circulate in such a sense as to oppose the field. This has the effect of partially shielding the nucleus from feeling the full value of the external field. Therefore, either the frequency or the field will have to be changed slightly to bring the shielded nucleus into resonance. In most NMR instruments this is accomplished by an adjustment of the magnetic field by means of direct current passed through an auxiliary winding. This auxiliary current is varied linearly to sweep the field over a narrow span. The electronic circuitry converts the value of the added field to its frequency equivalent for presentation to the recorder.

Chemical shift is defined as nuclear shielding or applied magnetic field. Chemical shift is a function of the nucleus and its environment. It is measured relative to a reference compound. This chemical shift, designated by δ is the ratio of the change of field necessary to achieve resonance to the field strength that resonates with a standard.

$$\delta = \frac{H \text{ sample} - H \text{ ref}}{H \text{ ref}} \qquad \textit{... eq. 8.9}$$

An important feature of high resolution NMR spectra is chemical shift. In different chemical environments the same type of nucleus is shielded slightly from the applied field in a manner that depends on the distribution of the surrounding electrons. For a fixed external field Ho different screening factors cause slightly different resonant frequencies. The magnitude of the effective field felt by each group of nuclei is expressed as follows:

$$H \text{ ref} = Ho(1 - \sigma) \qquad \textit{... eq. 8.10}$$

Where σ a non-dimensional shielding constant is may be either a positive or negative number.

The value of the shielding constant depends on several factors among which are the hybridization and electronegativity of the groups attached to the atom that contains the nucleus being studied. Shielding effects seldom extend beyond one bond length except with very strong electronegative groups.

NMR spectrometers with different field strengths are in use and it is desirable to express the position of resonance in field independent units and with respect to the resonance of a reference compound. For proton spectra in non-aqueous media, the reference material is tetramethylsilane, $(CH_3)_4Si$ (TMS) (Fig. 8.6) whose position is assigned as exactly 0.0 on the δ scale.

TMS contains 12 protons but these are all chemically equivalent and therefore give rise to a single sharp signal at *Ho* (applied field) higher than most other proton resonances. The magnitude of the chemical shift is expressed in parts per million:

$$\delta = \frac{Hr - Hs}{Hr} \times 10^6 = \frac{vs - vr}{vr} \times 10^6 \qquad \textit{... eq. 8.11}$$

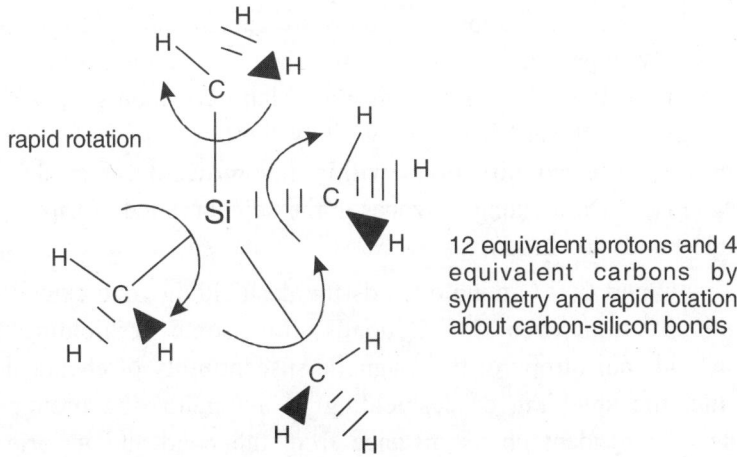

12 equivalent protons and 4 equivalent carbons by symmetry and rapid rotation about carbon-silicon bonds

Fig. 8.6: Tetramethylsilane (TMS)

Where *Hr* and *Hs* are the positions of the absorption lines for the reference and sample, v_r and v_s are the corresponding frequencies expressed in hertz. A positive δ value represents a greater degree of shielding in the sample than in the reference. Recommended reference materials for other nuclei include CS_2 or TMS for ^{13}C trichlorofluoromethane (CCL_3F) for ^{19}F, NH_3 liquid for ^{15}N, TMS for ^{29}Si, and 85% phosphoric acid for ^{31}P. The numbers on the dimensionless scale downfield from the reference are designated positive.

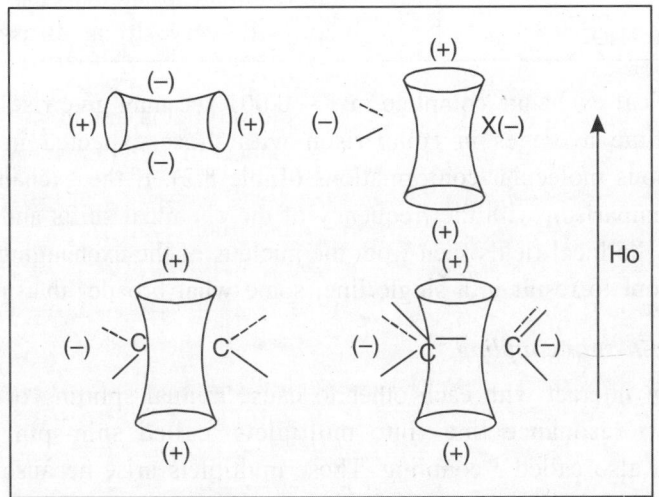

Fig. 8.7: Shielding (+) and Deshielding (–) zones in the neighborhood of triple, double, and single bonds to carbon and aromatic rings.

Proton resonances from C-H bonds are located in the range $\delta = 0.9 - 1.5$ when only aliphatic groups are substituents. Protons in CH_3 groups usually appear at $\delta = 0.9 - 1.0$, when the bonds on the adjacent carbon atom are to H, CH or CH_2 groups. Many common groups produce special shielding effects because they allow the circulation of electrons in only certain preferred directions within the molecule. Fig. 8.7 shows shielding (+) and Deshielding (–) zones in the neighborhood of triple, double, and single bonds to carbon.

In $C = C$ and $C = O$ double bonds the deshielding zone extends along the bond direction, even $C - C$ bonds show some deshielding in this direction. This anisotropy of the magnetic susceptibility of chemical bonds means that the shielding or deshielding of a neighboring proton in the molecule is dependent on its distance from the bond and its orientation with respect to that bond (Fig. 8.7). Aromatic rings exhibit a strong anisotropic effect. When such compounds are placed in a magnetic field the six π electrons circulate in two parallel doughnut shaped orbits on each side of the ring. The resulting magnetic field opposes Ho in a cone shaped zone of excess shielding that extends along the hexad axis.

Table 8.2: Proton chemical shifts

S.No.	Substituents group	Methyl protons (δ)	Methylene protons (δ)
1.	HC–Cl	3.05	3.45
2.	HC–OH	3.20	3.40
3.	HC–F	4.25	4.50
4.	HC–NO2	4.30	4.35

Chemical exchange complete in $1 - 0.001$ sec may give rise to spectra that are time averages in comparison with those expected in terms of instantaneous molecular conformations (Table 8.2). If the exchange rate is high in comparison with the frequency of the chemical shifts and spin-spin couplings the local fields seen from the nucleus of the exchanging atom are averaged out to result in a single line, some what broader than normal.

8.3.2.2. Spin-spin coupling

Nuclei can interact with each other to cause mutual splitting of the other wise sharp resonance lines into multiplets, called spin-spin coupling, sometimes also called J coupling. These multiplets arise because magnetic moments of nuclei interact with each other through the strongly magnetic electrons in the intervening bonds. The strength of the coupling, denoted by J is given by the spacing of the multiplets and is expressed in hertz. Proton-proton couplings in aliphatic organic compounds are transmitted through only two or three bonds, although weak couplings are transmitted further. When unsaturated systems occur between the protons, long range coupling is enhanced. In allylic systems, four bond couplings reach a maximum of about 3 Hz when the angle between the plane that contains the olefinic protons and the C–H bond of the allylic carbon atom is about 90°. In H–C–C = C–C–H systems, five bond couplings of about 3Hz are observed.

The ^{1}H NMR spectrum of ethanol shows the methyl peak has been split into three peaks (a triplet) (Fig. 8.8) and the methylene peak has been split into four peaks (a quartet) (Fig. 8.9). This occurs because there is a small interaction (coupling) between the two groups of protons. The spacing between the peaks of the methyl triplet is equal to the spacing between the peaks of the methylene quartet (Fig. 8.9). This spacing is measured in Hertz and is called the coupling constant, J.

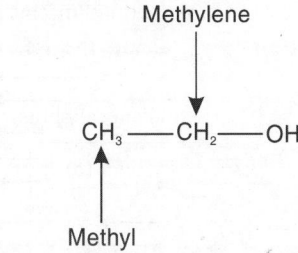

Fig. 8.8: The structure of Ethanol

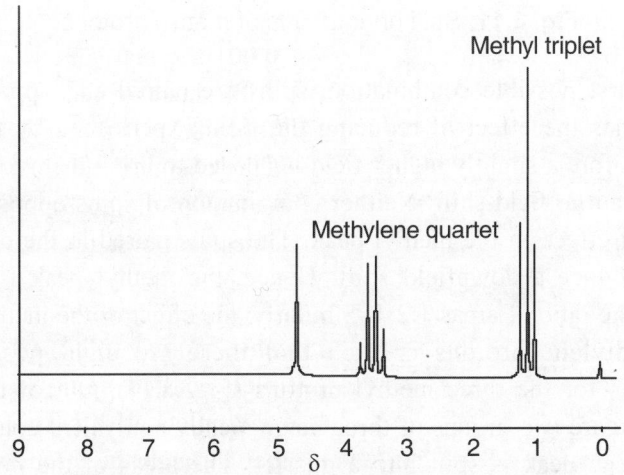

Fig. 8.9: The ^{1}H NMR spectrum of ethanol

Methyl peak can split into a triplet because of the methylene protons, each can have one of two possible orientations i.e aligned with or opposed against the applied field, this give a total of four possible states (Fig. 8.10).

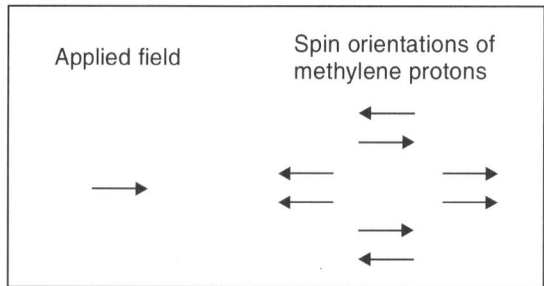

Fig. 8.10: Spin orientations of methylene protons

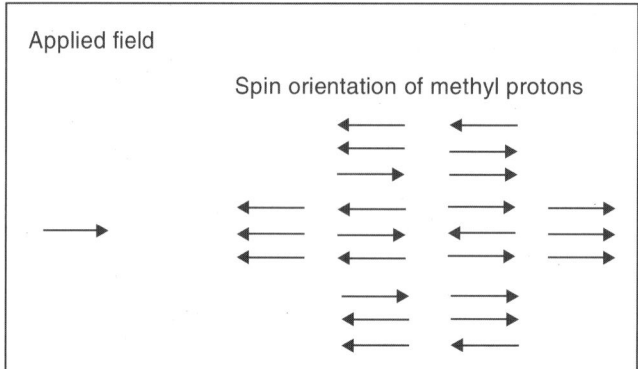

Fig. 8.11: Spin orientations of methyl protons

In the first possible combination, spins are paired and opposed to the field. This has the effect of reducing the field experienced by the methyl protons therefore a slightly higher field is needed to bring them to resonance resulting in an up field shift. Neither combination of spins opposed to each other has an effect on the methyl peak. The spins paired in the direction of the field produce a downfield shift. Hence, the methyl peak is split into three, with the ratio of areas 1:2:1. Similarly, the effect of the methyl protons on the methylene protons is such that there are eight possible spin combinations for the three methyl protons (Fig. 8.11). Out of these eight groups, there are two groups of three magnetically equivalent combinations. The methylene peak is split into a quartet. The areas of the peaks in the quartet have the ration 1:3:3:1. Couplings depend on geometry; the dihedral angle between planes determines the coupling of protons on adjacent carbon atoms. Adjacent axial-axial protons, displaying a dihedral angle of 180° are strongly coupled whereas axial-equatorial and equatorial-equatorial protons are coupled only moderately.

8.3.3. Instrument components

The important components of either type of spectrometer are given below (Fig. 8.12):

8.3.3.1. Magnet

Magnet is used to supply the magnetic field (H) in NMR. The magnet is most commercial NMR spectrometers may be either an electromagnetic or a permanent magnet. The field produced homogenous to 1 part in 10^8 with in the sample area and must be stable to similar degree for short periods of time. The suitable field strength gives a convenient frequency for H^1 resonance measurement. Frequencies of 100×10^6 CPS and fields of 23490 gauss are actually used in NMR spectrometers.

8.3.3.2. Magnetic field sweep

An alternation over a small range in the applied field may be made by making use of a pair of coils located parallel to the magnet face. These coils superimpose on the main field of the magnet, the additional field required to bring the total to the resonance condition. By varying a direct current through these coils the effective field can be changed by a few hundred mill gauss without any loss in field homogeneity.

8.3.3.3. Radio frequency source

The signal from a ratio frequency oscillator or transmitter is fed into a pair of coils mounted at right angles to the path of the field. In this manner a plane polarized beam of radiation has been found to be obtained.

8.3.3.4. Signal detector and recording system

The coil has been used to direct the radio frequency signal produced by the resonating nuclei. The electrical signal generated into the coil must be amplified before it can be recorded.

8.3.3.5. Sample holder

NMR sample cell consists of a 5 mm OD glass tube which has a capacity of about 0.4 ml of liquid. The sample must be in the liquid or solution state for high resolution spectra.

8.3.3.6. Sample probe

The sample probe is a device that holds the sample tube in a fixed position in the field and it is also provided with an air driven turbine for rotating the sample tube along its longitudinal axis at several hundred RPM. Probe may be either a single coil or a turn coil system. The single coil serves as both transmitter and receiver coils. In turn coil system the two coils are orthogonal to each other.

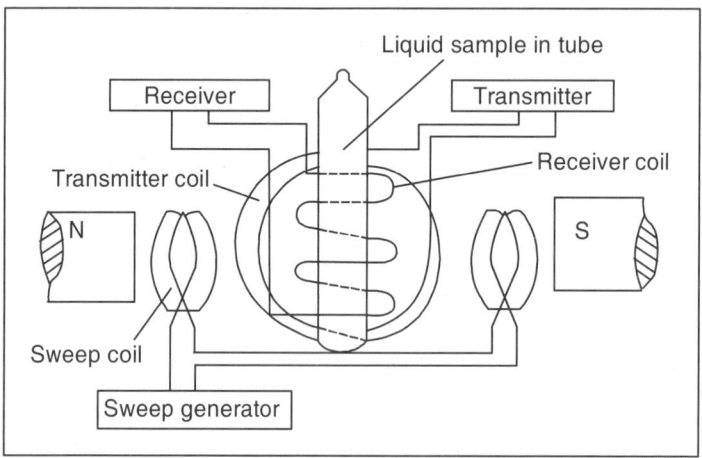

Fig. 8.12: Essential components of a double coil NMR Spectrometer

8.4. PRINCIPLE

Purcell and Bloch (1946) were the first who developed independently the various methods for studying nuclear magnetic resonance, the schematic diagram of one such method is shown in Fig. 8.13.

A is a radio frequency oscillator which is adjusted to generate a certain preset or definite frequency v which passes through a coil D located between the magnetic poles. Surrounding the sample S is a receiver coil. The receiver coil is used to pickup the broadcast signal. These signals are carried to a receiver and amplifier B and then to a cathode ray oscillograph recorder C. The strength of the magnetic field can be varied with the help of suitable device M known as Magnetic field sweep. The frequency of the magnetic field is gradually varied in such a manner that the frequency emitted by the nuclei of the samples equal to the radio frequency of the oscillator A. Under these conditions the same will be in resonance with the applied frequency. As a result absorption occurs, and stops again when H is raised further. A plot of the signal from the receiver versus magnetic field strength are called NMR spectrum.

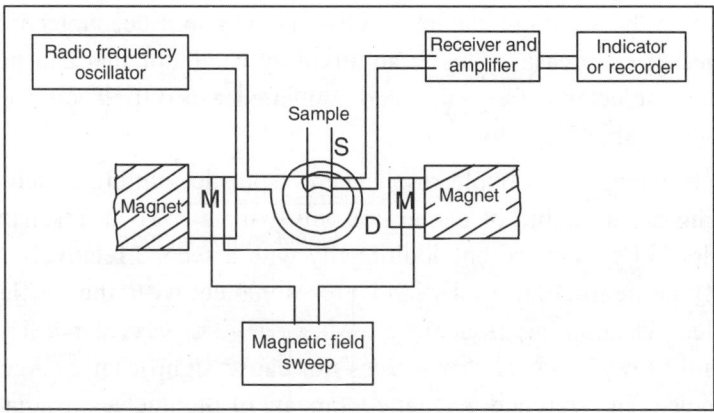

Fig. 8.13: NMR Spectrometer

8.5. FREQUENCY LOCK

An automatic circuit to maintain a constant field to frequency ratio is essential in a high resolution NMR Spectrometer. It uses a frequency controlling feedback circuit that locks onto some specific nuclear resonance and continually adjusts the RF oscillator to keep the reference signal maximized. Without this automatic control the performance of the instruments would be adversely affected by extraneous magnetic events in the vicinity such as operation of electrical machinery.

In most instruments the reference signal is taken from a nucleus in the sample itself (internal lock) often TMS which then serves a dual role being a standard for both chemical shifts and frequency locking. Some instruments provide for an external lock in which the reference material is located in a separate probe mounted close to that of the analytical sample. The external lock has the advantage that control is not interrupted during change of samples, and the disadvantage that the reference and analytical samples may not sense exactly equal magnetic fields it provides greater convenience at the expense of precision. Another lock is between a homo-nuclear and a hetero-nuclear lock. The lock is established on a different nuclear species than the one being examined. In ^{13}C NMR, the lock can be taken on the deuterium nuclei in a solvent such as $CDCl_3$.

8.6. DOUBLE RESONANCE OR SPIN DECOUPLING

Spin-spin coupling is sometimes of considerable in identifying resonances but in relatively complex molecules it may complicate the spectrum to the point where the structure cannot be elucidated, a spin decoupler is of great assistance in such cases. This is an auxiliary oscillator that can produce a signal at a selectable frequency and can impose this field on the sample with considerable intensity.

Nuclear magnetic double resonance or spin decoupling is achieved by irradiating an ensemble of nuclei not only with a rf H_1 at resonance with the nuclei to be observed but additionally with a second relatively strong rf field H_2 perpendicular to H_0 and at resonance with the nuclei to be decoupled. Decoupling is achieved when $\gamma H_2/2\pi >> /J_{AX} /$ but H_2 is still sufficiently low in power that it does not cause significant changes in the other nuclei. The spin and magnetic moments of the nucleus irradiated with v_2 are effectively averaged out over the possible spin states so that other nuclei do not interact with it, thus, the coupling is removed and any multiplets involved collapse into a single peak. Decoupling experiments are carried out to convert homo-nuclear ($^1H - {}^1H$) or hetero-nuclear ($^{19}F - {}^1H$, $^{13}C - {}^1H$) multiplets into singlets or less complex multiplets.

In the field swept technique of decoupling the spectrum is swept while a fixed frequency difference Δv is maintained between the observing v_1 and irradiating v_2 frequencies. Coupled resonances separated by the chosen frequency difference are observed. ^{13}C work usually all ^{13}C-^1H multiplets are decoupled for sensitive and simplicity reasons. This is achieved when the decoupling field, H_2, covers the range of all proton Larmor frequencies. This is at least 1 kHz at about 90 MHz in a magnetic field, H_0, of 23 kG. Decoupling fields with large frequency ranges are realized either by a very large rf power H_2 so that $\mu H_2/2\pi I = \Delta = 1$ kHz or by application of broad-band decoupling in which the coherent proton radio frequency is modulated with white noise.

8.7. NUCLEAR OVERHAUSER EFFECT (NOE)

Nuclear Overhauser Effect (NOE) depends on the disruption of the normal relaxation mechanism. Relaxation is the process by which a nucleus, after absorbing energy, returns to its normal state. The most important of such processes involves dipole interaction between spinning nuclei. This interaction depends strongly on the distance separating the nuclei concerned. If one of these nuclei is saturated by radiation from an auxiliary oscillator it becomes more difficult for the other nucleus to lose its excess energy. The result is a larger NMR signal. If the nuclei are both protons, the enhancement can be as much as 1.5 times.

If a ^{13}C resonance is observed while a nearby proton is irradiated, the signal can be increased by nearly 3 times. Thus, a spin excited nucleus may undergo spin relaxation via the transfer of its spin energy to that of an adjacent nucleus and the efficiency of this energy transfer is directly related to the distance between the two nuclei. This spin transfer is the basis of Nuclear Overhauser Effect (NOD). The interaction of magnetic nuclei through space does not lead to coupling and this NOE effect is observable over short distances, generally 2 to 4 A°. This effect increases the signal to noise ratio to a useful degree and is of great importance in studying the molecular geometry of the compounds. This effect also helps in showing which peaks are due to nuclei in close proximity to each other in the molecule. The effect can often distinguish between conformational or other stereoisomer's. Irradiation at the frequency of the methyl group protons produced an enhancement of the signal from the two protons in position 4 but not that at 2. This shows that the methyl is much closer to the carbon number 4 and therefore it Trans with respect to –COOH.

8.8. CARBON-13 NMR SPECTROSCOPY

^{1}H NMR spectroscopy is a tool for structural analysis when significant portions of a molecule lack C–H bonds no information is forthcoming. Examples include polychlorinated compounds such as chlordane, polycarbonyl compounds such as croconic acid and compounds incorporating triple bonds (Fig. 8.14).

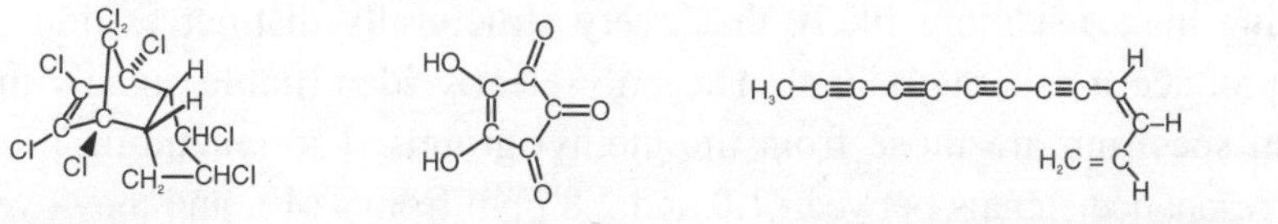

| chlordane | Croconic acid | a polyacetylene from Dahilia. |

Fig. 8.14: Structures of chloradan, croconic acid and polyacetylene

Even when numerous C–H groups are present an unambiguous interpretation of a proton NMR spectrum may not be possible. The following Fig. 8.15 depicts three pairs of isomers (A & B) which display similar proton NMR spectra. Although a careful determination of chemical shifts should permit the first pair of compounds to be distinguished the second and third cases might be difficult to identify by proton NMR alone.

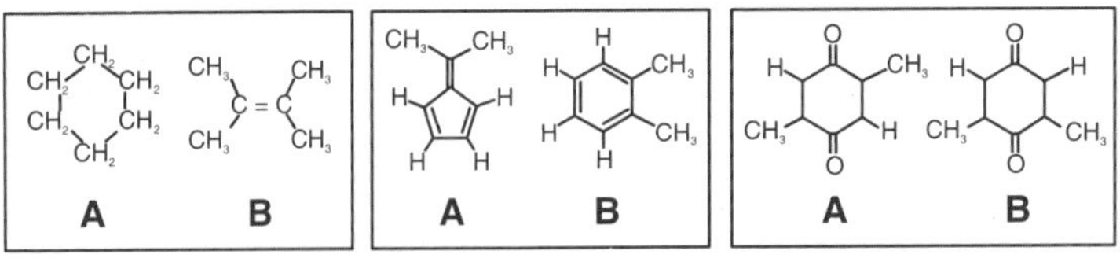

Fig. 8.15: Pairs of isomers

These difficulties would be largely resolved if the carbon atoms of a molecule could be probed by NMR in the same fashion as the hydrogen atoms. Since the major isotope of carbon (^{12}C) has no spin, this option seems unrealistic. Fortunately, 1.1% of elemental carbon is the ^{13}C isotope which has a spin I = 1/2 so in principle it should be possible to conduct a carbon NMR experiment. If much higher abundances of ^{13}C were naturally present in all carbon compounds proton NMR would become much more complicated due to large one-bond coupling of ^{13}C and ^{1}H.

The most important operational technique that has led to successful and routine ^{13}C NMR spectroscopy is the use of high field pulse technology coupled with broad band hetero-nuclear decoupling of all protons (Fig. 8.16). The results of repeated pulse sequences are accumulated to provide improved signal strength. Also, for reasons that go beyond the present treatment the decoupling irradiation enhance the sensitivity of carbon nuclei bonded to hydrogen. The carbon NMR spectrum of a compound displays a single sharp signal for each structurally distinct carbon atom in a molecule. The spectrum of camphor is typical furthermore a comparison with the ^{1}H NMR spectrums on the right illustrates some of the advantageous characteristics of carbon NMR (Fig. 8.17). The dispersion of ^{13}C chemical shifts is nearly twenty times greater than that for protons, and this together with the lack of signal splitting makes it more likely that every structurally distinct carbon atom will produce a separate signal. The only clearly identifiable signals in the proton spectrum are those from the methyl groups. The remaining protons have resonance signals between 1.0 and 2.8 ppm from TMS, and they overlap badly thanks to spin-spin splitting. Unlike proton NMR spectroscopy the relative strength of carbon NMR signals is not normally proportional to the number of atoms generating each one. Because of this, the number of discrete signals and their chemical shifts are the most important pieces of evidence delivered by a carbon spectrum.

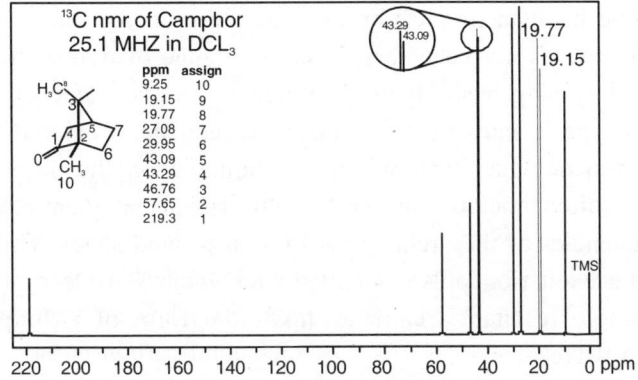

Fig. 8.16: ^{13}C NMR of Camphor

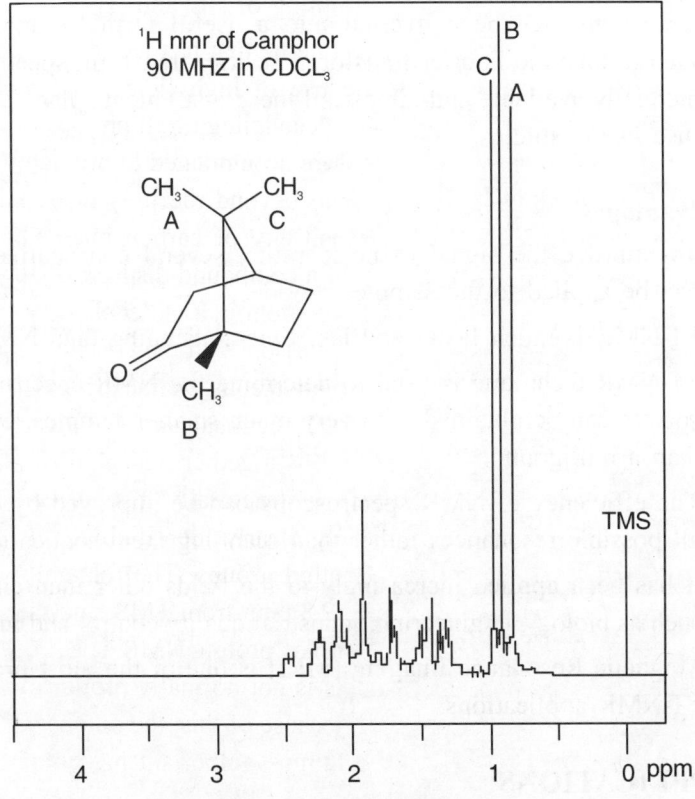

Fig. 8.17: ^{1}H NMR of Camphor

8.9. FOURIER TRANSFORM (FT) NMR

In NMR the sample is flooded with radiation of all possible frequencies in the range required followed b time to frequency (Fourier) conversion. These signals can be handled directly because they are on the time scale and so the interferometer is not necessary. The wide range of frequencies can best be obtained by using a short burst or pulse or radiofrequency energy. A pulse of this type is equivalent to a band of frequencies entered around the oscillator frequency the band width depending upon the duration of the pulse. The excited nuclei continue to emit signals at their characteristic Larmor frequencies as they relax back to their ground states. This emission following the excitation pulse is called Free Induction Decay (FID). The FID of a single resonant frequency takes the form of an exponentially decaying sin wave of $v = |v_L - c|$.

Where v_L = Larmor frequency, v_C = Carrier frequency

The FID shows well defined interference boats at 185 Hz, and because of this effect it is called as interferogram. The information gained by interferogram can then be converted into a useful form by a computer programmed to take the Fourier transform. Fourier transform spectrometers are commercially available and almost all these instruments place emphasis on their use in ^{13}C studies.

8.9.1. Advantages

1. To improve the signal to noise ration several consecutive pulses can be applied to the sample.

2. FT-NMR is much faster and less time consuming than NMR.

3. FT-NMR technique is used to determine the NMR spectrum good spectra can be obtained with very much smaller samples, even less than a milligram.

4. The efficiency of NMR spectroscopy can be improved by exciting all possible resonances rather than scanning them sequentially.

5. It has been applied increasingly to the fields other than chemistry such as biology, engineering, industrial quality control and medicine.

6. Magnetic Resonance Imaging (MRI) is one of the most prominent FT-NMR applications.

8.10. APPLICATIONS

1. **Qualitative analysis:** The utility of NMR in the identification of pure substances. For substances of unknown structure NMR provides a valuable diagnostic tool. Chemical shifts as well as spin coupling

and decoupling observations are all useful in this connection. One can establish how many different environments for hydrogen atoms exist in the molecule. Study of fine structure of peaks, it is possible to determine which hydrogen types are closest to each other.

2. **Quantitative analysis:** NMR absorption peak is directly proportional to the number of nuclei responsible for that peak. NMR spectra, the quantitative determination of a specific compound need not require pure samples of the compound for calibration. The concentration of species can be determined directly by making use of signal area per proton and the area of that identifiable peak of one of the constituents. The signal are per proton can be determined by making use of a known concentration of an internal standard. **Example:** The solvent, present in a known amount were benzene, cyclohexane or water, the areas of the single proton peak for these compounds could be used in order to set the required information, but the peak of the internal standard should not overlap with any of the sample peaks.

   2.1. **Analysis of multi-component mixtures:** NMR use of in the analysis of many multi-component mixtures. **Ex:** Analysis of benzene, heptane, ethylene glycol, water in mixture and determination of aspirin, phenacetin and caffeine. It has also been used for quantitative determination of water in food products, pulp and paper and in agricultural materials.

   2.2. **Elemental analysis:** The total concentration of a given kind of magnetic nucleus in a sample can also be determined by making use of NMR spectroscopy. **Ex:** Determination of organic compounds.

   2.3. **Identification of compounds:** An NMR spectrum provides an important tool for the characterisation of a pure compound when used in conjugation with other observations such as elemental analysis. The structure of an unknown compound from its NMR spectrum can be easily decided by making use of certain principles.

      a. The number of main NMR signals should be equal to the number of equivalent protons in compound of interest.

      b. The type of methylene hydrogen atoms, methyl group

hydrogen, ether hydrogen atom etc. is indicated by the chemical shift.

c. The possible arrangement of groups in the molecule is indicated by the multiplicity or spin splitting.

d. The area under an NMR absorption peak is directly proportional to the number of nuclei present in each group may be determined by making use of area of peaks.

2.4. Hydrogen bonding: The hydrogen bonding in organic compounds as well as in metal chelates has successfully been studied by NMR spectra. Proton signal is shifted towards low field, if hydrogen bonding is present in the molecule.

2.5. Structure determination: NMR spectroscopy is very helpful in studying and establishing the structures of complexes, organic and inorganic compounds. Ex: Structure of SOF_4, CIF_3, HF_2, WF_6, structure of bis (dipivaloyl methanido) mercury (II) complexes and polyethylene.

2.6. Determination of activation energy: NMR spectra bas been applied in determining the activation energy. Ex: N, N^1-dimethyl acetamide (DMA)

3. **Applications of NMR in biology:** In recent years the technology of NMR has also been used in various special fields that include industrial quality control, biology, engineering and medicine. One of the most important applications of NMR in medical sciences is that of Magnetic resonance imaging (MRI), in which data from pulsed FR excitation of solid or semisolid objects are subjected to Fourier transformation and converted to three dimensional images of the interior of the objects.

The linear variation or field gradient in B is created by auxiliary coils in the magnet bore that are under the control of the computer of the NMR instrument. Protons in different locations in the subject experience different magnetic field strengths B1 and B2 for two of

the regions. According to Larmor equation, $v = \dfrac{\gamma Bo}{2\pi}$ these nuclei are expected to exhibit different resonance frequencies $v1$ and $v2$. Each consecutive RF pulse produces an FID signal that encodes the concentration of protons at each 1 cm position along the direction of the field gradient.

The position of the slice along the z-axis is changed by adding a dc offset to the auxiliary coils as shown by the dotted field gradient. It is possible to probe each slice by changing the pulse width of the RF pulse to tune its frequency to the precession frequency of the protons in the slice. If the applied field gradient is made larger the thickness of the slice to be examined is made smaller. The ability to generate high resolution images from true three dimensional data is an important feature of MRI technique. MRI has now become an important tool in medical diagnosis, but its applications in other areas have not been tapped much. MRI has used in other areas of science and technology.

8.11. SUMMARY

Radio waves are considered to be the lowest form of electromagnetic radiation. At a particular combination of fields, energy is absorbed by the sample, and the absorption can be observed as a change in the signal developed by a radio frequency detector and amplifier. This energy absorption can be related to the magnetic dipolar nature of spinning nuclei. This technique, known as Nuclear Magnetic Resonance (NMR) .

Odd atomic number and odd mass number Ex: ^{1}H, $^{15N, 19F, 31P}$ etc.

Odd atomic number and even mass number Ex: ^{1}H, ^{14N} etc.

Even atomic number and odd mass number Ex: ^{13}C

The NMR spectroscopy is most often concerned with I = 1/2, the best example of which is proton, ^{1}H with a spin of 1/2. So NMR spectroscopy is also known as Proton Magnetic Resonance (PMR) Spectroscopy.

Protons and neutrons possess magnetic properties, they are associated with a magnetic moment µ for the nuclei. The relation between this magnetic moment and nuclear spin quantum number is given by ...

$$\mu = k\sqrt{[I(I+1)]}$$

The quantum character of atomic particles limits to a few number of possible energy levels. For a spin quantum number I and magnetic quantum number m, the energy of a quantum level is given by:

$$E = -\sqrt{\frac{m\mu}{I}}\beta H_o$$

The proton has magnetic quantum numbers of +1/2 and −1/2. The energies of these two states in the magnetic field are given by:

$$M = +1/2 \qquad E = -\frac{1/2(\mu\beta Ho)}{1/2} = -\mu\beta Ho$$

$$M = -1/2 \qquad E = -\frac{-1/2(\mu\beta Ho)}{1/2} = +\mu\beta Ho$$

For the lower energy state (m = 1/2), the vector of the magnetic moment is aligned with the field and for the higher energy state (m = −1/2), the alignment is reversed. The very small energy difference between the two levels is given by:

$$\Delta E = 2\mu\beta Ho$$

Excitation to a higher nuclear magnetic quantum level can be brought about by absorption of a photon with energy hv (= ΔE). Thus above equation can be written as:

$$hv = \frac{\mu\beta Ho}{I}$$

Chemical shift is defined as nuclear shielding / applied magnetic field. Chemical shift is a function of the nucleus and its environment. This is designated by δ, is the ratio of the change of field necessary to achieve resonance to the field strength that resonates with a standard:

$$\delta = \frac{H\ sample - H\ ref}{H\ ref}$$

In C = C and C = O double bonds, the deshielding zone extends along the bond direction, even C − C bonds show some Deshielding in this direction. This anisotropy of the magnetic susceptibility of chemical bonds means that the shielding or Deshielding of a neighboring proton in the molecule is dependent on its distance from the bond and its orientation with respect to that bond.

Nuclei can interact with each other to cause mutual splitting of the otherwise sharp resonance lines into multiplets, called spin-spin coupling, or also called as J coupling.

Frequency Lock - An automatic circuit to maintain a constant field to frequency ration is essential in a high resolution NMR Spectrometer. It uses a frequency controlling feedback circuit that locks onto some specific nuclear resonance and continually adjusts the RF oscillator to keep the reference signal maximized.

Nuclear magnetic double resonance or spin decoupling, is achieved by irradiating an ensemble of nuclei not only with a rf H_1 at resonance with the nuclei to be observed but additionally with a second, relatively strong rf field H_2 perpendicular to H0 and at resonance with the nuclei to be decoupled. Decoupling is achieved when $\gamma H_2/2\pi >>/J_{AX}/$, but H_2 is still sufficiently low in power that it does not cause significant changes in the other nuclei.

Nuclear Overhauser Effect (NOE) depends on the disruption of the normal relaxation mechanism. Relaxation is the process by which a nucleus, after absorbing energy, returns to its normal state. The most important of such processes involves dipole dipole interaction between spinning nuclei. This interaction depends strongly on the distance separating the nuclei concerned. If one of these nuclei is saturated by radiation from an auxiliary oscillator, it becomes more difficult for the other nucleus to lose its excess energy.

GLOSSARY

Atomic absorption spectrometry: The concentration of the element in the sample can be quantified by the amount of light absorbed by the element present within by the flame.

Absolute error: The difference between a measured value and the true or accepted value.

Absorbance: A measure of the amount of light absorbed by (usually) a solution, equal to the negative logarithm of the ratio of initial and final light intensity.

Absorption: The process by which a substance acquires energy by promotion to excited electronic, vibrational or rotational states.

Absorption spectrum: The spectrum of radiation intensity vs. wavelength or frequency produced when atoms, ions or molecules are excited from low energy states to higher energy states.

Absorption coefficient: A measure of the ability of particles or gases to absorb photons, a number that is proportional to the number of photons removed from the sight path by absorption per unit length.

Acid precipitation: Typically is rain with high concentrations of acids produced by the interaction of water with oxygenated compounds of sulfur and nitrogen which are the by-products of fossil fuel combustion.

Acid: A substance that ionizes in aqueous solution to yield hydronium ion as one of the ions or a substance that yields protons or accepts electron. The strength of an acid depends on the degree of ionization.

Amorphous: A substance having no definite shape or size and which is coarse in nature.

Anhydrous: A substance containing no water.

Atomic number: The number of protons in the nucleus of an atom.

Atomic weight: The weight of an atom with reference to the oxygen atoms as 16.00.

Accuracy: The agreement between a mean measured value and a true or accepted value as quantified by error. Related to determinate errors and quantified by bias.

Analyte: Species of interest in a determination.

Anion: A negative ion, such as sulfate, nitrate, or chloride.

Apparent contrast: Contrast at the observer of a target with respect to some background, usually an element of horizon sky directly above the target.

Attenuation: The diminution of quantity. In the case of visibility, attenuation or extinction refers to the loss of image-forming light as it passes from an object to the observer.

Anode: Place where oxidation reactions occur in an electrochemical cell. Electrons flow away from the anode in a cell and the anode becomes positively charged.

Band spectra: Spectra (absorption or emission) are showing broad peaks due to combination of electronic, vibrational and rotational transitions in molecules.

Bandwidth: The width of a band. Units of nm etc. The effective bandwidth is the width of the band at half height.

Base: A substance that ionizes in an aqueous solution to yield hydroxide ion as one of the ions or a substance that possesses at least a lone pair of electrons or which accepts a proton. The strength of a base depends on the degree of ionization.

Buffer solution: The resistance of a solution to changes in hydrogen ion concentration upon addition of small amount acid or alkali.

Base peak: The most intense m/z value in a mass spectrum, assigned 100 % intensity.

Blackbody radiation: A continuous emission spectrum produced by many small energy transitions (electronic and vibrational) in a solid when heated. Most emission is in the IR part of the EM spectrum but moves further into the visible (and UV) with increasing temperature.

Blank: A blank contains all matrix components except the analyte and is used to determine the instrumental signal associated with non-analyte components, or with determinate or systematic errors.

Bonded chromatographic column: Stationary phase covalently attached to support material rather than by simple physical adsorption, reduces stationary phase leaching.

Boundary potential E_b: Potential developing across a permeable membrane due to different positions of surface equilibrium when solutions of different activities are in contact with opposite sides of the membrane.

Background luminance: A measure of light power reflected or emitted from the background of an object within a solid angle of one steradian per unit area projected in a given direction.

Back trajectory: A trace backwards in time showing where an air mass has been.

Brightness: A measure of the light received from an object, adjusted for the wavelength response of the human eye, so as to correspond to the subjective sensation of brightness. For visually large objects, the brightness does not depend on the distance from the observer.

Brightness contrast: The ratio of the difference in brightness between two objects to the brightness of the brighter of the two. It varies from 0 to -1.

Calibration: The process of submitting samples of known value to an instrument, in order to establish the relationship of value to instrumental output.

Color: A qualitative sensation described by hue, brightness, and saturation

Color contrast or difference: Contrast between two adjacent scene element colors, any difference in color hue, saturation, or brightness, between two perceived objects.

Calorimetric analysis: Chemical analysis based on the colors of dyes formed by the reaction of the analysis with reagents.

Condensation: A process by which molecules in the atmosphere collide and adhere to small particles.

Condensation counter nuclei: An instrument that counts nucleation mode particles by causing them to grow in a humid atmosphere, and observing light reflections from the individual enlarged particles.

Condensation nuclei: The small nuclei or particles with which gaseous constituents in the atmosphere (water vapor) collide and adhere.

Contrast: Relative difference in light coming from a target compared to the surrounding background, usually the horizon sky and any difference in the optical quality of two adjacent images.

Contrast change threshold: Minimum change in contrast perceptible to an observer.

Contrast threshold: Minimum apparent contrast at which a target is just perceptible.

Contrast transmittance: Ratio between apparent and inherent spectral contrast. When the object is darker than its background, it has a value between 0 and -1. For objects brighter than their background, the value varies from 0 to infinity. When the contrast transmittance is equal to zero, the object cannot be seen.

Catalyst: A substance which alters the rate of reaction (increase or decrease) and which remains unchanged and can be recovered at the end of the reaction.

Cation: An ion or radical having a positive charge.

Chelation: Bond formation between a metal and two or more polar grouping of a single molecule.

Carrier gas: Mobile phase in gas chromatography usually helium, hydrogen or nitrogen.

Cathode: Place where reduction reactions occur in an electrochemical cell. Electrons flow towards from the cathode in a cell and the cathode becomes negatively charged.

Chemiluminescence: An exoergic chemical reaction that produces a product molecule in an electronically excited state that relaxes via fluorescence or phosphorescence. The intensity of light produced is proportional to the rate of production of the excited state molecule. Chemiluminescent reactions are relatively rare but often involve strong oxidizers (e.g. O_3 or H_2O_2).

Chromophore: An atom, ion or molecule that absorbs radiation and refer to a particular part or functional group of a molecule for example, the carbonyl group or halogen atom.

Concomitants: All species in a matrix excluding the analyte.

Cross-linked chromatographic column: Individual stationary phase molecules polymerized (cross-linked) after attaching to support material and reduces stationary phase leaching.

Detector: Any device which quantitatively responds to a change in its environment Ex: temperature, pressure, light intensity.

Deuterium lamp: A continuous source for UV part of the electromagnetic spectrum (180-340 nm). Some lines in visible.

Dipole moment (mu): A measure of the overall electric field associated with a molecule. Depends on (i) charges (or partial charges) and (ii) their separation. Units of Debye (D). 1 D = 3.33E-30 Cm. A dipole moment of ~10 D corresponds to a +1 and -1 electron equivalent charge separated by 1 Angstrom. Typical molecular dipole moments range from zero to a few Debye.

Direct probe: Inlet systems in mass spectrometry allow insertion of a small heated vial, capillary or plate into the ionization chamber through an air-lock system. Heating of low vapor pressure analytes causes vaporization inside the instrument.

Dispersion (1): Change in refractive index with frequency. In normal dispersion, the refractive index increases with frequency. In anomalous dispersion, the refractive index decreases with frequency indicative of energy absorption.

Dispersion (2): A measure of the angular spread of different wavelengths emerging from a prism or grating. More precisely, the reciprocal dispersion is the rate of change of wavelength with distance across a monochromator exit slit. It has units of nm/mm etc.

Dispersion (3): Measure of "dilution" of sample by solvent (usually in a tube or column) due to diffusion, migration and convection. Quantified is the sample concentration divided by the peak sample concentration measured at the detector. Influenced by many physical parameters including flow rate, temperature, tube inside diameter, tube length and sample volume.

Dissociation: Process of bond-breaking. For photo dissociation, the absorption of a photon produces an excited state which is directly repulsive for example an antibonding orbital or in a sufficiently high vibrational state that the molecule is above the dissociation limit.

Double-focusing mass analyzers: These are widely used in high resolution mass spectrometry. They consist of an electrostatic deflection- and magnetic sector-type analyzer in series to simultaneously reduce ion energy distribution and angular distribution aberrations.

Diffraction: Modification of the behavior of a light wave resulting from limitations of its lateral extent by an obstacle for example, the bending of light into the "shadow area" behind a particle.

Diffusion: A process by which substances, heat, or other properties of a medium are transferred from regions of higher concentration to regions of lower concentration.

Direct effects: The optical effects of aerosols on climate modification referring to absorption and scattering of solar radiation by airborne particles.

Discoloration: Any change in the apparent color of an image. Often refers to the loss of blue sky color due to air pollution.

Dose-response: The relationship between the dose of a pollutant and its effect on a biological system.

Dry deposition: Also known as dry fall includes gases and particles deposited from the atmosphere to water and land surfaces. This dry fall can include acidifying compounds such as nitric acid vapor, nitrate and sulfate particles, and acidic gases.

Dynamic dipole moment: A dipole moment whose magnitude varies with time, usually as a result of vibrational motion.

Dynamic range: The range of concentrations or other chemical/physical property able to be quantified by a particular instrumental technique typically 2 to 8 orders of magnitude or more.

Dynode: An electrode between the photocathode and anode in a photomultiplier device. May be multiple dynodes acting to give an electron avalanche effect.

Detection limit: The smallest concentration or amount of a component of interest that can be detected by a single measurement with a stated level of confidence, generally regarded as the concentration level with a precision of ±100%.

Efficiency (chromatographic): A measure of the zone or band broadening in chromatography expressed as plate height or number of plates.

Electromagnetic (EM) spectrum: Range of wavelengths or frequencies of electromagnetic radiation ranging from cosmic and gamma rays (high energy/short wavelength) to radio waves (low energy/long wavelength). The UV part of the EM spectrum has wavelengths extending from about 180-350 nm, the visible from about 350 (violet)-700 (red) nm and the IR from about 700 to >3000 nm.

Electron capture detector (ECD): Gas chromatography detector based on measurement of current produced by capture of electrons from a radioactive source by carrier or analyte molecules essentially non-destructive, poor dynamic range only moderate sensitivity for many species.

Electron impact (EI) ionization: Widely used in mass spectrometry to produce positive analyte ions by bombarding molecules with 70-100 eV electrons. EI excited spectra are characterized by multiple fragment ions.

Electro spray ionization (ESI): Widely used in mass spectrometry to ionize large mw molecules. A solution of the analyte passes through a high-voltage capillary tube whereby individual droplets pick up multiple charges. As the solvent evaporates, the drops shrink and the charge density increases. At some critical size, charged analyte begin to desorb from the droplet.

Elute/eluent: The mobile phase in chromatography.

Elution chromatography: The process of transporting material through a stationary phase by the continuous addition of solvent (mobile phase) causing separation.

Elution order in chromatography: The order in which various components elute on a particular column can often be approximately determined by the following rules of thumb: volatile components (low boiling point) elute before non-volatile components (ii) small molecules elute before large molecules (iii) in a polar mobile phase, polar molecules elute before non-polar molecules (iv) for a polar stationary phase, non-polar molecule elute before polar molecules. Exceptions to these rules frequently occur.

Emission spectrum: The spectrum of radiation intensity vs. wavelength or frequency produced when atoms, ions or molecules relax from high energy states to lower energy states.

Excitation: The process of raising the energy of an atom, ion or molecule from some low energy state to some higher energy state.

Extinction coefficient: A measure of the ability of particles or gases to absorb and scatter photons from a beam of light; a number that is proportional to the number of photons removed from the sight path per unit length.

Excited state: An atomic, ionic or molecular state with higher energy than the ground state.

External conversion: Occurs when a molecule in an excited state relaxes non-radioactively by interactions (energy transfer) with neighboring molecules (usually solvent).

Equilibration: A balancing or counter balancing to create stability, often with a standard measure or constant.

Equivalent contrast: Any scene can be Fourier decomposed into light and dark bars of various frequencies and intensities modulated in accordance with a sine wave function. Equivalent contrast is the average contrast of those sine waves within a specified range of spatial frequencies.

f **Number:** A measure of the light gathering efficiency of an optical element (mirror, lens). Defined as the focal length, F, divided by the diameter of the optic, d. Small *f* numbers imply better light gathering efficiency.

Faraday: The charge carried by one mole of electrons or 96,485 C.

Flame ionization detector (FID): Gas chromatography detector based on measurement of ion current in a flame.

Fluorescence: Relaxation route producing radiation from short lived (<10 microseconds) excited states.

Fluoro-meter: An instrument for measuring luminescence (fluorescence, phosphorescence) spectra. Contains filters for excitation and emission but no wavelength scanning. Light is collected at 90 degrees from the incident light (excitation) direction.

Focal length: For parallel incident rays, the distance between an optical element (lens, mirror) and the focal point where all rays converge.

Fourier transforms (FT): A mathematical technique for converting time dependant data into frequency data. Most often used in Fourier transform infrared spectrometry (FTIR). The conversion from frequency to time is performed by the inverse Fourier transform function.

Fourier Transformation Infrared Spectroscopy (FTIR): An instrumental method of analysis used for qualitative identification of organic and inorganic compounds. The technique can sometimes be used for quantitative analysis of mixtures of materials.

Frequency: Number of peaks of a wave passing a point per second. Units are /s or Hz (Hertz).

Gas chromatography: Usually refers to gas-liquid chromatography where the mobile phase is a gas (helium, hydrogen or nitrogen) and the stationary phase is an immobilized liquid bound to a support material.

General elution problem: Compromise between resolution and experiment time, optimized by gradient elution in liquid chromatography and temperature programming in gas chromatography.

Glass electrode or pH electrode: Electrode with porous non-crystalline glass membrane sensitive to changes in hydrogen ion activity used with standard reference electrode if in same housing called a combination electrode.

Gradient elution in liquid chromatography: Controlled change of solvent (mobile phase) composition to reduce retention time of strongest retained species.

Grating: A diffraction/refraction dispersion element used to separate radiation into component wavelengths. Consist of a series of grooves of lines blazed onto a piece of quartz or silicon. These lines are parallel and closely spaced (10-2000 per mm). Constructive interference at certain incidence angles produces monochromatic radiation at specific diffraction angles. Multiple order effects also produce radiation at one half, one third, one quarter of the primary order (longest wavelength) radiation at the same diffraction angle.

Ground state: The lowest energy electronic, vibrational or rotational state. A species may be in the ground electronic state but excited vibrational or rotational states or any other combination of ground and excited states.

Hooke's law: Relationship between restoring force, F, and displacement, d, for a harmonic oscillator: $F=-kd$. The proportionality constant k is called the spring constant.

High performance liquid chromatography (HPLC): An instrumental method of analysis used to separate mixtures of compounds based on their chemical or physical properties.

Humidity: Water in air as a gas. Often measured as a percentage compared to the maximum amount of water vapor the air can contain at that temperature.

Hydrocarbons: Compounds containing only hydrogen and carbon. Ex: methane, benzene, decane.

Hydrophobic: Lacking affinity for water, or failing to adsorb or absorb water.

Hygroscopic: Readily absorbing moisture, as from the atmosphere.

Hydration: Association of one or more molecules of water with an ion or molecule or compound.

Hydrolysis: A reaction involving double decomposition in which water is one of the reactants.

Hygroscopic: A solution that has the ability to absorb moisture or water from atmosphere.

Hypertonic: A solution having osmotic pressure higher than the other solution with which it is compared.

Hypotonic: A solution having osmotic pressure lower than the other solution with which it is compared.

Interference: Potential or actual species whose presence in a matrix affects the quantitative measurement of an analyte.

Internal conversion: Occurs when a molecule in one electronic state non-radioactively converts to another state of same multiplicity (without spin change).

Intersystem crossing (ISC): Occurs when a molecule in one electronic state non-radioactively converts to another state of different multiplicity (with spin change, for example S_1 to T_1).

Isocratic elution: The use of a single solvent composition in liquid chromatography.

Isothermal elution: The use of a single column temperature in gas chromatography.

Illumination: Application of visible radiation to an object.

Inherent contrast: Contrast of the target against the horizon sky background when viewed at the target same as intrinsic contrast. The contrast would be seen between two adjacent scenic elements if there were no intervening atmosphere.

Inherent spectral contrast: Percent difference in radiant energy associated with an object and its background at an observer distance equal to zero.

Ion: A charged molecular group or atom.

Ion chromatography: A method of separating ions by their different speeds of passage through an ion-exchange resin. The ions are usually detected by their conductivity.

Indicator paper: A paper impregnated with a solution of indicator which shows a change in color at a certain hydrogen ion concentration.

Ionization constant: It is the ratio of the product of concentration in moles of the anion and cation of an electrolyte divided by the concentration in moles of the un-dissociated electrolyte. The constant remains same under standard conditions of temperature and pressure.

Ionization potential: It is the energy required to convert neutral atom from its lowermost energy level to form a positive ion.

Isotonic solution: A solution having same osmotic pressures as that of body fluids.

Isopleths: A line drawn on a map through all points having the same numerical value.

Isotropic: A situation where a quantity (or its spatial derivatives) are independent of position or direction.

Isotropic scattering: The process of scattering light equally in all directions.

J (quantum number): Total orbital angular momentum quantum number (simply) given by L+S.

Junction potential E_j: Potential developing across any permeable material with different activity solutions on either side caused by different diffusion rates through the material for different ions.

Koschmeider constant: The constant in the reciprocal relationship between standard visual range and the extinction coefficient.

Litmus: A pigment prepared from *Rochella tinctoria* and other species used as acid base indicator.

Limit of linearity (LOL): The maximum concentration of analyte that can be quantitatively determined using an instrumental technique.

Liquid chromatography: Usually refers to chromatography where the mobile phase is a liquid and the stationary phase is an immobilized liquid bound to a support material (partition) or a silica/alumina surface (adsorption). High performance liquid chromatography (HPLC) refers to the recent trend towards efficient separations using relatively high column inlet pressures, small particle sizes and immobilized liquid stationary phases.

Level of quantization (LOQ): The minimum detectable concentration of analyte at which quantitative measurements can be confidently conducted, defined as 10x the standard deviation of repeated measurements on a blank sample.

Molar solution: A solution containing one mole (molecular weight in gm) of solute per 1000 gm of the solvent.

Magnetic sector: A mass spectrometric analyzer based on the deflection of charged particles (ions) by a magnetic field. The field lines are perpendicular to the ion direction.

Mass to charge ration (m/z): The quantity measured in mass spectrometry. Units are usually Daltons (Da) or atomic mass units.

Matrix: All chemical components in a mixture to be analyzed including the analyte.

Mobile phase: A liquid or gas used in chromatography used to transport an analyte through a stationary phase.

Matrix filter: A filter that is formed of a mat or matrix of fibers. It is physically thick, and particles are trapped deep in its structure.

Membrane filter: A thin filter usually made of a synthetic polymer, with microscopic holes in it. Particles are collected only on the surface facing the air flow.

Micron: Units of length equal to one millionth of a meter the unit of measure for wavelength.

Molecular peak (M$^+$): In mass spectrometry is the m/z value corresponding to the singly-charged molecular ion. Often $(M-1)^+$ or $(M+1)^+$ peaks are observed rather than the molecular peak depending upon the ionization method.

Multichannel transducer: A transducer sensitive to spatial position of incident stimulus for example, a photodiode array consists of a series of p-n Si junctions arranged on a chip. If each junction is wired individually, current is produced only on those elements upon which light falls. As in all transducers, the magnitude of the current is proportional to the intensity of the radiation (up to saturation).

Multiplicity (spin multiplicity): Given by 2S+1 where S is the sum of individual electron spins, s. When 2S+1=1 (anti-parallel spin electron pair) the state is called a singlet state, when 2S+1=2 (single unpaired electron) the state is called a doublet state and when 2S+1=3 (parallel spin electron pair) the state is called a triplet state.

Normal phase liquid (-liquid) chromatography: Is performed using a non-polar mobile phase (hexane) and a polar stationary phase (polyethylene glycol).

Nephelometer: An instrument used to measure the light scattering component of light extinction.

Neutron activation: A method of chemical analysis in which the sample is bombarded with analysis neutrons in a nuclear reactor. The nuclei of various elements in the sample are modified to radioactive forms, and the concentrations of the elements are then determined by the intensities and wavelengths of the radiation emitted.

Nucleation: Process by which a gas interacts and combines with droplets.

Nuclei mode: A size range of particles below about 0.1 micrometer in diameter these particles are the nuclei around which larger particles grow.

Neutralization: A chemical reaction in which an acid and base reacts to form water and salt.

Normal solution: A solution containing one gram equivalent weight of substance per 1000 ml of solution.

Open tubular column: Have smaller internal diameters than regular packed columns but are more efficient. Often coated on the inside with stationary phase (wall-coated open tubular WCOT) or coated with stationary phase-coated small particles on the inside (support-coated open tubular SCOT).

Object luminance: A measure of light power reflected or emitted from an object itself within a solid angle of one steradian per unit area projected in a given direction.

Optical depth: The degree to which a cloud or haze prevents light from passing through it. It is a function of physical composition, size distribution and particle concentration often used interchangeably with turbidity.

Optical monitoring: Optical monitoring refers to directly measuring the behavior of light in the ambient atmosphere.

Optical particle: An instrument which measures the size of individual particles by the counter amount of reflected light from a microscopic illuminated volume.

Organic carbon: Aerosols composed of organic compounds, which may result from emissions from incomplete combustion processes, solvent evaporation followed by atmospheric condensation, or the oxidation of some vegetative emissions.

Organic compounds: Chemicals that contain the element carbon.

Oxidation: The process which results in the loss of one or more electrons by atoms or ions.

Parts per billion (ppb): a weight unit of measurement. 1% is equivalent to 10,000,000 ppb 1 ppb is equivalent to 0.001 ppm or 0.0000001%.

Parts per million (ppm): a weight unit of measurement. 1% is equivalent to 10,000 ppm 1 ppm is equivalent to 1,000 ppb or 1,000,000 ppt.

Parts per trillion (ppt): a weight unit of measurement. 1% is equivalent to 10,000,000,000 ppt 1 ppt is equivalent to 0.000001 ppm or 0.0000000001%.

Packed column: Chromatographic (gas) column filled with small particles to provide a porous phase with large surface area for partitioning.

Partition ratio or partition coefficient (K): The equilibrium constant describing the concentration distribution of analyte between stationary and mobile phases in chromatography.

Parallel digital data: Digital data encoding scheme using multiple data pathways usually in binary representation in which each transmission line represents a different power of two.

Period (of a wave): Time required for one wave oscillation (peak to adjacent peak for example) to pass a stationary point.

Peristaltic pump: Variable speed (flow rate) pump consisting of 8-12 rotating rollers pinching a flexible tube. Multiple channel pumps can pump several tubes at one time. Some pulsation used in flow injection and continuous flow techniques.

Phosphorescence: Relaxation route producing radiation from a long-lived (triplet) state >10 microseconds to hours.

Photomultiplier tube (PMT): A radiation transducer consisting of a photo emissive cathode that produces electrons when irradiated by light of appropriate wavelength usually visible followed by a series of dynodes and an anode. The dynodes are biased increasingly positive. Electron collision with the first dynode produces a cascade of secondary electrons. These travel to the second dynode, producing more electrons and so on until the last dynode releases many electrons which travel to the anode. Current traveling between the anode and cathode is proportional to the light intensity falling on the cathode for weak light intensities. Gains of more than one million are easily possible (one incident photon producing one million electrons at the anode).

Phototube: A radiation transducer consisting of a photocathode that produces electrons (photoelectric effect) when irradiated by light of appropriate wavelength (usually visible), and an anode. The current traveling between the cathode and anode is proportional to the light intensity for moderate intensities. Not as sensitive as a photomultiplier tube.

Plate height (height of an equivalent theoretical plate) (H): A term used to describe the efficiency of a chromatographic column and defined as [sigma]^2/L where [sigma] is the standard deviation of the Gaussian-shaped peak and L the column length. More plates produce narrower bands and more efficient separation. The number of plates, N, in a column is given by L/H. It should be noted that physical "plates" do not exist in a column!

Polarity index P': Number representing the average polarity of a molecule in a variety of solvents ranges from about -2 (very non-polar) to >10 (polar), n-Hexane has a value of 0.1 and water has a value of 10.2.

Precision: The reproducibility of a measurement made multiple times in identical ways caused by indeterminate or random errors usually quantified by absolute or relative standard deviation.

Pressure drop: The difference in pressure between the inlet and outlet of any tube. The pressure drop represents the resistance of an open tubular or packed column to mobile phase flow. As a consequence, while the amount of material entering and exiting the column per unit time is constant, the flow rate at the inlet is lower than at the outlet.

Pyroelectric transducer: A device based on a piezoelectric material used to sense temperature changes. The piezoelectric material has a permanent dipole moment due to the arrangement of ions in the crystal structure. A deformation of the material (thermal or mechanical) changes the magnitude of this dipole moment which can be measured electronically as a small current flow.

Phase function: Relationship of scattered to incident light as a function of scattering angle; volume scattering function.

Phase shift: A change in the periodicity of a wave-form such as light.

Photochemical: Any chemical reaction which is initiated by light such processes is process important in the production of ozone and sulfates in smog.

Photometer: Instrument for measuring photometric quantities such as luminance, luminous intensity and luminous flux. An instrument for measuring the brightness of an object it has been suggested that this name be reserved for those instruments which have been adjusted to match the wavelength response of the human eye but established usage is not yet this consistent and radiometers are sometimes called photometers.

Photometry: Instrumental methods, including analytical methods employing measurement of light intensity.

Photon: A bundle of electromagnetic energy that exhibits both wave-like and particle-like characteristics.

Photonic: Vision or wavelength response of the cones of a normal eye when exposed to a luminance of at least 3.4 candelas per square meter.

Polarization: A property of light, light can be linearly polarized in any direction perpendicular to the direction of travel, circularly polarized (clockwise or counterclockwise) or mixtures of the above.

Quadrupole: A mass analyzer consisting of four parallel rods. DC and variable frequency RF voltages are applied to opposite pairs of the rods. Ions entering this field are forced into helical trajectories and for a particular set of DC potentials and RF frequencies, only ions of a narrow range of m/z values can exist in stable trajectories and reach the detector.

Quantum mechanics: Mathematical formalism developed in the first part of the 20th century based on treating atoms and molecules as possessing specific energy states. Discrete transitions are only allowed between two states.

Quantum numbers: Are used in quantum mechanics to describe electron wave function. Principle quantum number n (values: 1, 2, 3...), angular momentum quantum number l *or* L (values: 0 (=s), 1 (=p), 2 (=d)...), magnetic quantum number m *or* M, spin quantum number s *or* S (values: +1/2, -1/2). See also term symbols. A vibrational quantum number (nu) (values: 0, 1, 2, 3) is used in the description of the energy of molecular vibrational levels.

Quartz tungsten halogen lamp (QTH): A high intensity tungsten lamp continuous (band) source producing ~200-3000 nm radiation. The quartz envelope to resist high temperatures and pass UV is filled with a small concentration of iodine vapor. This reacts with any W evaporating from the filament to produce gaseous tungsten iodide that is decomposed upon contact with the hot filament to redeposit metallic tungsten. Consequently, QTH lamps have a long lifetime.

Reactor: Small coil of tightly-wound tubing used in flow injection analysis used to thoroughly mix reactants.

Reciprocating pump: Piston pump used in HPLC for mobile phase pressurization up to 10,000 psi produces very pulsating flow.

Refractive index (RI): The ratio of the speed of EM radiation in a medium divided by the speed of EM radiation in a vacuum, c. c=2.99 million km/s (2.99E8 m/s). The refractive index varies with wavelength according to dispersion. Glass has a refractive index of about 1.50 at 590 nm, while air is about 1.00 at 590 nm.

Relaxation: Return to lower energy state in atoms, ions or molecules accompanied by release of heat (non-radioactive relaxation) or radiation.

Resolution R_s **(1):** A measure of the separation of two chromatographic peaks defined as twice the separation of the peaks divided by the sum of the width of the peak bases. Values of less than 1.0 are considered unresolved peaks.

Resolution R **(2):** A measure of the separation of two mass spectrometric peaks. The peaks are usually considered separated if the intensity between the two peaks is <10% of the intensity of the weakest peak. R is defined as the average m/z values of the two peaks divided by their separation. For example, if two peaks at m/z=400 and 401 are just separated, the resolution of the instrument is 400.5.

Retention time t_R: The time after injection for a retained analyte to elute (be detected) in a chromatographic separation.

Retention volume V_R: The volume of mobile phase eluted before the elution of an analyte in a gas chromatographic separation defined as the retention time t_R multiplied by the average mobile phase flow rate u.

Radiometer: A name for light measuring instruments which do not match the wavelength response of the human eye.

Reconstructed light extinction: The relationship between atmospheric aerosols and the light extinction coefficient can usually be approximated as the sum of the products of the concentrations of individual species and their respective light extinction efficiencies.

Reflectance: Ratio of reflected to incident light.

Reflection: Return of radiation by a surface without a change of frequency.

Refraction: The change of direction of a ray of light in passing obliquely from one medium into another in which the speed of propagation differs.

Selectivity coefficient $h_{x,y}$: Ratio of selectivity of membrane material to interfering ions. A selectivity coefficient of 0.00 means the membrane reacts only with the ion of interest x, a selectivity coefficient of 1.00 means that the membrane is as selective to interfering ion y as ion x. A selectivity coefficient of much greater than 1.00 implies the electrode is selective for species y and might be useful as an ion selective electrode (ISE) for species y.

Selectivity factor: The partition ratio K_B of species B divided by the partition ratio K_A of species A, A measure of the relative interactions strengths of species A and B with a stationary phase material in chromatography.

Serial digital data: Digital data encoding scheme using one data pathway (usually in binary representation in which each successive bit represents an increasing power of two) more efficient than simple counting.

Sensor: A device able to quantitatively and reversibly detect chemical species and produce an electrical signal.

Snell's law: Relationship between the sines's of incident and refracted angles for a beam passing between two media of different refractive indices.

Soft ionization sources: Deposit small excess quantities of energy into a molecule to ionize it and cause relatively little subsequent fragmentation.

Spectrofluorometer: An instrument for measuring luminescence (fluorescence, phosphorescence) spectra and contains two monochromators: one for varying excitation and one for varying emission wavelengths. Only one monochromator is scanned during spectral acquisition (excitation fluorescence spectra are collected with fixed emission wavelength and emission fluorescence spectra are collected with fixed excitation wavelength). Light is collected at 90 degrees from the incident light (excitation) direction.

Spring constant k: A measure of the stiffness of a spring, the proportionality constant in Hooke's Law. Units of N/m and typical effective molecular spring constants are 1-10 N per meter.

Standard hydrogen electrode: Standard reference electrode assigned a value of 0.000 V. Made by dipping an inert Pt wire into a acid solution with hydrogen ion activity of 1.000 M and bubbling H_2

gas around the wire at a partial pressure of 1.000 atm. Difficult to operate so alternatives are often used (calomel or Ag/AgCl for example).

Stationary phase: A solid or immobilized liquid used in chromatography that has some interaction with an analyte in a mobile phase, and so can partition the analyte between phases. Common stationary phase materials for gas chromatography include polysiloxanes and polyethylene glycol (PEG). Stationary phase polarity is an important component affecting capacity factors and selectivity factors.

Support coated open tubular (SCOT) column: A capillary chromatography column of about 150-400 micron diameter with the inside wall coated with a thin layer (10-50 microns) of solid particles each coated with a thin layer (1-10 microns) of stationary phase.

Scattering: An interaction of light wave with an object, which causes the light to be redirected in its path in elastic scattering, no energy is lost to the object.

Solute: A substance (the component of the solution) which is dissolved in the solvent.

Solvent: A substance (component of the solution) which dissolves solute.

Temperature programming in gas chromatography: Controlled change of column temperature during elution to reduce retention time of strongest retained species.

Thermal conductivity detector (TCD): Gas chromatography detector based on measurement of a heated filament temperature when surrounded by gas of a given thermal conductivity. Sensitive and non-destructive but variable sensitivity thermocouple a pair of junctions of two fine wires (dissimilar materials) that produces a small voltage proportional to the difference in temperature between the two junctions. A number of thermocouples arranged in series form a thermopile.

Transducer: A device able to quantitatively convert non-electrical domain data into electrical domain data (e.g. photocell)

Tracer elements: An element which is emitted most strongly by a specific source or class of sources, and can therefore be used as evidence for an impact by such a source when the element is detected in an air pollution sample.

Transmission gauge: A device for determining the amount of particles collected on a filter by the attenuation of light passing through the filter. Beta rays are sometimes used in place of visible light, and the resulting instrument is called a beta gauge.

Transmittance: The fraction of initial light from a light source that is transmitted through the atmosphere. Light is attenuated by scattering and absorption from gases and particles.

Turbidity: A condition that reduces atmospheric transparency to radiation, especially light, the degree of cloudiness or haziness, caused by the presence of aerosols, gases and dust.

Valence electrons: The electrons present in the outermost shell of element which are gained, lost or shared in chemical reactions.

Vibrational relaxation: Can occur as a vibrational excited molecule non-radioactively loses energy by internal or external conversion processes, rapid - 10^{-13} to 10^{-11} s.

Wall coated open tubular (WCOT) column: A capillary chromatography column of about 150-400 micron diameter with the inside wall coated with a thin layer of stationary phase.

Wavelength: Distance between two adjacent peaks (or troughs) of a wave.

Wave number: Reciprocal of wavelength measured in cm. Units of per cm.

Wave velocity: The number of peaks on a wave passing per second (frequency) multiplied by the distance between two adjacent peaks (wavelength).

Zone or band: The resolved Gaussian-shaped concentration profile of analyte emerging from a chromatographic column.

INDEX